Basic Mechanical Engineering

AUTHORS

Dr. Dharmendra Singh
Assistant Professor,
Department of Mechanical Engineering,
Govt. Engineering College Bikaner,
Rajasthan, India

Dr. Chandra Shekhar Rajoria
Assistant Professor,
Department of Mechanical Engineering,
Govt. Engineering College, Bikaner,
Rajasthan, India

REVIEWERS

Dr. R P Saini
Professor,
Alternate Hydro Energy Centre, IIT,
Roorkee-247667, (Uttarakhand), India

Dr. S. P. Harsha
Professor,
Department of Mechanical and Industrial
Engineering, IIT, Roorkee-247667,
(Uttarakhand), India

KHANNA BOOK
PUBLISHING CO. (P) Ltd.
4C/4344, Ansari Road, Darya Ganj,
New Delhi, India

All India Council for
Technical Education
Nelson Mandela Marg, Vasant Kunj,
New Delhi, India

Dear Readers,

To prevent the piracy, this book is secured with HIGH SECURITY HOLOGRAM on the front title cover. In case you don't find the hologram on the front cover title, please write us to at contact@khannabooks.com or whatsapp us at +91-99109 09320 and avail special gift voucher for yourself.

Specimen of Hologram on front Cover title:

Moreover, there is a SPECIAL DISCOUNT COUPON for you with EVERY HOLOGRAM.

How to avail this SPECIAL DISCOUNT:

Step 1: Scratch the hologram

Step 2: Under the scratch area, your "coupon code" is available

Step 3: Logon to www.khannabooks.com

Step 4: Use your "coupon code" in the shopping cart and get your copy at a special discount

Step 5: Enjoy your reading!

ISBN: 978-93-5538-753-0
Book Code: DIP447EN

Basic Mechanical Engineering
by Dr. Dharmendra Singh and
Dr. Chandra Shekhar Rajoria

First Edition: 2025

Originally Published by:
All India Council for Technical Education, Ministry of Education, GOI

Exclusively Published & Distributed by:
Khanna Book Publishing Co. (P) Ltd.
4C/4344, Ansari Road, Daryaganj,
New Delhi (INDIA)

Visit us at: **www.khannabooks.com**
Write us at: **contact@khannabooks.com**

To view complete list of books,
Please scan the QR Code:

KPH

Printed in Bharat.

प्रो. टी. जी. सीताराम
अध्यक्ष
Prof. T. G. Sitharam
Chairman

सत्यमेव जयते

आज़ादी का अमृत महोत्सव

अखिल भारतीय तकनीकी शिक्षा परिषद्
(भारत सरकार का एक सांविधिक निकाय)
(शिक्षा मंत्रालय, भारत सरकार)
नेल्सन मंडेला मार्ग, वसंत कुंज, नई दिल्ली–110070
दूरभाष : 011–26131498
ई–मेल : chairman@aicte-india.org

ALL INDIA COUNCIL FOR TECHNICAL EDUCATION
(A STATUTORY BODY OF THE GOVT. OF INDIA)
(Ministry of Education, Govt. of India)
Nelson Mandela Marg, Vasant Kunj, New Delhi-110070
Phone : 011-26131498
E-mail : chairman@aicte-india.org

FOREWORD

Engineers are the backbone of any modern society. They are the ones responsible for the marvels as well as the improved quality of life across the world. Engineers have driven humanity towards greater heights in a more evolved and unprecedented manner.

The All India Council for Technical Education (AICTE), have spared no efforts towards the strengthening of the technical education in the country. AICTE is always committed towards promoting quality Technical Education to make India a modern developed nation emphasizing on the overall welfare of mankind.

An array of initiatives has been taken by AICTE in last decade which have been accelerated now by the National Education Policy (NEP) 2020. The implementation of NEP under the visionary leadership of Hon'ble Prime Minister of India envisages the provision for education in regional languages to all, thereby ensuring that every graduate becomes competent enough and is in a position to contribute towards the national growth and development through innovation & entrepreneurship.

One of the spheres where AICTE had been relentlessly working since past couple of years is providing high quality original technical contents at Under Graduate & Diploma level prepared and translated by eminent educators in various Indian languages to its aspirants. For students pursuing 2nd year of their Engineering education, AICTE has identified 87 books, which shall be translated into 12 Indian languages - Hindi, Tamil, Gujarati, Odia, Bengali, Kannada, Urdu, Punjabi, Telugu, Marathi, Assamese & Malayalam. In addition to the English medium, books in different Indian Languages are going to support the students to understand the concepts in their respective mother tongue.

On behalf of AICTE, I express sincere gratitude to all distinguished authors, reviewers and translators from the renowned institutions of high repute for their admirable contribution in a record span of time.

AICTE is confident that these outcomes based original contents shall help aspirants to master the subject with comprehension and greater ease.

(Prof. T. G. Sitharam)

ACKNOWLEDGEMENT

The authors are grateful to the authorities of AICTE, particularly Prof. T. G. Sitharam, Chairman; Dr. Abhay Jere, Vice-Chairman; Prof. Rajive Kumar, Member-Secretary, Dr. Ramesh Unnikrishnan, Advisor-II and Dr. Sunil Luthra, Director, Training and Learning Bureau for their planning to publish the book on Basic Mechanical Engineering. We sincerely acknowledge the valuable contributions of the reviewer of the book Dr. R P Saini, Professor, Alternate Hydro Energy Centre, Indian Institute of Technology, Roorkee and Dr. S P Harsha, Professor, Department of Mechanical and Industrial Engineering, Indian Institute of Technology, Roorkee for making it students' friendly and giving a better shape in an artistic manner.

We also wish to express gratitude to Dr. Jaiprakash Bhamu, Dr. Manoj Kuri, Dr. Vineet Kumar, Dr. Mohd. Yunus Sheikh, Mr. Rajendra Singh Shekhawat and Dr. Ganesh Prajapat for their valuable suggestions and continuous support. We would like to extend heartiest gratitude to our family members for their motivation and support during the period of book writing.

This book is an outcome of various suggestions of AICTE members, experts and authors who shared their opinion and thought to further develop the engineering education in our country. Acknowledgements are due to the contributors and different workers in this field whose published books, review articles, papers, photographs, footnotes, references and other valuable information, which enriched us at the time of writing the book.

Dr. Dharmendra Singh

Dr. Chandra Shekhar Rajoria

PREFACE

The book titled "Basic Mechanical Engineering" is an outcome of the rich experience of our teaching of Mechanical Engineering courses. The initiation of writing this book is to expose basic science to the engineering students, the fundamentals of Mechanical Engineering systems as well as enable them to get an insight of the subject. Keeping in mind the purpose of wide coverage as well as to provide essential supplementary information, we have included the topics recommended by AICTE, in a very systematic and orderly manner throughout the book. Efforts have been made to explain the fundamental concepts of the subject in the simplest possible way.

During the process of preparation of the manuscript, we have considered the various standard text books and accordingly we have developed sections like critical questions, solved and supplementary problems etc. While preparing the different sections, emphasis has also been laid on definitions and laws and also on comprehensive synopsis of formulae for a quick revision of the basic principles. The book covers all types of problems and these have been presented in a very logical and systematic manner. The gradations of those problems have been tested over many years of teaching to a wide variety of students.

Apart from illustrations and examples as required, we have enriched the book with numerous solved problems in every unit for proper understanding of the related topics. Under the common title, "Basic Mechanical Engineering" there are two parts in book covering different aspects. Out of those, the first one covers the basic thermodynamic principles including heat transfer, thermal power plants systems, steam turbines, IC engines and refrigeration. Whereas, the second one covers the basics of materials, manufacturing process, machine tools and their mechanisms.

As far as the present book is concerned, "Basic Mechanical Engineering" is meant to provide a thorough grounding in Mechanical Engineering systems on the topics covered. This book will prepare engineering students to develop abilities and capabilities to apply the acquired knowledge of Mechanical Engineering to tackle the engineering challenges and address the related aroused questions towards sustainable development. The subject matters are presented in a constructive manner so that engineering students are prepared to work in different sectors or in national laboratories at the very forefront of technology.

We sincerely hope that the book will inspire the students to learn and discuss the ideas behind basic principles of Mechanical Engineering systems and will surely contribute to the development of a solid foundation of the subject. We would be thankful to all beneficial comments and suggestions that will contribute to the improvement of the future editions of the book. It gives us immense pleasure to place this book in the hands of the teachers and students. It was indeed a big pleasure to work on different aspect covered in the book.

Dr. Dharmendra Singh

Dr. Chandra Shekhar Rajoria

OUTCOME BASED EDUCATION

For the implementation of an outcome based education the first requirement is to develop an outcome based curriculum and incorporate an outcome based assessment in the education system. By going through outcome based assessments, evaluators will be able to evaluate whether the students have achieved the outlined standard, specific and measurable outcomes. With the proper incorporation of outcome based education there will be a definite commitment to achieve a minimum standard for all learners without giving up at any level. At the end of the programme running with the aid of outcome based education, a student will be able to arrive at the following outcomes:

Programme Outcomes (POs) are statements that describe what students are expected to know and be able to do upon graduating from the program. These relate to the skills, knowledge, analytical ability, attitude and behaviour that students acquire through the program. The POs essentially indicate what the students can do from subject-wise knowledge acquired by them during the program. As such, POs define the professional profile of an engineering diploma graduate.

National Board of Accreditation (NBA) has defined the following seven POs for an Engineering diploma graduate:

PO-1. Basic and Discipline specific knowledge: Apply knowledge of basic mathematics, science and engineering fundamentals and engineering specialization to solve the engineering problems.

PO-2. Problem analysis: Identify and analyses well-defined engineering problems using codified standard methods.

PO-3. Design/development of solutions: Design solutions for well-defined technical problems and assist with the design of systems components or processes to meet specified needs.

PO-4. Engineering Tools, Experimentation and Testing: Apply modern engineering tools and appropriate technique to conduct standard tests and measurements.

PO-5. Engineering practices for society, sustainability and environment: Apply appropriate technology in context of society, sustainability, environment and ethical practices.

PO-6. Project Management: Use engineering management principles individually, as a team member or a leader to manage projects and effectively communicate about well-defined engineering activities.

PO-7. Life-long learning: Ability to analyse individual needs and engage in updating in the context of technological changes.

COURSE OUTCOMES

By the end of the course the students are expected to learn:

CO-1: To understand the fundamental concepts of elements of mechanical engineering.

CO-2: To provide the knowledge of basic thermodynamic principles and laws with their applications.

CO-3: Apply the acquired knowledge heat transfer and thermal power plants systems and apply them to solve real life problems.

CO-4: To provide the knowledge in relating the concepts to steam turbines, IC engines and refrigeration.

CO-5: To understand the various materials, manufacturing process, machine tools and their mechanisms.

CO-6: To develop abilities and capabilities to apply the acquired knowledge towards sustainable development.

Mapping of Course Outcomes with Programme Outcomes to be done according to the matrix given below:

Course Outcomes	EXPECTED MAPPING WITH PROGRAMME OUTCOMES (1- Weak Correlation; 2- Medium correlation; 3- Strong Correlation)						
	PO-1	PO-2	PO-3	PO-4	PO-5	PO-6	PO-7
CO-1	3	1	1	1	2	1	3
CO-2	3	3	3	3	2	1	3
CO-3	3	3	3	3	2	1	3
CO-4	3	3	3	3	2	1	3
CO-5	3	2	2	1	2	1	3
CO-6	1	2	1	1	3	1	3

GUIDELINES FOR TEACHERS

To implement Outcome Based Education (OBE) knowledge level and skill set of the students should be enhanced. Teachers should take a major responsibility for the proper implementation of OBE. Some of the responsibilities (not limited to) for the teachers in OBE system may be as follows:

- Within reasonable constraint, they should manoeuvre time to the best advantage of all students.
- They should assess the students only upon certain defined criterion without considering any other potential ineligibility to discriminate them.
- They should try to grow the learning abilities of the students to a certain level before they leave the institute.
- They should try to ensure that all the students are equipped with the quality knowledge as well as competence after they finish their education.
- They should always encourage the students to develop their ultimate performance capabilities.
- They should facilitate and encourage group work and team work to consolidate newer approach.
- They should follow Blooms taxonomy in every part of the assessment.

Bloom's Taxonomy

Level	Teacher should Check	Student should be able to	Possible Mode of Assessment
Creating	Students ability to create	Design or Create	Mini project
Evaluating	Students ability to Justify	Argue or Defend	Assignment
Analysing	Students ability to distinguish	Differentiate or Distinguish	Project/Lab Methodology
Applying	Students ability to use information	Operate or Demonstrate	Technical Presentation/ Demonstration
Understanding	Students ability to explain the ideas	Explain or Classify	Presentation/Seminar
Remembering	Students ability to recall (or remember)	Define or Recall	Quiz

GUIDELINES FOR STUDENTS

Students should take equal responsibility for implementing the OBE. Some of the responsibilities (not limited to) for the students in OBE system are as follows:

- Students should be well aware of each UO before the start of a unit in each and every course.
- Students should be well aware of each CO before the start of the course.
- Students should be well aware of each PO before the start of the programme.
- Students should think critically and reasonably with proper reflection and action.
- Learning of the students should be connected and integrated with practical and real life consequences.
- Students should be well aware of their competency at every level of OBE.

ABBREVIATIONS AND SYMBOLS

List of Abbreviations

General Terms			
Abbreviations	**Full form**	**Abbreviations**	**Full form**
BDC	Bottom dead centre	KE	Kinetic energy
BHP	Brake horse power	PCB	Printed circuit board
CI engine	Compression ignition engine	PE	Potential energy
COP	Coefficient of performance	PMM	Perpetual motion machine
CV	Calorific value	R	Refrigerator
HE	Heat engine	SFEE	Steady flow energy equation
HP	Heat pump	SI engine	Spark ignition engine
IC engine	Internal combustion engine	TDC	Top dead centre
IHP	Indicated horse power	TR	Tons of refrigeration

List of Symbols

Symbols	Description	Symbols	Description
A	Area	R	Universal gas constant
C	Velocity, constant	R_t	Thermal resistance
c_p	Specific heat at constant pressure	s	Specific entropy
c_v	Specific heat at constant volume	T	Temperature
d	Bore diameter	U	Internal energy
E	Total energy	V	Volume
g	Gravitational acceleration	V_c	Clearance volume
h	Specific enthalpy	V_s	Swept volume
k	Thermal conductivity	W	Work done
L	Stroke length	R_t	Thermal resistance
m	Mass	z	Datum
n	Polytropic index, number of molecules in a gas, number of cylinders	α	Explosion ratio
N	Engine speed	γ	Adiabatic index

Symbols	Description	Symbols	Description
p	Pressure	δ	Inexact differential, thickness
Q	Heat supplied	η	Efficiency
r	Compression ratio	ρ	Density, cut-off ratio

LIST OF FIGURES

LIST OF TABLES

CONTENTS

1 Introduction to Thermodynamics

UNIT SPECIFICS

Through this unit we will discuss the following aspects:

The following aspects of thermodynamics will be discussed in this unit:

- Various type of systems like open, closed and isolated system along with the concept of control volume and control mass;
- The understanding of property, state, process, cycle, point and path functions, thermodynamic equilibrium, reversible and irreversible processes;
- Various gas laws applied in thermodynamics;
- Overview of concept of work and heat, internal energy, enthalpy, specific heats;
- Zeroth law, first law and second law of thermodynamics in addition to the hypothesis of PMM-1 and PMM-2;
- Application of first law to non-flow (closed) and flow (open) systems together with the concept of steady flow energy equation (SFEE);
- Work and heat equations in steady flow processes;
- Second law of thermodynamics, Clausius and Kelvin-Plank statements and their equivalence;
- Heat engine, refrigeration and heat pump, COP, Carnot cycle and Carnot theorem with its corollary;
- Concept of entropy and Clausius inequality and problems related to application of first law, second law and entropy.

The concepts have been explained in a simple language to develop a better understanding of the various terms and concepts of thermodynamics. The content has been carefully designed with simple day to day examples so as to facilitate the learning in anticipation to its correlation with thermodynamics. A number of practical applications of thermodynamics are presented for developing interest and curiosity in conjunction to enhanced problem solving capability. Aside from a large number of multiple-choice questions as well as questions of short and long answer types marked in two categories following the lower and higher order of Bloom's taxonomy, the unit includes assignments through a number of numerical problems, and a list of references and suggested readings for practise. It is important to note that some QR codes have been provided in various sections for more information on various topics of interest, which can be scanned for relevant supportive knowledge. There is a "Know More" section following the content-based related practical. This section has been thoughtfully constructed to maximise the value of the supplemental data offered to the readers of this book.

RATIONALE

The unit of thermodynamics will help the readers to understand the behaviour of physical systems with an axiomatic approach and a minimum of assumptions. Thermodynamic laws govern all the processes occurring in nature and daily life and its applications are everywhere be it like refrigeration and air conditioning, I C engines, power plants, heat engines, industrial processes etc., the basic terms and concepts of thermodynamics used in the unit will help the students to gain an insight to get started with this subject. The knowledge of gas laws will be useful while doing the thermodynamic analysis. Thermodynamic laws give a mathematical framework to understand how energy transfer occurs inside a system, hence help us to design a better and efficient system. The first law tells us about the conservation of energy and the second law tells whether a system is possible or not.

The heat and work interactions and their equations are derived to develop an understanding of various processes of thermodynamics and will be used for solving the numerical problems in this unit. Applications related to flow and non-flow processes along with steady flow energy equation is discussed in length to apply the laws of thermodynamics and to develop a deep understanding of the subject. The second law helps us understand the spontaneity of some process and is often described as arrow of *time. Hence thermodynamics is one of the few basic branches of science which gives us some notion* of time. Perpetual motion machines of first kind (PMM-1) and second kind (PMM-2) are discussed to demonstrate that the machines which violate the first and second law are not possible.

The concept of entropy is also discussed which is the basic outcome of the second law and describes the possibility of a process to occur. Entropy lays the foundation of energy which is available for useful work. The unit covers almost all the aspects of thermodynamics and the concepts have been explained with the help of a number of numerical problems to make the content more interesting. Since thermodynamics applies to all the domains of our day-to-day life, it is essential that everyone should learn to conserve energy and use it judiciously. In conclusion, a strong basic understanding of thermodynamics is essential in a variety of disciplines, particularly mechanical and chemical engineering, chemistry, physics, and the life sciences.

PRE-REQUISITES

- Mathematics: Class XII
- Physics: Class XII

UNIT OUTCOMES

At the end of the unit, the students will be able to:

U1-O1: Identify and explain the basic concepts of thermodynamics such as system, properties, state, process cycle point and path function, Zeroth law of thermodynamics.

U1-O2: Calculate and analyse the thermodynamic properties and processes on p-V, T-s diagrams to solve various engineering problems.

U1-O3: Understand the concept of heat and work applied to different thermodynamic cycles for open and closed systems.

U1-O4: Demonstrate and apply the first law of thermodynamics flow and non-flow systems with their application to various systems like nozzles and diffusers, compressors and boilers.

U1-O5: Understand Kelvin Plank and Clausius statements of second law of thermodynamics and prove the equivalence. Evaluate the feasibility of a cycle while applying second law for various engineering problems.

U1-O6: Establish Clausius inequality and apply it to explain the concept of entropy. Apply the principle of increase of entropy to evaluate the feasibility of a thermodynamic process.

Unit-1 Outcomes	EXPECTED MAPPING WITH COURSE OUTCOMES (1- Weak Correlation; 2- Medium correlation; 3- Strong Correlation)					
	CO-1	CO-2	CO-3	CO-4	CO-5	CO-6
U1-O1	3	3	1	-	-	2
U1-O2	3	3	-	1	-	3
U1-O3	3	3	-	1	-	2
U1-O4	3	3	-	1	-	2
U1-O5	3	3	-	-	2	3
U1-O6	3	3	-	-	-	3

1.1 INTRODUCTION

The topic of thermodynamics, like other disciplines, arose through the systematisation of empirical facts. Data has no value until it is examined inside a sufficiently well-defined conceptual framework. The origins of such a framework can be found in everyday sense impressions and experiences. The concept of hotness and coolness must have existed for a long time. A key aspect of this experience is the realisation that when two bodies come into touch, the hotter body becomes colder and the colder body becomes hotter.

Thermodynamics is a key scientific area that has extensive applicability in all fields of science and engineering. Because thermodynamic rules are generic, the emphasis, substance and technique vary significantly among scientific and engineering fields. A general definition of thermodynamics is the science that studies the rule that controls how energy transforms from one form to another. It is established on observations from everyday life that were made into thermodynamics laws. These laws control the idea of energy conservation.

Any thermodynamic system can be described by four basic laws of thermodynamics namely first law, second law, third law and the Zeroth law. The concept of thermal equilibrium is described by the Zeroth law and thus forms the foundation for defining temperature.

1.2 ROLE OF THERMODYNAMICS IN ENGINEERING AND SCIENCE

Thermodynamics is a science related with the matter and conversion of energy from one form to another, mainly into heat and work. All physical objects contains matter, therefore thermodynamic is associated with analysis of all engineering problems. The principle of thermodynamics depends on the laws of conservation of mass and conservation of energy. Thermodynamics also defines how systems respond to changes in their environment. Engines, pumps, chemical reactions, human body, corrosion engineering, aerospace engineering, mechanical engineering, biomedical engineering, materials science, cell biology, black hole or entire universe are the system distinct from surrounding on which these laws may be applied to define engineering and science. These systems are free to transfer the work, heat and other forms of energy with the surrounding.

Thermodynamics is having physical significance from ancient times where man was harvesting mechanical work from nature and wind and water wheels were being used for irrigation. Combustion of wood and other fuel were used for cooking food, melting and other mechanical work. In 1769 James Watt patented steam engine to convert heat into work triggered industrial revolution. In 1804, Richard Trevithick built the first steam locomotive, whereas in 1814, George Stephenson invented first engine locomotive for passengers in railway. In 1824, Sadi Carnot, the father of thermodynamics published "Reflections on the Motive Power of Fire" was the time to bring up thermodynamics as a modern science.

It also affects our daily lives and we apply the concepts and laws to design and operation of power plants, chemical reactions, refrigerators, and many more important concepts which makes our life comfortable. The engine we use to drive our vehicles, refrigerators and air conditioners used in houses, heating and cooling systems, all are based on laws and information from thermodynamics for their design achieve most efficient performance. The design of houses and buildings to have proper heating and cooling arrangements to use energy efficiently, and minimize the wastes is another significance of these laws. According to the first law of thermodynamics, when we walk on road, we transform the chemical energy of the food into kinetic energy. Engineering courses such as power generation, heat transfer, and fluid mechanics make use of this law. Likewise, second law of thermodynamics is related with reversible and irreversible processes is the basis for all the power generation cycles. Therefore, this subject is having practical significance to deal with all engineering problems.

1.3 BASIC CONCEPTS AND DEFINITIONS

1.3.1 Types of Systems

A *system* may be described as a region or space containing a matter or quantity where attention is focused in the examination of a problem. Everything which lies outside the system is referred as *surroundings*. The *system* and *surroundings* are separated by a real or fictitious envelope called as *boundary* as depicted in Fig. 1.1. Furthermore, this boundary may either be moving or fixed and a system along with its surrounding together constitute a *universe.*

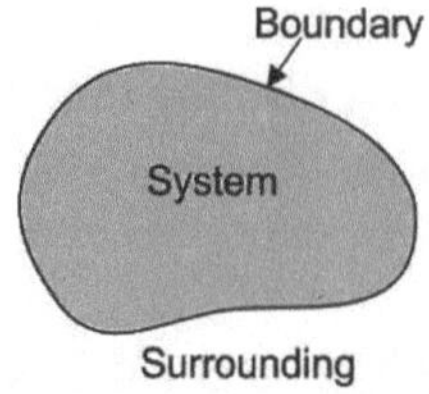

Fig. 1.1: System, boundary and surrounding

A system may be further classified into three types namely: (a) closed system (b) open system and (c) isolated system. In a *closed* system (Fig. 1.2 a) no mass transfer takes place across the system boundary but energy may cross into or out of the system. A closed lid container and a piston cylinder arrangement without valve are some of the examples of close system with fixed and moving boundaries respectively. In an *open* system (Fig. 1.2 b) both energy and mass can cross the system boundary. The confinement of mass within the boundary may change or remain constant depending upon the rate at which the mass enters and leaves the system. An air compressor and a turbine are some examples of an open system. In an *isolated* system (Fig. 1.2 c) neither mass nor the energy crosses the system boundary which implies that it is a system of fixed mass and energy. A thermos flask may be closely related to an isolated system because it is a closed container of fixed mass and does not allow energy transfer.

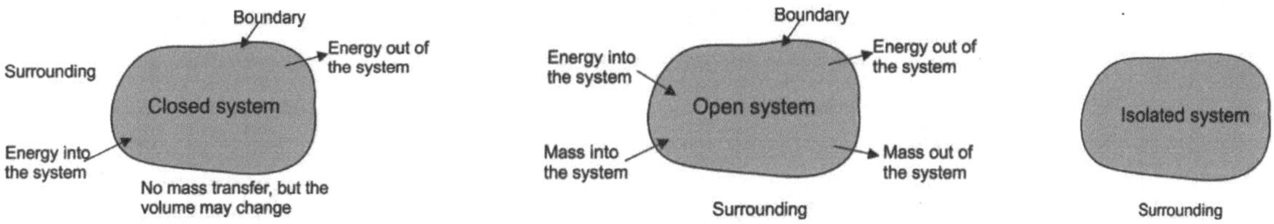

(a) (b) (c)

Fig. 1.2: Types of systems: (a) Closed system (b) Open System (c) Isolated system

1.3.2 Concept of Control Volume and Control Mass

A *thermodynamic system* comprises of a fixed quantity of matter upon which we focus the attention for the analysis of the system. A *control volume* is chosen in a thermodynamic system to bound the matter and devices under study inside a *control surface*. Everything lying outside a control volume is considered as surroundings which are separated by the control surface. A control surface may have energy interaction in the form of heat transfer and work transfer while it may be open or closed to mass flows and the boundaries may be either movable or stationary. If the control surface is such that it does not allow mass to escape or enter the control volume, or in other words if the control surface is closed to mass interaction, it is known as *control mass*.

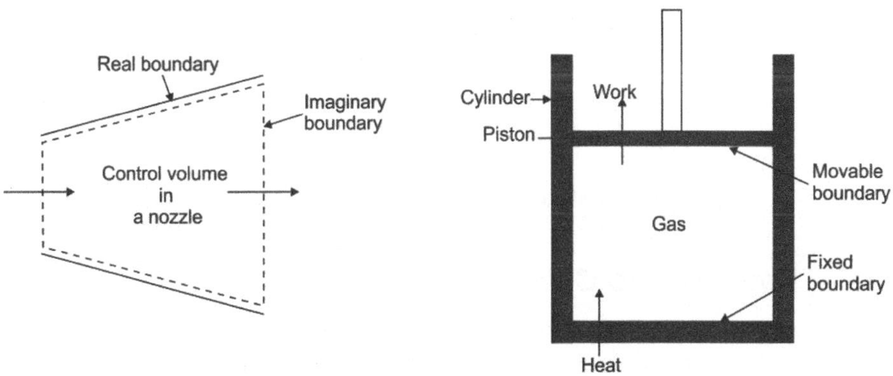

(a) (b)

Fig. 1.3: Control volumes with (a) real and imaginary; (b) fixed and moving boundaries

A control volume is related with open system while control mass is related with closed system, although there is no specific rule for selecting a control volume, but appropriate selection makes the analysis simple. A control volume may be of fixed shape and size or may have a moving boundary. The boundaries in contact with the physical surfaces of the system are referred as real boundaries while those involving no physical surfaces are called as imaginary boundaries as shown in Fig. 1.3 (a). The boundaries which cannot move are the fixed boundaries while moving boundaries can change its shape and size as illustrated in Fig. 1.3 (b).

1.3.3 Property

The measurable and observable characteristic of any system is called the *property* of the system. Some of the familiar properties are Temperature *T*, Pressure *P*, mass *m*, volume *V* etc. Any change in the property is governed by their end states and is independent of the path followed. Therefore, a thermodynamic property has a fixed value at a fixed state. Properties are classified into two categories as;

(i) Intensive properties

(ii) Extensive properties

The intensive properties are those which are independent of mass of the system e.g., density, temperature, pressure etc. The values of extensive properties depend on the size and mass of the system e.g., total mass, total volume, total momentum etc. A very simple approach to identify whether a property is intensive or extensive is seeing that whether its value in a system change or remains same at a given state when we divide it into two equal parts. To understand this let us assume a system involving mass, temperature, pressure, volume and density as depicted in Fig. 1.4.

Now divide this system into two halves, we can see that each part of the system will have half the value of mass and volume, while the value of temperature, pressure and density will remain the same for each half. Therefore, mass and volume are extensive properties and pressure, temperature and density are intensive properties. When the extensive properties are expressed on per unit mass basis, it is called as specific properties for e.g., specific volume, specific entropy etc. Specific properties are intensive properties and are generally denoted by small letters.

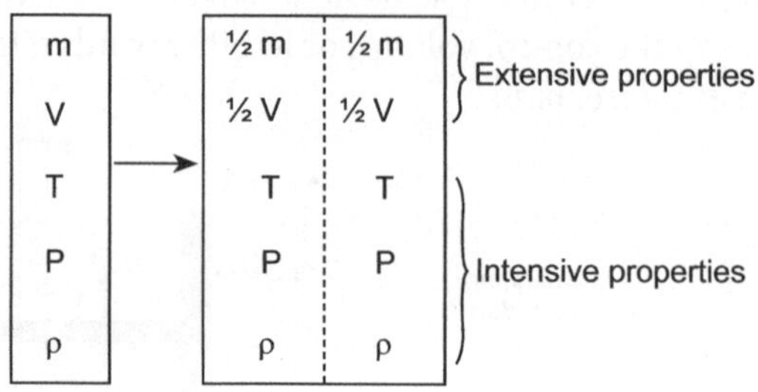

Fig. 1.4: Differentiation of intensive and extensive properties

1.3.4 State, Process and Cycle

When a system is not undergoing any change, then all the properties are measurable and carries a definite or fixed value which describes the condition of the system known as *state*. The change in value of even one property changes the state of the system therefore, properties may be considered as state variable of a system. The method by which a system passes from one equilibrium state to another is called the *process*. The successive series of states through which a system passes during a process is called as *path* of the given process (Fig. 1.5).

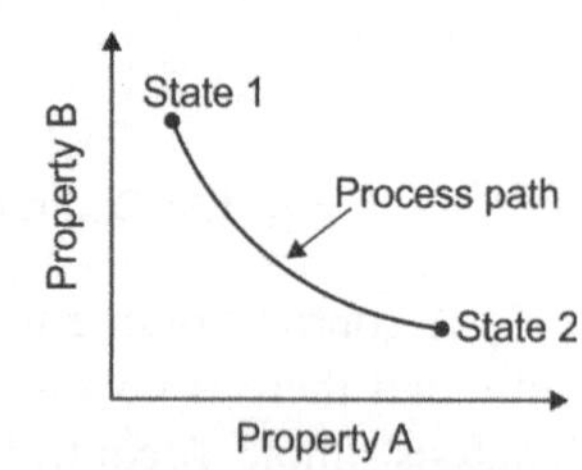

Fig. 1.5: A process between two states and its path

When a process progresses in such a manner that the system always remains infinitesimally close to an equilibrium state, it is referred as *quasi-static* process, where quasi means almost. Quasi-static process may be considered as a 'reversible process' and is characterized by process of infinite slowness (Fig. 1.6)

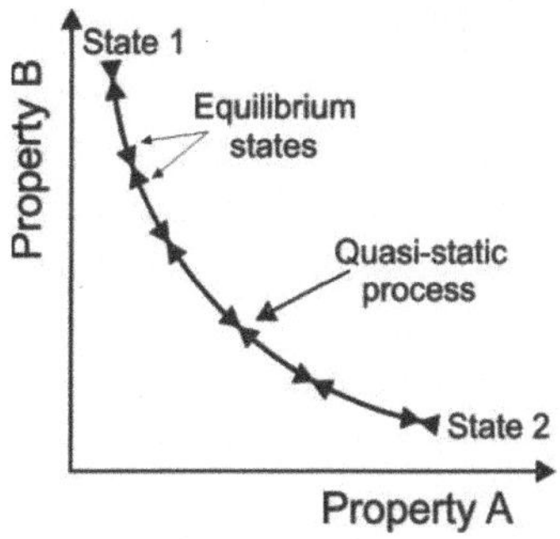

Fig. 1.6: Quasi-static process

When a system undergoing a process passes over a series of states in such a manner that its final state is identical with its initial state is called as thermodynamic *cycle* (Fig. 1.7). At the end of the cycle all the properties have the same values as they had at the beginning.

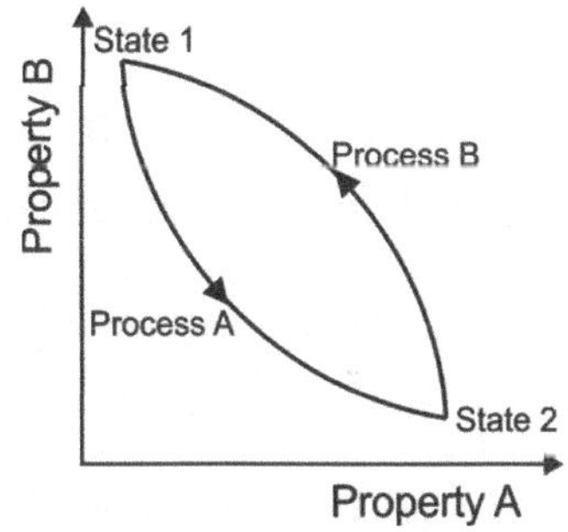

Fig. 1.7: Thermodynamic cycle

1.3.5 Point Function and Path Function

A *point function* is also called as *state function* is a function whose value depends on the initial and final conditions of a process. The properties can be located as a point on the graph and are independent of the path followed in a thermodynamic process. All properties of a system are point function hence, for a cyclic process the change in property is zero. Some examples of point function are internal energy, entropy, pressure, temperature, density, volume etc. The point functions are exact differentials.

$$\int_1^2 dV = V_2 - V_1 \tag{1.1}$$

The value of a path *function* depends on the path followed; they cannot be located on a graph by a point. Since, the process is a function of path therefore they are computed as area on that graph. The examples of path function are heat and work and their change cannot be represented on the basis of their end states therefore they are inexact differentials.

$$\int_1^2 \delta Q \neq Q_2 - Q_1 \text{ rather it is represented as } {}_1Q_2 \text{ or } Q_{1-2} \tag{1.2}$$

Also, $$\int_1^2 \delta W \neq W_2 - W_1 \text{ and it is represented as } {}_1W_2 \text{ or } W_{1-2} \tag{1.3}$$

1.3.6 Thermodynamic Equilibrium

The thermodynamic analysis is carried out at equilibrium states where equilibrium refers to a balanced state. Equilibrium can be of many types, such as thermal equilibrium, mechanical equilibrium, phase equilibrium and chemical equilibrium. A system cannot be in thermodynamic equilibrium unless all the relevant conditions pertaining to the given equilibrium are satisfied.

For example, when we talk of heat flow, its driving force is the temperature differential. Now, if the temperature of the system is same throughout, there will not be any temperature differential and the system is said to be in *thermal equilibrium*. Similarly, when we talk of mechanical work, the driving force is pressure differential.

Therefore, a system is in *mechanical equilibrium* if there is no pressure variation at any point of system with respect to time. A system is said to be in *phase equilibrium* if the mass of each phase reaches an equilibrium level. Finally, when the chemical composition of a given system does not alter with time and the system becomes inert to chemical reaction then the system is said to be in *chemical equilibrium*. Therefore, if a system maintains thermal, mechanical phase and chemical equilibriums then the system is said to be in *thermodynamic equilibrium*.

1.3.7 Reversible and Irreversible Processes

A *reversible process* is a process in which the system and surroundings can be brought back to its initial state without changing the thermodynamic properties of the universe (system + surroundings) when the process is reversed. In an *irreversible process* the system and surroundings cannot be restored to its initial state when the process is reversed. No process is truly reversible but it may approach to reversibility to a close approximation. Quasi-static processes are reversible processes, some examples of reversible processes are compression and expansion of spring, electrolysis, frictionless movement etc.

The irreversible processes may be identified into two forms namely *internal irreversibility* which is due to the dissipating effect within the working fluid example, viscosity and inertia of gas. The *external irreversibility* occurs due to the dissipating effect outside the working fluid example, friction due to some external source.

1.4 GAS LAWS

Gas is a substance which cannot be liquified isothermally with the application of pressure while vapour is a substance which can be liquified with the application of pressure isothermally. There are certain gas laws which are important with reference to thermodynamics.

1. **Boyle's law:** It states that if a gas is heated isothermally, then the product of pressure and volume always remains constant.

   PV = constant (at constant temperature)

   or, $P_1V_1 = P_2V_2$ 1.4

2. **Charle's law:** This law states that when heating a gas at constant pressure the volume of gas is directly proportional to the temperature or the ratio of volume to time is constant.

   $$\frac{V}{T} = \text{constant} \quad \text{(at constant pressure)}$$

   $$\text{or, } \frac{V_1}{T_1} = \frac{V_2}{T_2} \qquad 1.5$$

The gas which strictly follows the Boyle"s and Charle's law are called as perfect (ideal) gas and this law does not apply to vapours. Both the laws become ineffective at higher temperatures and pressures.

3. **Avagadro's law:** This law states that if different perfect gases of equal volume are put under same conditions of pressure and temperature, they all will contain equal number of molecules. It can be mathematically expressed as,

$$\frac{V}{n} = \text{constant (at constant pressure and temperature)}$$

$$\text{or, } \frac{V_1}{n_1} = \frac{V_2}{n_2} \qquad 1.6$$

where, V is the volume of the perfect gas and n is the number of molecules in the gas.

4. **Ideal gas law:** The combination of above three laws will yield the *ideal gas law.*

From Boyle's law, $V \,\alpha\, \frac{1}{P}$ 1.7

From Charle's law, $V \,\alpha\, T$ 1.8

From Avagadro's law, $V \,\alpha\, n$

Combining the three laws, $V \,\alpha\, \frac{nT}{P}$ 1.9

$$PV = nRT \qquad 1.10$$

Where R is the constant of proportionality called as the *universal gas constant* and its value is 8.314J/mol-K and n is the number of moles

5. **Gay Lussac's law:** According to this law the pressure of a gas is directly proportional to the temperature if the gas is held at constant volume.

$$\frac{p}{T} = \text{constant (at constant volume)}$$

$$\text{or, } \frac{p_1}{T_1} = \frac{p_2}{T_2} \qquad 1.11$$

6. **Joule's Law:** The Joule's law states that for a perfect gas the internal energy of a system is a function of temperature only.

i.e., $u = f(T)$ 1.12

It is evident that for a non-flow process the change in internal energy of a system is equal to the difference of net heat interaction and the work interaction

$$du = dQ - dW \qquad 1.13$$

Let us consider 1 kg of perfect gas is heated at constant volume, therefore the work interaction of the system will be zero *i.e.,* $dW = 0$, therefore Eq. 1.13 will become

$$dQ = du \qquad 1.14$$

$$dQ = du = c_v dT$$

on integration $\int du = \int c_v dT$ 1.15

$$u = c_v T + C \qquad 1.16$$

As we see from Eq. 1.12 that for a perfect gas internal energy is the function of temperature only implying a linear relation between u and T.

For a perfect gas it can be assumed that $u = 0$ when $T = 0$, which gives the value of $C = 0$. So, the internal energy,

$$u = c_v T \qquad 1.17$$

For a perfect gas of mass m the internal energy can be written as,

$$U = mc_v T \qquad 1.18$$

For any process the gain in internal energy between the two states 1 and 2 is given as,

$$U_2 - U_1 = mc_v(T_2 - T_1) \qquad 1.19$$

1.5 ZEROTH LAW OF THERMODYNAMICS

The *Zeroth law of thermodynamics* establishes the foundation of temperature measurement which states that 'if two thermodynamic systems or bodies are in thermal equilibrium with a third body, they are also in thermal equilibrium with each other'. For example, if a body A is in thermal equilibrium with another body B, and also separately with a third body C, then A, B and C will be in thermal equilibrium with each other (Fig. 1.8). The temperature differential is the basic driving force for heat to flow and also internal energy is a function of temperature.

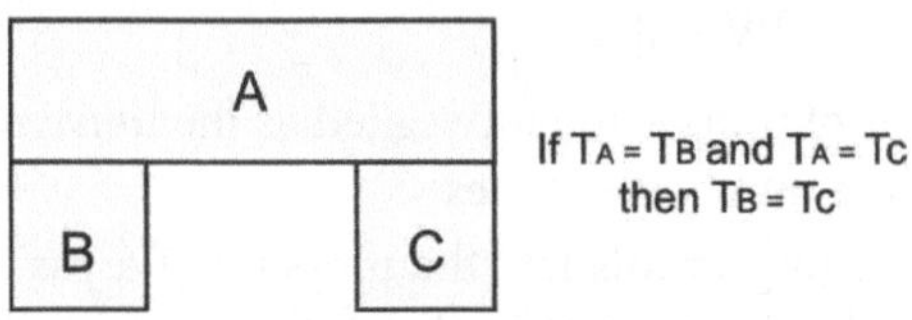

Fig. 1.8: Zeroth law of thermodynamics

Scan to know more about temperature scale

1.6 WORK AND HEAT

Energy can be transferred from one system to another either in the form of work or heat. Both have the same unit (Joule) and both carries equal importance in thermodynamic analysis. Work is related with the transfer of mechanical energy and heat is related to the transfer of thermal energy between the two systems. Heat is energy flow due to the random intermolecular collisions of the particles within a body while work is the energy transfer associated with an orderly motion in one direction. Both heat and work are mutually convertible i.e., heat can be transformed to work and vice-versa, but complete conversion of heat into work is not possible due to the violation of second law.

On the contrary, work can be completely transformed into heat therefore, work is considered as *high grade energy'* and heat as *'low grade energy'*. Work is connected with mechanical equilibrium while heat is related thermal equilibrium. Mechanical equilibrium is said to be achieved when there is no pressure differential in the system while in case of thermal equilibrium there is no temperature differential between the system and the surrounding.

For a non-flow system such as piston cylinder arrangement the reversible work done is given by,

$$W_{1-2} = \int_1^2 \delta W = \int_1^2 pdv = p(v_2 - v_1) \qquad 1.20$$

The solution of the given equation is possible when the path of the process is known form the p-V diagram where the area under the p-V curve gives the work done. The work done by the system is taken as positive while the work done on the system is taken as negative.

1.7 INTERNAL ENERGY

The energy accumulated into the system after the heat and work interaction is called as the *internal energy (U)* of the system. In other words, it is the sum of all forms of energy within the system like internal kinetic energy, internal potential energy, energy due to magnetism and chemical energy. The total energy *E* of the system is thus, the sum of the energies described below.

Scan to know more about energy

$$E = PE + KE + U + (magnetism,\ chemical\ energy,\ etc.) \qquad 1.21$$

For the sake of simplicity only PE, KE and U are considered in engineering thermodynamics.

$$E = PE + KE + U$$

$$= mgz + \frac{1}{2}mV^2 + U \qquad 1.22$$

On per unit mass basis

$$e = gz + \frac{1}{2}V^2 + U \qquad 1.23$$

For a stationary system without gravity (*PE=0* and *KE=0*)

Therefore,

$$E = U \text{ or } e = u \qquad 1.24$$

1.8 ENTHALPY

Enthalpy simply represents the total heat content in a system. When we supply heat to a system with a fixed amount of gas it will increase the internal energy of the gas which will lead to expansion of the system. Therefore, the system will have the energy interaction in the form of heat plus the work done for the expansion of system. Since, work done is the product of pressure and volume the enthalpy can be mathematically expressed as,

$$H = U + PV \qquad 1.25$$

Between the states 1 and 2, the change in enthalpy is given as,

$$\Delta H = H_2 - H_1 = \Delta U + \Delta(PV) \qquad 1.26$$

$$= (U_2 - P_2V_2) - (U_1 - P_1V_1) \qquad 1.27$$

1.9 SPECIFIC HEATS

Different substances require different amount of heat energy to raise its temperature. For example, if we take 1 kg of copper and 1 kg of water and heat them both, so that their temperature is raised by 1 °C. We observe that the amount of heat required for water is nearly 9 time than that required for copper. One such property that shows the energy storage capacity of different substances is the specific heat.

Specific heat is defined as the amount of heat required to raise the temperature of a substance of unit mass by one degree temperature. There are two types of specific heats which plays a significant role in thermodynamic analysis: *specific heat at constant pressure* c_p and *specific heat at constant volume* c_v.

Solids and liquids have only one value of specific heats and it is interesting to note that c_p is always greater then c_v because at constant pressure the system is allowed to expand and extra energy is required for this expansion work. The heat flow for reversible non-flow process at constant pressure is given by;

$$dQ = mc_p dT$$

Therefore, for a process occurring between two states 1 and 2 at constant pressure the expression for heat flow is given as,

$$Q_2 - Q_1 = mc_p(T_2 - T_1) \quad 1.28$$

Similarly, for processes occurring at constant volume the expression for heat flow is as follows,

$$Q_2 - Q_1 = mc_v(T_2 - T_1) \quad 1.29$$

1.9.1 Relationship between the two specific heats

Let us consider perfect gas which is heated at constant pressure from a temperature T_1 to T_2.

For a non-flow process the heat flow is equal to the sum of energy stored within the system and the work interaction from state 1 to 2 which is given as,

$$Q = (U_2 - U_1) + W \quad 1.30$$

Also, $U_2 - U_1 = mc_v(T_2 - T_1)$ $\qquad Q = mc_v(T_2 - T_1) + W$ 1.31

The work done in a constant pressure process is given by,

$$W = p(V_2 - V_1) = mR(T_2 - T_1) \qquad \because pV = mRT \text{ and } pv = RT$$

On substituting the above values in Eq. 1.30,

$$Q = mc_v(T_2 - T_1) + mR(T_2 - T_1)$$

$$Q = m(c_v + R)(T_2 - T_1) \quad 1.32$$

But for a constant pressure process $Q = mc_p(T_2 - T_1)$ 1.33

Equating Eq. 1.32 and Eq. 1.33

$$m(c_v + R)(T_2 - T_1) = mc_p(T_2 - T_1)$$

Therefore; $(c_v + R) = c_p$ or $c_p - c_v = R$ 1.34

On dividing both sides by c_v, we get

$$\frac{c_p}{c_v} - 1 = \frac{R}{c_v} \quad 1.35$$

$$\therefore \; c_v = \frac{R}{\gamma - 1} \quad 1.36$$

Where $\gamma = c_p/c_v$ is called the *adiabatic index* and its value for air is taken as 1.4. Also, when we divide Eq. 1.34 by c_p we get,

$$c_p = \frac{\gamma R}{\gamma - 1} \quad 1.37$$

1.10 THE FIRST LAW OF THERMODYNAMICS

The first law of thermodynamics is the law of conservation of energy which states that "energy can neither be created nor be destroyed but can be transformed from one form to another". We often encounter a number of situations in our daily life witnessing the effect of work and heat.

For example, when we use a bicycle pump to inflate the tyre, we observe that the pump becomes hot due to friction. Similarly, if we supply heat to a turbine or an IC Engine, we get mechanical work as an output. From the above examples it can be suggested the heat and work are mutually convertible and that there must be some relationship between heat and work.

According to first law of thermodynamics "when a system undergoes a thermodynamic cycle then the cyclic integral of heat is proportional to the cyclic integral of work". Symbolically it can be represented as:

$$\oint \delta Q = \oint \delta W \qquad 1.38$$

Where $\oint$ represents the sum for a complete cycle and $\oint \delta Q$ and $\oint \delta W$ have the same units.

The operator 'δ' is used for *inexact differential* and is applied to the quantities which are path dependant. Since, heat and work are path dependant therefore, they are inexact differential and are not the *property* of a system. *Exact differentials* use the operator '*d*' and is used for point functions i.e., they depend only on initial and final states and therefore, they are the *property* of a system. The first law of thermodynamics is also called as law of nature and cannot be proven. However, the experimental evidence has repeatedly confirmed its validity.

1.10.1 Perpetual Motion Machine of First Kind (PMM-1)

The machines that violate the first law of thermodynamics are called as *perpetual motion machine of* first kind (PMM-1). The PMM-1 are impossible because they are the hypothetical devices that can either produce continuous work without the supply of any other form of energy or continuously emits heat without any other form of energy supplied to the device as shown in Fig. 1.9 (a) and (b), which is not feasible.

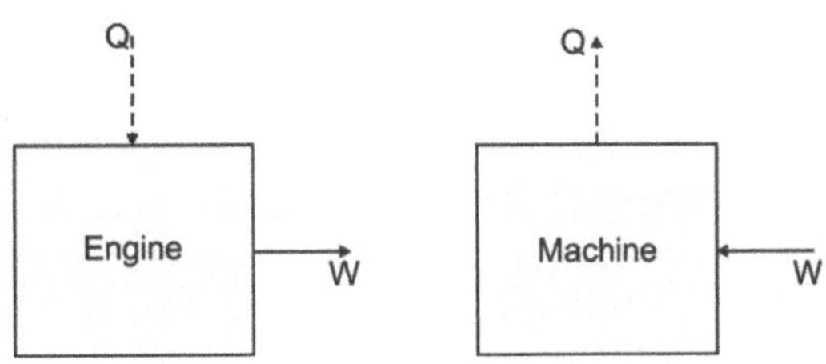

Fig. 1.9: PMM-1

1.10.2 Application of First Law to a Non-Flow System or Closed System

1. Constant pressure (isobaric) process (p = constant)

To achieve a constant pressure process the boundary must be movable against the external resistance so that the additional pressure rise by the application of heat is compensated by the moving boundary. The piston cylinder arrangement is shown in Fig. 1.10 (a) allows the piston to move freely inside the cylinder when heat is supplied to it; the piston moves away due to the force exerted by the heated gas, it is said that the work is done by the gas on the surroundings. The p-V plot of the initial and final states achieved by the piston is shown in Fig. 1.10 (b).

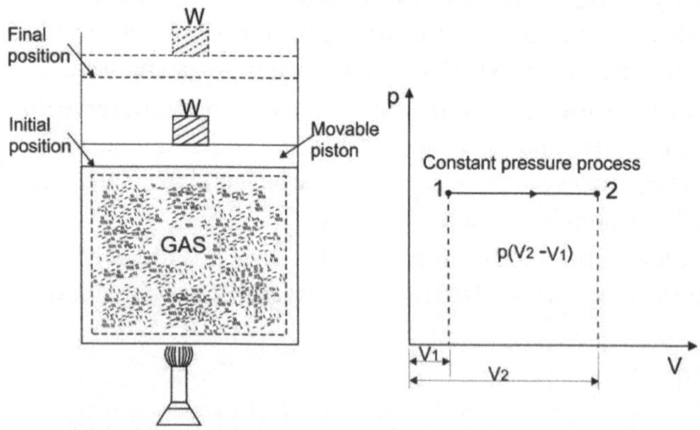

Fig. 1.10: Constant pressure process and its p-V plot

Applying first law to the given process for a unit mass of gas

$$(u_2 - u_1) = Q - W \quad 1.39$$

The work done under the given condition is given as,

$$W = \int_1^2 pdv = p(v_2 - v_1) \quad 1.40$$

Therefore,

$$(u_2 - u_1) = Q - p(v_2 - v_1) \quad 1.41$$

or, $$Q = u_2 - u_1 + pv_2 - pv_1 = (u_2 + pv_2) - (u_1 + pv_1) \quad 1.42$$

we know that specific enthalpy $h = u + pv$, therefore

$$Q = h_2 - h_1 = c_p(T_2 - T_1) \qquad [\because \Delta h = c_p \Delta T] \quad 1.43$$

Where, c_p is the specific heat at constant pressure

2. Constant volume (isochoric) process (V = constant)

When the piston is constrained in the cylinder by putting stopper at the inside wall of the cylinder such that the boundaries of the system are immovable then it is considered as constant volume process (Fig. 1.11 a). Under such condition no work is done on the system or by the system. The processes with zero work are often implied as constant volume processes. When heat is added to such a system, then only pressure rises from initial state 1 to final state 2 as depicted in the p-V plot in Fig. 1.11 b.

Applying first law to the working substance of unit mass

$$Q = (u_2 - u_1) + W \quad 1.44$$

$W = 0$, (change in volume $dV = 0$)

$$\therefore \; Q = u_2 - u_1 = c_v(T_1 - T_2) \quad 1.45$$

If the system has working fluid of mass *m*, then heat added will be

$$Q = U_2 - U_1 = mc_v(T_1 - T_2) \tag{1.46}$$

where, $mu = U$

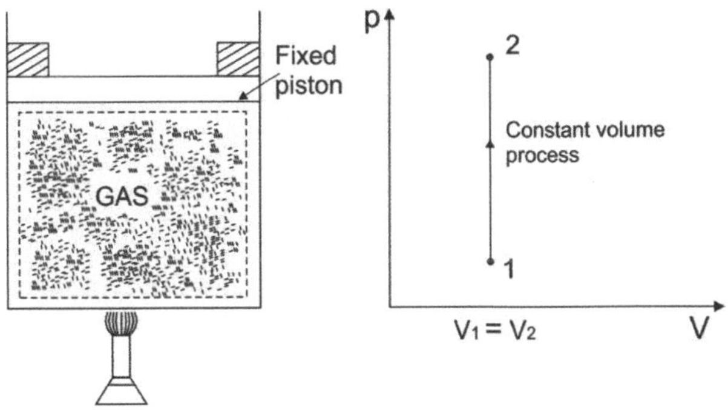

Fig. 1.11: Constant volume process and its p-V plot

3. Constant temperature (isothermal) process (T = constant)

When a process is executed in such a way that its temperature remains the same at each state of the process then it is called as constant temperature or isothermal process. When the working fluid is allowed to expand in a piston-cylinder arrangement form high pressure to low pressure then its temperature will eventually decrease therefore a heat source is employed to supply heat in order to maintain a constant temperature (Fig. 1.12 a). Similarly, when the working fluid is compressed isothermally then its temperature will increase therefore heat must be removed continuously from the working fluid to maintain a constant temperature. The p-V plot of isothermal expansion between states 1 and 2 is shown in Fig. 1.12 b

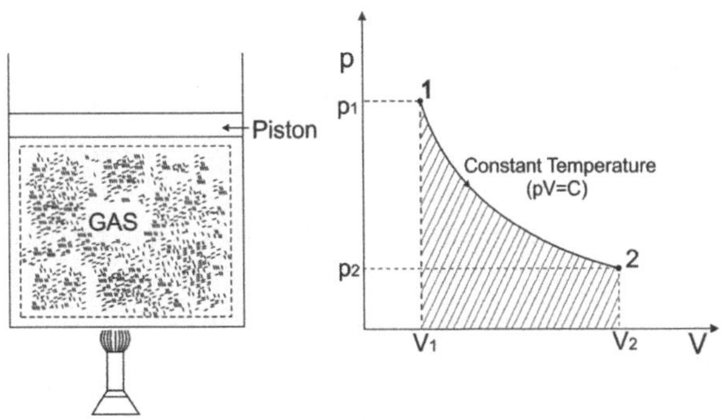

Fig. 1.12: Constant temperature process and its p-V plot

Writing the energy balance equation for the above system for a unit mass of working fluid (first law)

$$Q = (u_2 - u_1) + W$$

$$= c_v(T_1 - T_2) + W$$

$$= 0 + W = 0 + \int_1^2 pdv \qquad [\because T_1 = T_2]$$

Referring to the ideal gas equation $pv = mRT$ for the isothermal process we can write

$$pv = C \text{ (Where; } C = \text{constant) or } p = \frac{C}{v}$$

$$\therefore \ W = \int_{v_1}^{v_2} C\frac{dv}{v} = C\ln\frac{v_2}{v_1} \quad 1.46$$

For a unit mass of working fluid $p_1v_1 = p_2v_2 = C$ which can be replaced in Eq. 1.46

$$Q = W = p_1v_1 \ln\frac{v_2}{v_1} = p_2v_2 \ln\frac{v_2}{v_1} \quad 1.47$$

If mass m of the working fluid is considered then $V = mv$ and the expression of heat flow is equal to

$$Q = p_1V_1 \ln\frac{V_2}{V_1} \quad 1.48$$

or $$Q = p_1V_1 \ln\frac{p_2}{p_1} \quad \left[\because \frac{V_2}{V_1} = \frac{p_2}{p_1}\right] \quad 1.49$$

4. Reversible adiabatic (isentropic) process (pv^γ = constant)

In an adiabatic process there is no heat interaction to and from the system undergoing a process (i.e., $Q = 0$). Applying first law to such system,

$$Q = (u_2 - u_1) + W \quad 1.50$$

$$0 = (u_2 - u_1) + W$$

or $$W = (u_1 - u_2) \quad 1.51$$

Eq. 1.51 is equally valid for both reversible and irreversible process. The internal energy of the fluid reduces in an adiabatic expansion and increases in case of adiabatic compression. The system needs to be perfectly insulated for an adiabatic process to occur. Fig. 1.13 (a) and (b) shows the reversible adiabatic process and its p-V diagram respectively.

(i) Derivation of $pv^\gamma = C$

The governing equation for a reversible adiabatic process can be derived with the help of first law and the ideal gas law.

The energy equation can be put in the differential form as

$$dQ = du + dW \quad 1.52$$

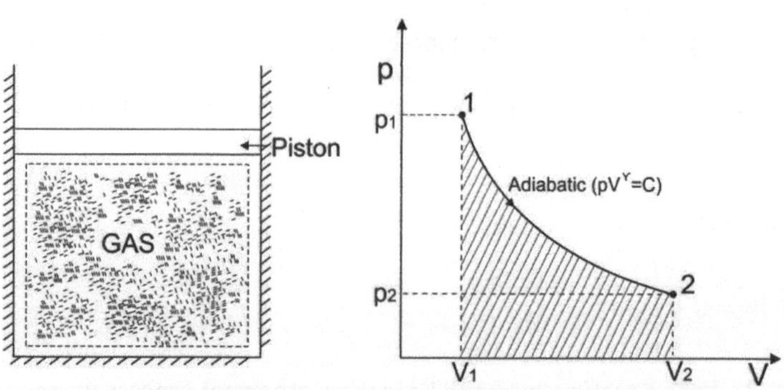

Fig. 1.13 Reversible adiabatic process and its p-V plot

We know that work done for a reversible process is

$$dW = pdv$$

Since, $dQ = 0$ Eq. 1.52 will be rewritten as

$$0 = du + pdv \tag{1.53}$$

From perfect gas relation; $p = \frac{RT}{v}$

Substituting the value in Eq. 1.53 will yield

$$du + \frac{RT}{v} dv = 0$$

Also, $du = c_v\, dT$

Therefore, $c_v dT + \frac{RT}{v} dv = 0$

Dividing both the sides by T

$$\frac{c_v dT}{T} + \frac{R}{v} dv = 0 \tag{1.54}$$

Integrating Eq. 1.54 will give

$$c_v \ln T + R \ln v = C$$

On substituting $T = pv/R$ and dividing by c_v *we get*

$$\ln \frac{pv}{R} + \frac{R}{c_v} \ln v = C \tag{1.55}$$

From Eq. 1.36, $c_v = \frac{R}{\gamma - 1}$ or $\frac{R}{c_v} = \gamma - 1$

Hence, Eq. 1.55 will become

$$\ln \frac{pv}{R} + (\gamma - 1) \ln v = C \tag{1.56}$$

or $\ln \frac{pv}{R} + \ln v^{\gamma-1} = C$

$$\ln \frac{pv \times v^{\gamma-1}}{R} = C \tag{1.57}$$

i.e., $\ln \frac{pv^{\gamma}}{R} = C$ 1.58

or, $\frac{pv^{\gamma}}{R} = e^C$ = Constant

or pv^{γ} = Constant 1.59

(ii) Work done W

The area under the *p-V* diagram (Fig. 1.13 b) indicates the work done in the process

i.e., $W = \int_{v_1}^{v_2} p dv$ 1.60

putting, $p = \frac{C}{v^\gamma}$ in Eq. 1.60

then; $W = \int_{v_2}^{v_1} C \frac{dv}{v^\gamma}$

$$= C\left[\frac{v_2^{-\gamma+1} - v_1^{-\gamma+1}}{1-\gamma}\right] = C\left[\frac{v_1^{-\gamma+1} - v_2^{-\gamma+1}}{\gamma-1}\right] \quad 1.61$$

Since $p_1v_1 = p_2v_2 = C$, Eq.1.61 will become

$$W = \left[\frac{p_1v_1^\gamma \times v_1^{-\gamma+1} - p_2v_2^\gamma \times v_2^{-\gamma+1}}{\gamma-1}\right] = \frac{p_1v_1 - p_2v_2}{\gamma-1} \quad 1.62$$

or $W = \frac{R(T_1 - T_2)}{\gamma-1}$ 1.63

(iii) Relationship between the ratios of p, v and T

The relation between *T* and *v*, and *T* and *p* can be obtained by establishing relation between the ideal gas law and $pv^\gamma = C$. Following are the relations which are important form the point of view of problem solving:

(i) $\frac{p_2}{p_1} = \left(\frac{v_1}{v_2}\right)^\gamma$ 1.64

(ii) $\frac{T_2}{T_1} = \left(\frac{v_1}{v_2}\right)^{\gamma-1}$ 1.65

(ii) $\frac{T_2}{T_1} = \left(\frac{p_2}{p_1}\right)^{\frac{\gamma-1}{\gamma}}$ 1.66

5. Reversible polytropic process (pv^n = constant)

The polytropic process applies to the reversible process on any open or closed system of gas or vapour which involves both heat and work interaction. Almost all practical processes obey the law pv^n = *constant* and such processes are internally reversible and *n* is called the *polytropic index.*

The work done for the process involving pv^n = *constant (C)*

$$W = \int p dv$$

Also, $p = \frac{C}{v^n}$

Adopting the same methodology as applied in finding work done in an isentropic process we will get

$$W = \left[\frac{p_1 v_1^{\,n} \times v_1^{\,-n+1} - p_2 v_2^{\,n} \times v_2^{\,-n+1}}{n-1}\right] = \frac{p_1 v_1 - p_2 v_2}{n-1} \qquad 1.67$$

$$W = \frac{R(T_1 - T_2)}{n-1} \qquad 1.68$$

(i) Heat transfer equation for a polytropic process

Applying the first law for non-flow process the heat transfer can be found,

i.e., $Q = (u_2 - u_1) + W$

$$= c_v(T_2 - T_1) + \frac{R(T_1 - T_2)}{n-1} \qquad 1.69$$

i.e., $$Q = \frac{R(T_1 - T_2)}{n-1} - c_v(T_1 - T_2) \qquad 1.70$$

We know that, $c_v = \dfrac{R}{\gamma - 1}$

$$\therefore Q = \frac{R(T_1 - T_2)}{n-1} - \frac{R}{\gamma - 1}(T_1 - T_2) \qquad 1.71$$

$$= R(T_1 - T_2)\left(\frac{1}{n-1} - \frac{1}{\gamma - 1}\right) = \frac{(\gamma - n)\,R(T_1 - T_2)}{(\gamma - 1)\;(n-1)}$$

$$\therefore\; Q = \left(\frac{\gamma - n}{\gamma - 1}\right) W \qquad 1.72$$

It can be observed from Eq. 1.72 that the polytropic index n depends only on heat and work. It is also interesting to note that the various processes discussed earlier are the special cases of polytropic process for a perfect gas. For example,

$n = 0$ for a constant pressure (isobaric) process

$n = \infty$ for constant volume (isochoric)n process

$n = 1$ for constant temperature (isothermal) process

$n = \gamma$ for reversible adiabatic (isentropic) process

The illustration of above processes on p-V diagram is shown in Fig. 1.14.

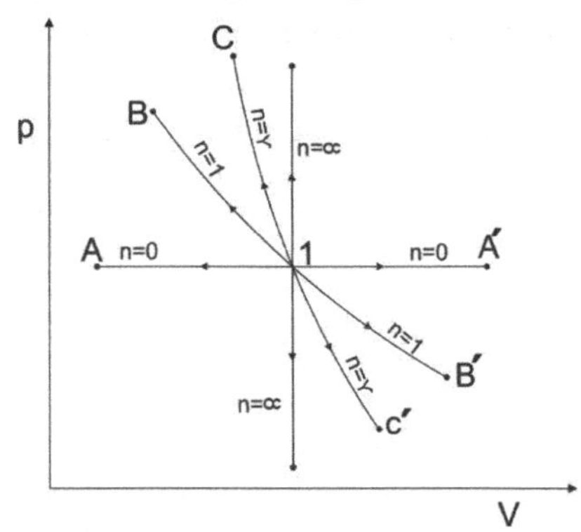

Fig. 1.14: Illustration of processes for different values of n

6. Joule expansion or free expansion

To explain free expansion let us consider two perfectly insulated vessels *A* and *B* connected with a small length pipe having a valve *P* (Fig. 1.15). Let the vessel *A* be initially filled with a gas at a certain pressure and vessel *B* be completely in vacuum. When the valve *P* is opened, the gas from vessel *A* rush to occupy space of vessel *B* without any external pressure. The gas thus expands freely from one vessel to another hence, it is called free expansion. This is a classic example of highly *irreversible* process due to continuous eddying of gas while passing through one vessel to another. The final pressure of the gas will be less than the initial pressure of the gas in vessel *A*.

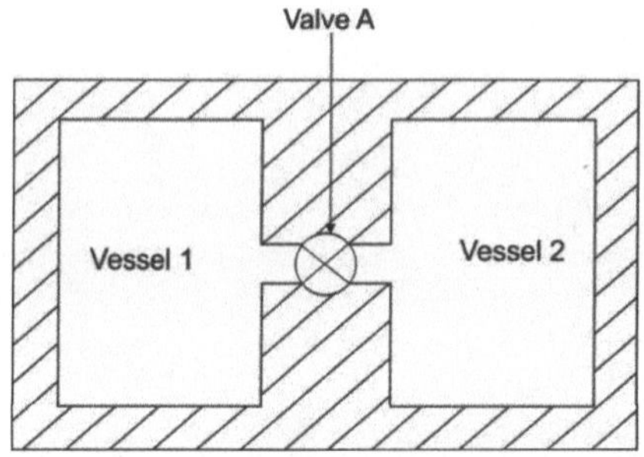

Fig. 1.15: Free expansion

Since, in this system the boundaries are fixed so work done on or by the gas is *zero* and the heat transfer is also *zero* because the system is perfectly insulated therefore no heat flows to or from the system. Under such condition applying the first law of thermodynamics between initial and final states.

$$Q = (u_2 - u_1) + W \tag{1.73}$$

or $$u_2 - u_1 = 0 \text{ or } u_2 = u_1 \tag{1.74}$$

Free expansion thus has same internal energy initially and finally i.e., the change in internal energy is zero and for a perfect gas we know that

$$u = c_v T$$

Therefore, for a free expansion $c_v T_1 = c_v T_2$

or $$T_1 = T_2$$

This implies that for a perfect gas going through free expansion the initial and final temperatures are equal.

Example 1.1: 1 litre of nitrogen gas at 0 °C is compressed to one fourth of the volume. Calculate the change in temperature of the gas, given the ratio of specific heat as 1.4.

Solution:

Given: V_1 = 1 litre; T_1 = 0 °C = 0+ 273 = 273 K; $V_2 = V_1/4$ = 0.25 litre

Let us consider T_2 as the final temperature of the gas.

$$\frac{T_2}{T_1} = \left(\frac{V_1}{V_2}\right)^{\gamma-1} = \left(\frac{1}{0.25}\right)^{1.4-1}$$

$$= (4)^{0.4} = 1.741$$

$$\frac{T_2}{273} = 1.741$$

$$T_2 = 273 \times 1.741 = 475.293 \text{ K}$$

$$T_2 = 475.293 - 273 = 202.293\ ^{o}C \quad \text{Ans}$$

Example 1.2: A closed vessel contains 1.5 kg of CO_2 at 20 °C and 1 bar pressure. Heat is supplied to the vessel till gas acquires a pressure of 2 bar. Find out;

(a) Final temperature of the gas

(b) Work done

(c) Heat supplied

(d) The change in internal energy

Consider specific heat of the gas at constant volume as 0.657 kJ/kg K.

Solution:

Given: m=2 kg; T_1= 20 °C = 273+ 20= 293 K; P_1 = 1 bar; P_2 = 2 bar

(a) Let us consider T_2 as the final temperature of the gas.

As thc gas is heated in a closed vessel, the volume of the gas will be constant. Therefore;

$$\frac{P_1}{T_1} = \frac{P_2}{T_2}$$

$$T_2 = \frac{P_2 \times T_1}{P_1} = \frac{2.0 \times 293}{1} = 586 \text{ K} \quad \text{Ans}$$

(b) Work done

As there is no change in volume (dv=0), therefore the work done is zero. Ans

(c) The heat is added at constant volume;

$$Q = m\, c_v (T_2 - T_1)$$

$$= 1.5 \times 0.657\,(586 - 293) = 288.75 \text{ kJ} \quad \text{Ans}$$

(d) Change in internal energy (dU)

$$Q_{1\text{-}2} = W_{1\text{-}2} + dU$$

$$Q_{1\text{-}2} = 0 + dU$$

$$dU = Q_{1\text{-}2} = 288.75 \text{ kJ} \quad \text{Ans}$$

Example 1.3: A gas at 10 bar pressure and 150 °C initial temperature having 0.336 m^3 volume is expanded adiabatically upto a pressure of 4 bar. It is then compressed isothermally to the original volume. Calculate the:

(a) Final temperature

(b) Final pressure

(c) The change in internal energy.

Assume, c_p = 0.996 kJ/kg K and c_v = 0.703 kJ/kg K;

Solution:

Given: V_1 = 0.336 m^3; P_1 = 10 bar = 10 x 10^5 N/m^2; T_1 = 150 °C = 273 + 150= 423 K; P_2= 4 bar = 4x10^5 N/m^2; V_3 = V_1 = 0.336 m^3;

Consider process 1-2 as adiabatic expansion of gas and
Process 2-3 as isothermal compression of gas to its original volume ($V_3 = V_1$) as shown in Fig. 1.16

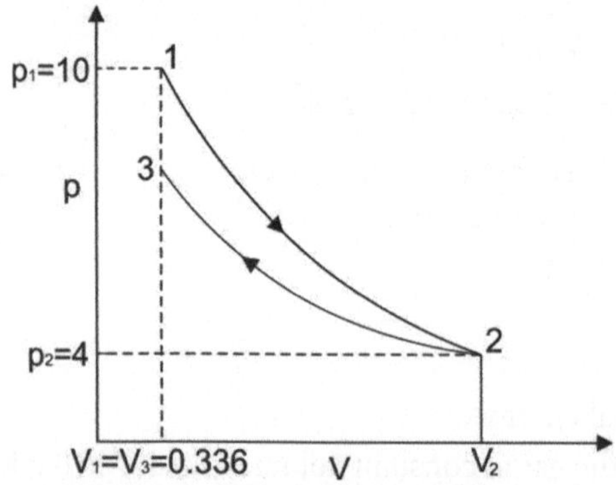

Fig. 1.16: p-V curve for example1.3

(a) Final temperature of the gas
For adiabatic process, let T_2 is the temperature after adiabatic expansion

$$\frac{T_1}{T_2} = \left[\frac{P_1}{P_2}\right]^{\frac{\gamma-1}{\gamma}} = \left[\frac{10}{4}\right]^{\frac{1.417-1}{1.417}} \qquad \left[\text{As } \gamma = \frac{c_p}{c_v} = \frac{0.996}{0.703} = 1.417\right]$$

$$= (2.5)^{0.294} = 1.31$$

$$T_2 = \frac{T_1}{1.31} = \frac{423}{1.31} = 323 \text{ K}$$

During process 2-3, Compassion is isothermal
Therefore, the final temperature of the gas (for isothermal compression)

$$T_3 = T_2 = 323 \text{ K} = 50\ ^\circ\text{C} \quad \text{Ans}$$

(b) Final pressure of the gas
P_3 is the final pressure of the gas after isothermal compression (process 2-3)
It is also mentioned that, $V_3 = V_1$ (process 3-1)
Therefore,

$$\frac{P_3}{T_3} = \frac{P_1}{T_1}$$

$$P_3 = \frac{P_1 \times T_3}{T_1} = \frac{10 \times 10^5 \times 323}{423} = 0.76 \times 10^6 \text{ N/m}^2$$

$$= 7.6 \text{ bar} \qquad \text{Ans}$$

(c) Change in internal energy
As we know that gas constant

$$R = c_p - c_v = 0.996 - 0.703$$

$$= 0.293 \text{ kJ/kg K}$$

$$= 293 \text{ J/kg K}$$

Also, $P_1V_1 = m R T_1$

$$m = \frac{P_1 V}{R T_1} = \frac{10 \times 10^5 \times 0.336}{293 \times 423}$$

$= 2.7$ kg

Now; $dU = U_3 - U_1 = m c_v (T_3 - T_1)$

$$= 2.7 \times 0.703 (323 - 423)$$

$$= -189.8 \text{ kJ} \quad \text{Ans}$$

The negative sign shows the decrease in internal energy

Example 1.4: An ideal gas is compressed in a reversible manner from 690 kPa and 277 K to a final specific volume of 0.47 m³/kg according to the relation; p= 561 +200 v+100 v², where p is the pressure in kPa, v is the specific volume in m³/kg. The specific heat at constant volume is 0.887 kJ/kgK and R=1.279 kJ/kgK. Determine;

(a) The initial specific volume
(b) The work
(c) The final temperature
(d) The heat

Solution:

Given: $P_1 = 690$ kPa; $T_1 = 277$ K; $c_v = 0.887$ kJ/kg K; $v_1 = 0.47$ m³/kg

(a) The initial specific volume

$$P_1V_1 = mRT_1$$

$$P_1v_1 = RT_1$$

$$v_1 = \frac{RT_1}{P_1} = \frac{1.279 \times 277}{690_1} = 0.5135 m^3 / kg$$

(b) Work done

$$W_{1-2} = \int_1^2 p dv = \int [561 + 200v + 100v^2] dv$$

$$= \left[561v + \frac{200v^2}{2} + \frac{100v^3}{3} \right]_{v_1=0.5135}^{v_2=0.47} = -29.73 kJ / kg$$

(c) The final temperature

$$P_2 = 561 + 200v + 100v^2$$

$$= 561 + 200(0.47) + 100(0.47)^2 = 677.1 \quad kPa$$

$$P_2v_2 = RT_2$$

$$T_2 = \frac{P_2v_2}{R} = \frac{677.1 \times 0.47}{1.279} = 248.8K$$

(d) Heat

$$\Delta u = c_v(T_2 - T_1) = 0.837(248.8 - 277) = -23.6kJ/kg$$

$$q_{1-2} = w_{1-2} + \Delta u = -29.73 - 23.6 = -53.33kJ/kg$$

Example 1.5: A gas of mass 2.57 kg with ratio of specific heat as 1.35 occupies a volume of 0.15 m^3 at 10 bar and 500 K. The gas undergoes an expansion of 0.6 m^3. Determine the heat transfer by the gas for each of the following methods of expansion.

(a) Constant pressure
(b) According to law $PV^{1.32}=C$
(c) Adiabatic

Solution:

Given: m=2.57 kg; γ =1.35; V_1=0.15 m^3; V_2=0.6 m^3; P_1= 10 bar; T_1=500 K

$$P_1V_1 = mRT_1$$

$$10 \times 100 \times 0.15 = 2.57 \times R \times 500$$

$$R = \frac{10 \times 100 \times 0.15}{2.57 \times 500} = 0.1167 \quad kJ/kgK$$

(a) Constant pressure (P=C)

$$\frac{V_1}{T_1} = \frac{V_2}{T_2}$$

$$T_2 = \frac{V_2 \times T_1}{V_1} = \frac{500 \times 0.6}{0.15} = 2000 \quad K$$

$$W_{1-2} = P(V_2 - V_1)$$

$$= 10 \times 100(0.6 - 0.15) = 450 \quad kJ$$

$$\Delta U = mc_v(T_2 - T_1)$$

$$= 2.57 \times 0.33(2000 - 500) = 1283.7 \quad kJ$$

$$Q_{1-2} = W_{1-2} + \Delta U$$

$$= 450 + 1283.7 = 1733.7 \quad kJ$$

(b) According to law $PV^{1.32}=C$

$$P_1V_1^n = P_2V_2^n$$

$$P_2 = P_1\left(\frac{V_1}{V_2}\right)^n$$

$$= 10\left(\frac{0.15}{0.6}\right)^{1.32} = 1.605bar$$

$$T_2 = T_1\left(\frac{V_1}{V_2}\right)^{n-1}$$

$$= 500\left(\frac{0.15}{0.6}\right)^{1.32-1} = 320.85kJ$$

$$W_{1-2} = \frac{P_1V_1 - P_2V_2}{n-1}$$

$$= \frac{100(10 \times 0.15 - 1.604 \times 0.6)}{1.32 - 1} = 168 \quad kJ$$

$$Q_{1-2} = W_{1-2} + \Delta U$$

$$= 168 - 153.3 = 14.7 \quad kJ$$

(c) According to Adiabatic process ($PV^{\gamma}=C$)

$$P_1V_1^{\gamma} = P_2V_2^{\gamma}$$

$$P_2 = P_1\left(\frac{V_1}{V_2}\right)^{\gamma}$$

$$= 10\left(\frac{0.15}{0.6}\right)^{1.35} = 1.5389 \quad bar$$

$$T_2 = T_1\left(\frac{V_1}{V_2}\right)^{\gamma-1}$$

$$= 500\left(\frac{0.15}{0.6}\right)^{1.35-1} = 307.78 \quad kJ$$

$$W_{1-2} = \frac{P_1V_1 - P_2V_2}{n-1}$$

$$= \frac{100(10 \times 0.15 - 1.5389 \times 0.6)}{1.35 - 1} = 164.75 \quad kJ$$

$$\Delta U = mc_v(T_2 - T_1)$$

$$= 2.57 \times 0.333(307.78 - 500) = -164.75 \quad kJ$$

$$Q_{1-2} = W_{1-2} + \Delta U$$

$$= 164.75 - 164.75 = 0$$

Example 1.6: A perfect gas is compressed reversibly from 1 bar, 17 °C to a pressure of 5 bar in a perfectly thermally insulated cylinder, the final temperature being 77 °C. The work done on the gas during the compression is 45 kJ/kg. Calculate **γ**, c_v, R and the molecular mass of the gas.

Solution:

Given: T_1= 17+273= 290 K; P_1= 1 bar; P_2= 5 bar; T_2= 77+273= 350 K; W=45 kJ/kg

Perfectly thermally insulated cylinder has no heat transfer, hence the process is adiabatic.

$$\frac{T_2}{T_1} = \left(\frac{P_2}{P_1}\right)^{(\gamma-1)/\gamma}$$

Taking log on both sides

$$\log\frac{T_2}{T_1} = \left(\frac{\gamma-1}{\gamma}\right)\log\frac{P_2}{P_1}$$

$$\log \frac{350}{290} = \left(\frac{\gamma-1}{\gamma}\right) \log\frac{5}{1}$$

$$\left(\frac{\gamma-1}{\gamma}\right) = \frac{\log\frac{350}{290}}{\log 5} = 0.1168$$

$$\gamma = \frac{1}{(1 - 0.1168)} = 1.132 \qquad \text{Ans}$$

$$Q_{1-2} = W_{1-2} + \Delta U$$

$$0 = -45 + c_v (T_2 - T_1)$$

$$45 = c_v (350 - 290)$$

$$c_v = \frac{45}{60} = 0.75 \text{ kJ/kgK} \qquad \text{Ans}$$

$$c_v = \frac{R}{\gamma-1}$$

$$R = c_v(\gamma - 1) = 0.75(1.132 - 1)$$

$$= 0.088 \text{ kJ/kgK} \qquad \text{Ans}$$

$$R_u = MR$$

$$M = \frac{R_u}{R} = \frac{8.3143}{0.099} = 84 \qquad \text{Ans}$$

1.11 FLOW PROCESSES

A flow process is the process which permits mass transfer to and from the system, it applies for open system in which fluid enters the system and leaves after doing useful work. A flow process may be of two types namely:

(i) Steady flow process
(ii) Unsteady flow process

1.11.1 Steady flow process

In a steady flow process the rate at which the fluid flows through the control volume remains constant. There is no accumulation of mass or energy within the control volume. The following assumptions are made for the analysis of steady flow systems:

(i) Uniform mass flow rate through the system
(ii) Uniform composition of fluid
(iii) The state of a fluid at any given time doesn't change over time.
(iv) Only work and heat interact with the surroundings through the system.

The schematic of flow process for an open system (steady flow system) is shown in Fig. 1.17. The fluid enters the system at section 1-1 and leaves at section 2-2.

The pressure, specific volume, cross-section area, mass flow rate and energy and velocity at section 1-1 are p_1, v_1, A_1, m_1, e_1, C_1 and at 2-2 are p_2, v_2, A_2, m_2, e_2, C_2 respectively. The stored energy of a system is simply the difference between heat and work interactions.

i.e., $$E = Q - W \tag{1.75}$$

Further, this stored energy is comprised of three major components of energy as internal energy (U), kinetic energy (KE = (1/2). mC²) and potential energy (PE = mgz).

The other components of energy are neglected for simplicity.

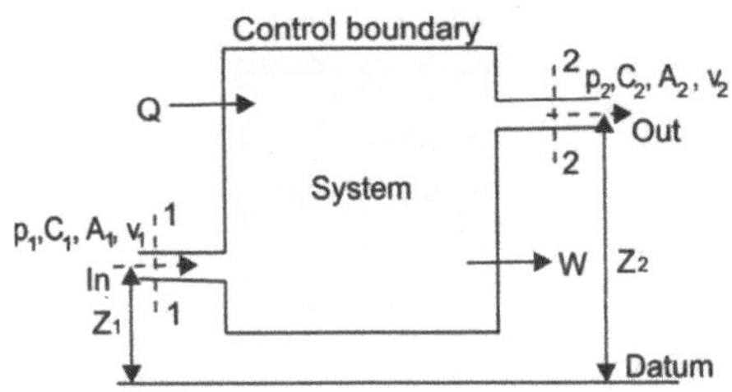

Fig. 1.17: Steady flow system

Therefore, the stored energy can be written as,

$$E = U + KE + PE \tag{1.76}$$

or, $$E = U + \frac{1}{2}mC^2 + mgz \tag{1.77}$$

on per unit mass basis;

$$e = u + \frac{C^2}{2} + gz \tag{1.78}$$

Let the heat Q be supplied to the steady flow system and the work W is being done by the system. The energy balance can be applied to the two sections which is given as,

(Heat supplied to the system + Stored energy of the fluid at inlet) = Stored energy of the fluid at outlet

The energy of the fluid at inlet and outlet are composed of stored energy and the flow energy which is given as,

at the inlet, $m_1(e_1 + p_1v_1)$

at the outlet, $m_2(e_2 + p_2v_2)$

Therefore, the energy balance can be mathematically written as,

$$Q + m_2(e_1 + p_1v_1) = m_2(e_2 + p_2v_2) + W \tag{1.79}$$

Where; $e_1 = u_1 + \frac{C_1^2}{2} + gz_1$ and $e_2 = u_2 + \frac{C_2^2}{2} + gz_2$

The quantity z_1 and z_2 are the datum levels at inlet and outlet respectively.

Substituting the values of e_1 and e_2

$$Q + m_1\left(u_1 + \frac{C_1^2}{2} + gz_1 + p_1v_1\right) = m_2\left(u_2 + \frac{C_2^2}{2} + gz_2 + p_2v_2\right) + W \tag{1.80}$$

The enthalpy at inlet and outlet can be given as,

$$h_1 = u_1 + p_1 v_1$$

$$h_2 = u_2 + p_2 v_2$$

Therefore Eq. 1.80 can be written as

$$Q + m_1\left(h_1 + \frac{C_1^2}{2} + gz_1\right) = m_2\left(h_2 + \frac{C_2^2}{2} + gz_2\right) + W \qquad 1.81$$

The above equation is known as *steady flow energy equation (SFEE)*. For same mass flow rate at the inlet and outlet i.e., $m_1 = m_2 = m$

then, $$Q + m\left(h_1 + \frac{C_1^2}{2} + gz_1\right) = m\left(h_2 + \frac{C_2^2}{2} + gz_2\right) + W$$

on per unit mass basis the *SFEE* will become;

$$q + h_1 + \frac{C_1^2}{2} + gz_1 = h_2 + \frac{C_2^2}{2} + gz_2 + w \qquad 1.82$$

where, $q = \frac{Q}{m}$, $w = \frac{W}{m}$

The *SFEE* forms the basis for solving all the thermodynamic problems and it is applied to different engineering systems like turbines, pumps, compressors, nozzles, boilers and condensers etc.

1.11.2 Work Done in a Steady Flow Process

We have already discussed in the earlier section that the work done in a non-flow reversible process is given by $\int_1^2 pdv$. Similarly, the work done in a flow process can be obtained using the SFEE.

We know that for a unit mass the SFEE is given as;

$$Q + h_1 + \frac{C_1^2}{2} + gz_1 = h_2 + \frac{C_2^2}{2} + gz_2 + W \qquad 1.83$$

or $$Q - W = \left(\frac{C_2^2}{2} - \frac{C_1^2}{2}\right) + (h_2 - h_1) + g(z_2 - z_1)$$

If the changes in kinetic and potential energies are neglected then the above equation reduces to

$$Q - W = h_2 - h_1$$

Converting the above equation to differential form

$$dQ - dW = dh \qquad 1.84$$

From the definition of enthalpy $h = u + pv$

After differentiation $dh = du + pdv + vdp$ 1.85

From the first law of thermodynamics for a close system, $dQ = du + pdv$

Substituting dQ in Eq. 1.85, $dh = dQ + vdp$

Substituting the value of dh in Eq. 1.84, $dQ - dW = dQ + vdp$

or $$dW = -vdp \qquad 1.86$$

The Eq. 1.86 is known as *flow work equation.*

1.11.3 Applications of steady flow energy equation (SFEE)

The SFEE finds a large number of applications in all of the flow devices. The solution to any such problem is achieved by applying the appropriate boundary condition.

1. Nozzles and diffusers

A nozzle is a device used to increase the kinetic energy of the fluid at the expense of its pressure drop, whereas diffuser is a device that increases the pressure of the fluid at the expense of its kinetic energy. Fig. 1.18 shows an insulated convergent-divergent nozzle used to increase the kinetic energy of steam.

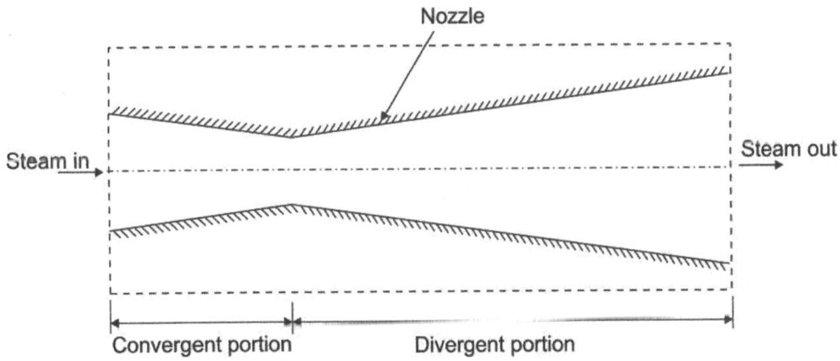

Fig. 1.18: Convergent-divergent nozzle

For the above system,

$PE = 0$

$Q = 0$

$W = 0$

Applying the SFEE to the system

$$h_1 + \frac{C_1^2}{2} = h_2 + \frac{C_2^2}{2} \qquad 1.87$$

or $$\frac{C_2^2}{2} - \frac{C_1^2}{2} = h_1 - h_2$$

or $$C_2 = \sqrt{C_1^2 + 2(h_1 - h_2)} \qquad 1.88$$

Where C is the velocity in m/s and h is the enthalpy in *Joules.*

The velocity C_1 is very small as compared to C_2 then,

$$C_2 = \sqrt{2(h_1 - h_2)}\ m/s$$

2. Turbine

In a turbine the energy of the working fluid such as steam or gas is converted into useful work which may be used to run a generator to produce electricity. A turbine is a power producing device and hence the work done is taken positive. Fig. 1.19 shows an insulated steam turbine so that there is no transfer of heat. The flow velocities may be considered negligible and also the datum level is fixed so KE and PE terms will be zero.

With the given conditions the SFEE becomes

$$h_1 = h_2 + W$$

or $$W = h_1 - h_2 \quad 1.89$$

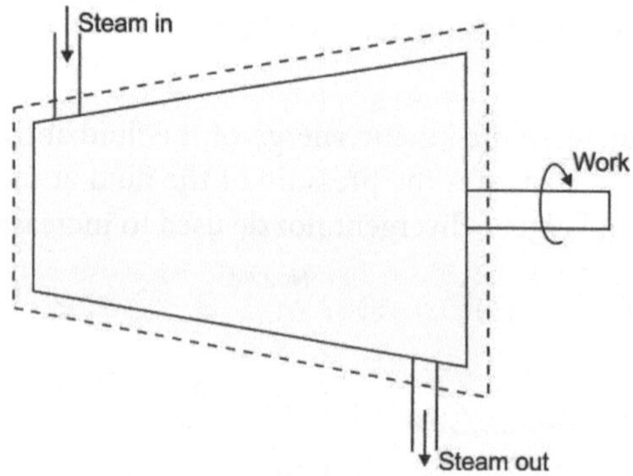

Fig. 1.19: Steam turbine

3. Reciprocating compressor

A reciprocating compressor is a device that draws air from the atmosphere and compresses it to a considerably higher pressure to be supplied in small quantities as shown in Fig. 1.20. A compressor is a power consuming device therefore the work done is taken negative.

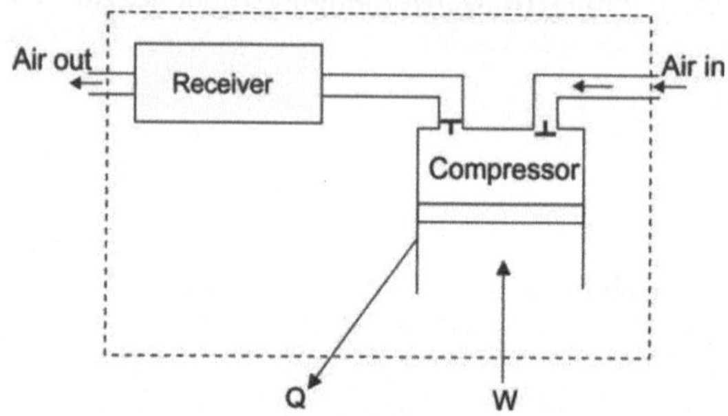

Fig. 1.20: Reciprocating compressor

If the control volume contains the receiver, which significantly reduces flow fluctuations, the reciprocating compressor can be regarded as a steady flow system. Since, in a reciprocating compressor the changes in KE and PE are negligible, so the SFEE becomes.

$$h_1 - Q = h_2 - W \quad 1.90$$

It may be noted that the both heat and work acquires negative value because heat is coming out of the system and work is supplied to the system.

4. Boiler

A boiler in a power plant supplies steam at high temperature and pressure to the turbine. It supplies heat to the incoming water and converts it into steam. Fig 1.21 shows the simplified system for analysis. For this system PE and KE are negligible and because the height of boiler is fixed and there is no velocity change either. Also, the work done $W = 0$ since neither any work is produced nor absorbed in the system. Therefore, the SFEE will become

$$h_1 + Q = h_2 \quad 1.91$$

or $$Q = h_2 - h_1$$

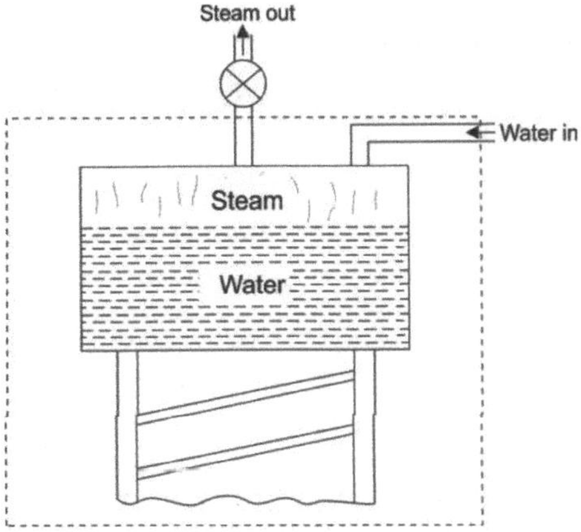

Fig. 1.21: Boiler

5. Throttling process (No heat and work transfer)

An appreciable drop in pressure occurs when a fluid passes through a restricted passage, such as a partially opened valve, orifice, or porous plug, and the flow is referred to as being throttled. The throttling process is shown in Fig. 1.22. The process is said to be adiabatic but irreversible.

Since the system is isolated so $Q = 0$ and there is no work interaction so $W = 0$. The inlet and outlet are at same level therefore $PE = 0$ also the pipe velocities in throttling process are so small that it can be neglected so $KE = 0$.

Applying these conditions, the SFEE will become

$$h_1 = h_2$$

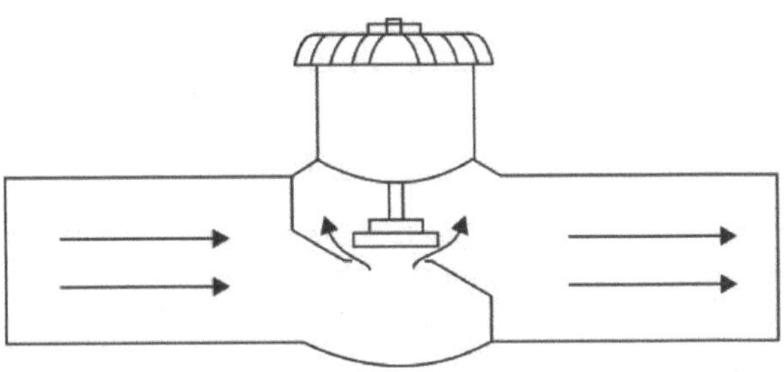

Fig. 1.22: Throttling process

Thus, throttling process is *constant enthalpy* process or enthalpy of the fluid before and after throttling remains same. Both free expansion and throttling are the examples of constant enthalpy processes, the difference being; in free expansion the gas leaves with a large velocity while in throttling the gas has negligible velocity.

Example 1.7: In a steady flow system, a substance flows at the rate of 300 kg/min. It enters at a pressure of 6 bar, a velocity of 300 m/s, specific internal energy of 2000 kJ/kg and specific volume of 0.4 m³/kg. It leaves the system at a pressure of 0.1MPa, a velocity of 150 m/s, specific internal energy of 1600 kJ/kg and specific volume of 1.2 m³/kg. The inlet is 10 m above the outlet. During its passage through the system, the substance has a work transfer of 3 MW to the surrounding. Calculate the heat transfer in kJ/s. State whether it is from or to the system.

Solution:

Given: $P_1 = 6$ bar $= 600$ kPa; $C_1 = 300$ m/s; $u_1 = 2000$ kJ/kg; $v_1 = 0.4$ m³/kg; $Z_1 = 10$m; W=3 MW

$P_2 = 0.1$ MPa $= 100$ kPa; $C_2 = 150$ m/s; $u_2 = 1600$ kJ/kg; $v_2 = 1.2$ m³/kg; $Z_2 = 0$; mass flow rate = 300 kg/min.

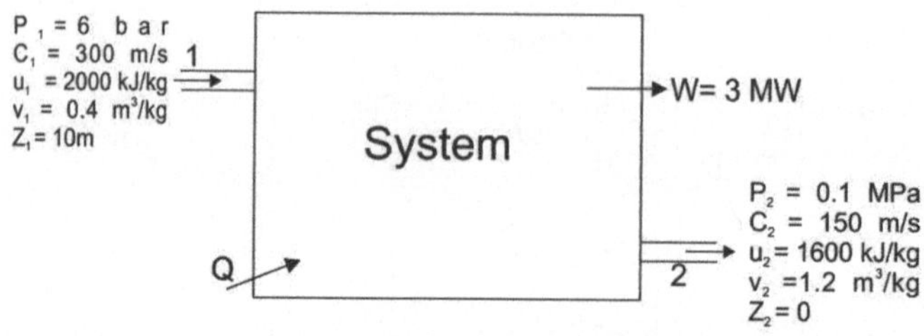

Fig. 1.23: System for example1.7

Applying the SFEE ($E_{in} = E_{out}$)

$$\dot{m} = 300 \text{ kg/min} = 5 \text{kg/s}$$

$$\dot{W} = \dot{m} w$$

$$3000 = 5 \times w$$

$$w = 600 \text{ kJ/kg}$$

$$p_1 v_1 + u_1 + \frac{C_1^2}{2} + gZ_1 + q = p_2 v_2 + u_2 + \frac{C_2^2}{2} + gZ_2 + w$$

$$600 \times 0.4 + 2000 + \frac{300^2}{2 \times 1000} + \frac{9.81 \times 10}{1000} + q = 100 \times 1.2 + 1600 + \frac{150^2}{2 \times 1000} + \frac{9.81 \times 0}{1000} + 600$$

$$q = 46.1519 \text{ kJ/kg}$$

$$Q = q \times \dot{m}$$

$$= 46.1519 \times 5 = 230.7595 \text{ kJ/s} \quad \text{Ans}$$

Example 1.8: The air compressor receives air from the ambient atmosphere where the pressure is 1.01 bar and the temperature is 300 K. The outlet pressure is 3.924 bar and the temperature is 480 K, and the velocity is 100 m/s. The mass flow rate is 15 kg/s. Determine the power required to operate the compressor. Take c_p= 1.005kJ/kgK

Solution:

Given: P_1= 1.01 bar; T_1= 300 K; C_1=0; P_2= 3.924 bar; T_2= 480 K; C_2= 100 m/s, mass flow rate= 15 kg/s.

Here we assume that the initial velocity (C_1=0) and the heat transfer as zero. The process is considered as steady flow, with no change in potential energy.

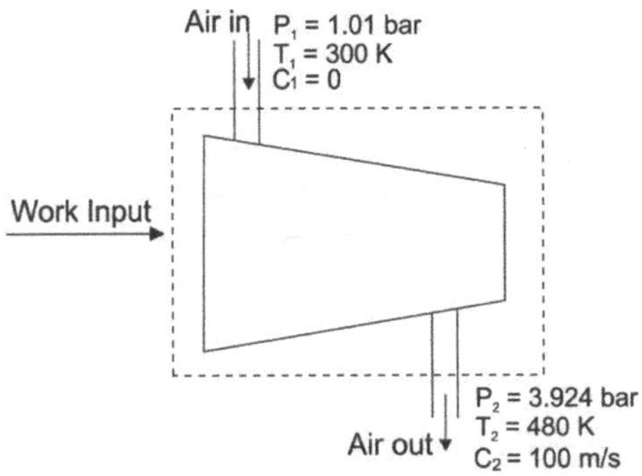

Fig. 1.24: Compressor for example 1.8

$$Q + m\left(h_1 + \frac{C_1^2}{2} + gz_1\right) = m\left(h_2 + \frac{C_2^2}{2} + gz_2\right) + W$$

$$0 + m\left(h_1 + \frac{0}{2} + gz_1\right) = m\left(h_2 + \frac{C_2^2}{2} + gz_2\right) + W$$

$$W = -m\left((h_2 - h_1) + g(z_2 - z_1) + \frac{C_2^2}{2}\right)$$

$$= -m\left((h_2 - h_1) + 0 + \frac{C_2^2}{2}\right)$$

$$= -m\left((c_p T_2 - c_p T_1) + \frac{C_2^2}{2}\right)$$

$$= -15\left((1.005 \times 480 - 1.005 \times 300) + \frac{100^2}{2 \times 1000}\right)$$

$$= -2782.5 kW$$

Example 1.9: A heat exchanger is supplied with air at 20 °C at a velocity of 40 m/s, where its temperature is raised to 700 °C. It then enters the turbine with the same velocity until the temperature falls to 550 °C. On leaving the turbine, the air is taken taken at a velocity of 70 m/s to a nozzle where it expands until the temperature has fallen to 400 °C . If the rate of air flow is 2 kg/s, Calculate

(a) The rate of heat transfer to the air in heat exchanger

(b) Power output from the turbine assuming no heat loss

(c) Velocity at exit from the nozzle assuming no heat loss.

Assume c_p for air =1.005 kJ/kg K

Solution:

Given: T_1= 20+273= 293 K; T_2= 273+700= 973 K; T_3= 273+550= 823 K; C_1= C_2= 40 m/s; C_3= 70 m/s; mass flow rate of air =2 kg/s

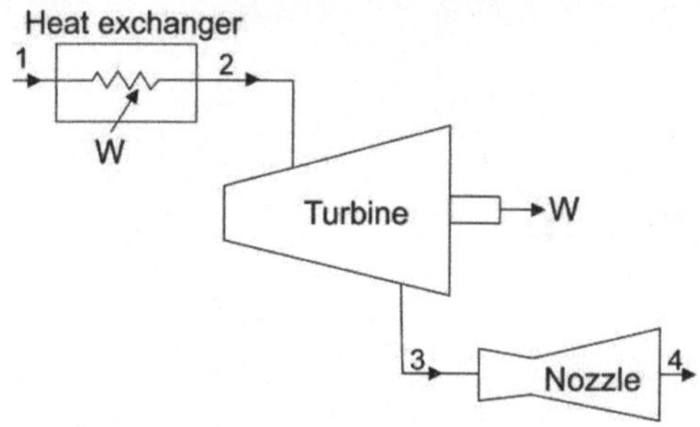

Fig. 1.25: Heat exchanger for example 1.9

(a) Heat exchanger

No work transfer

KE and PE neglected

$h_1 = c_p T_1 = 1.005 \times 293 = 294.465$ kJ/kg

$h_2 = c_p T_2 = 1.005 \times 973 = 977.865$ kJ/kg

Applying SFFFE

$Q = h_2 - h_1 = 977.865 - 294.465 = 683.4$ kJ/kg

Rate of heat transfer $= \dot{m} Q = 2 \times 683.4 = 1366.8$ kJ/s Ans

(b) Turbine

No heat transfer

PE neglected

$h_3 = c_p T_3 = 1.005 \times 823 = 827.115$ kJ/kg

Applying SFFFE

$$h_2 + \frac{C_2^2}{2} = h_3 + \frac{C_3^2}{2} + w$$

$$w = (h_2 - h_3) + \frac{C_2^2 - C_3^2}{2000}$$

$$= (977.865 - 827.115) + \frac{40^2 - 70^2}{2000}$$

$$= 150.75 - \frac{3300}{2000}$$

$$= 149.1 \text{ kJ/kg}$$

Power output $= \dot{m} w$

$= 2 \times 149.1 = 298.2$ kW Ans

(c) For Nozzle

Applying SFEE will give

$$h_3 + \frac{C_3^2}{2} = h_4 + \frac{C_3^2}{2}$$

$$\frac{C_4^2 - C_3^2}{2000} = c_p(T_3 - T_4)$$

$$C_4^2 - C_3^2 = 1.005 \times 2000\ (550 - 400)$$

$$= 301.50 \times 10^3\ m^2/s^2$$

$$C_4^2 = 30.15 \times 10^4 + 0.49 \times 10^4$$

$$= 30.64 \times 10^4 m^2/s^2$$

Exit velocity from the nozzle= 553.53 m/s

Power output = $\dot{m}$ w

= 2 × 149.1 = 298.2 kW   Ans

1.12 THE SECOND LAW OF THERMODYNAMICS

The first law tells us about conservation of energy but there is a limitation of this law that it fails to explain the non-occurrence of certain events, direction of the process etc. This law places no restriction on conversion of heat into work or conservation of heat because according to this law all the heat transferred could be completely converted into work, which is not true. The limitation of first law can be understood with the help of following examples.

Consider a block placed on a rough inclined plane as shown in Fig. 1.26. As a common understanding, the block will slide down and heat will be generated. The reverse of this process will be in agreement of the first law whereby allowing the block to slide up to the rough plane and become cooler. We know that this is not possible even though the principle of conservation of energy is not violated.

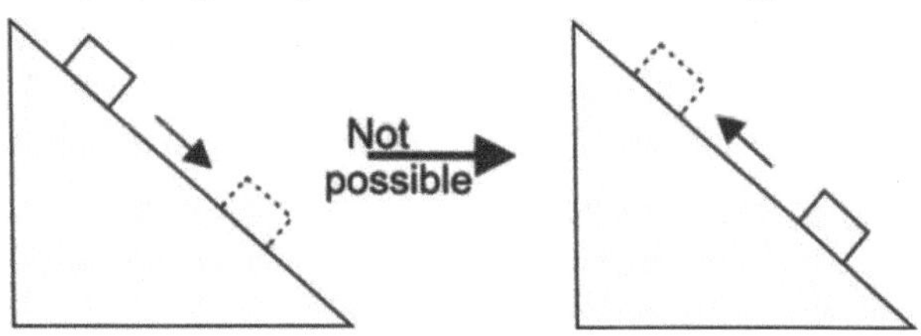

Fig. 1.26: Rough inclined plane

Consider the heating of a room caused by the passage of an electric current through a resistor as shown in Fig 1.27. Again, according to the first law, the amount of electric energy supplied to the resistance wires must be equal to the amount of energy transferred to the room air as heat. Let us now try to reverse this process.

It should come as no surprise that transferring some heat to the wires does not result in the generation of an equivalent amount of electric energy in the wires.

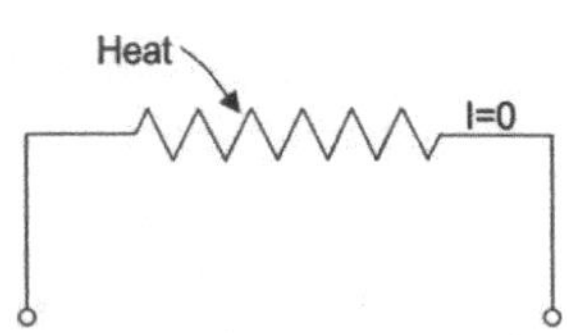

Fig. 1.27: Resistor

It is clear from the above examples that processes occur in a specific direction and reverse of it is not possible. As a result, it was felt that more thermodynamic laws were necessitated to deal with such complex situations. The second law emerged as an embodiment of actual events while retaining the fundamental nature of the first law of thermodynamics. The potential answers provided by the second law include process feasibility, process direction, and energy grades such as low and high. The second law of thermodynamics is capable of indicating the maximum possible efficiencies of heat engines, heat pump and refrigerator coefficients of performance, defining a temperature scale independent of physical properties, and so on.

It can be further stated that although energy is conserved but its quality is not, it is continuously degraded. The conservation and degradation of energy is a parallel process and second law provides a platform for measuring the degradation of energy and hence also known as *law of degradation of energy.*

1.12.1 The Second Law Statements

After studying Sadi Carnot's work, Rudolph Julius Emmanuel Clausius, a German physicist, presented the first general statement of the second law of thermodynamics in 1850. It was known as Clausius' statement of second law. Lord Kelvin and Max Planck also developed another statement of the second law, known as the Kelvin-Planck statement for the second law of thermodynamics. As a result, there are two statements of the second law of thermodynamics however both have an interconnection (the equivalence of both the statements are discussed in subsequent section).

1.12.2 Clausius Statement

It states that "*it is impossible to have a device that produces no effect other than the transfer of heat from a body at low temperature to a body at higher temperature while operating in a cycle*" (Fig. 1.28).

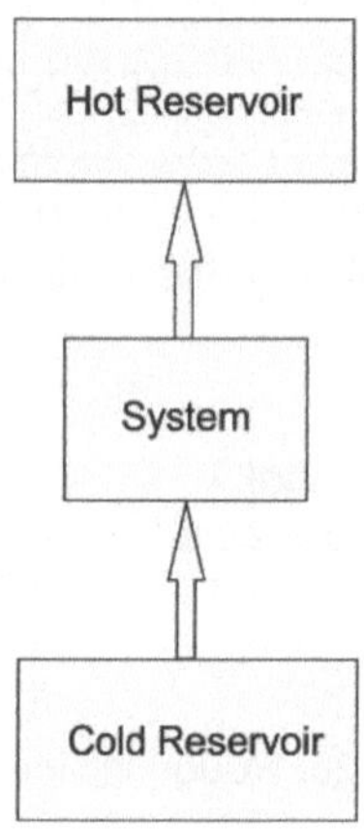

Fig. 1.28: Clausius statement

1.12.3 Kelvin-Plank Statement

It states that "*it is impossible to construct a device operating in a cycle that produce net work while exchanging heat with bodies at single fixed temperature*". It says that a device operating in cycle (such as a heat engine) must have heat interaction at two different temperatures or with body/reservoirs at two different temperatures in order to produce net work (i.e., source and sink).

Alternatively, the efficiency of an engine cannot be 100 percent and to get net work from a system, some heat must be rejected from a high temperature source to a low temperature sink. The illustration of Kelvin-Plank statement is depicted in Fig. 1.29.

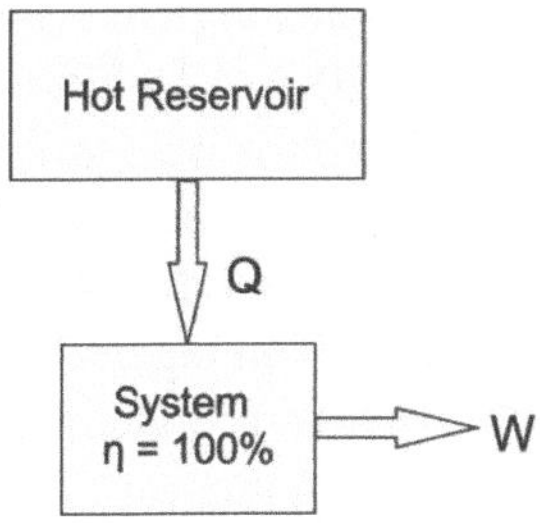

Fig. 1.29: Kelvin-Plank statement

1.12.4 Perpetual Motion Machine of Second Kind (PMM-II)

The above two statements refer to the feasibility of operation of heat pump/refrigerator and heat engine respectively. The PMM-II are devices that violates the second law of thermodynamics. Fig. 1.30 depicts an example of a PMM-II. It shows a hypothetical machine that is operating in a cycle and absorbing heat from a single reservoir and delivering equivalent amount of work.

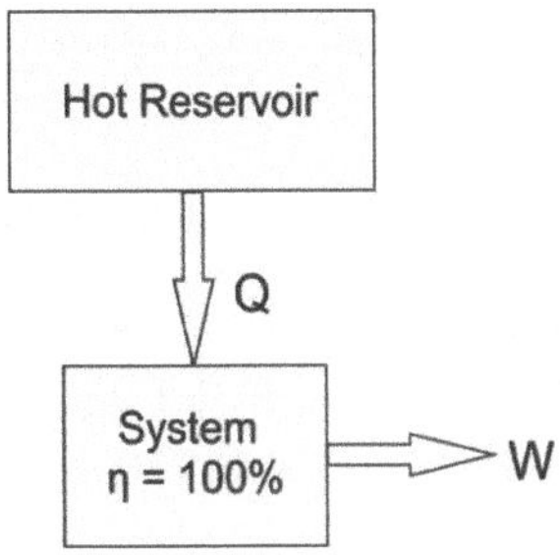

Fig. 1.30: A PMM-II

The Kelvin-Plank statement and Clausius statement both denies the possibility of PMM-II. The PMM-II does not violate the first law of thermodynamics, but it violates the second law. The existence of PMM-II would indicate that continuous work can be produced while extracting heat from an inexhaustible single heat reservoir, which is not possible.

1.12.5 Equivalence of Clausius Statement and Kelvin-Plank Statement

The Clausius and Kelvin-Plank statements are basically two different interpretations of the same basic fact. The violation of one statement will automatically be the violation of the second statement. Therefore, equivalence may be established based upon the violation of one statement resulting in the violation of second statement also.

(i) System based on violation of Clausius statement

Let us assume that a device (heat pump) may be constructed which violates the Clausius statement i.e., the device will transfer heat from cold reservoir to hot reservoir without the expense of any work ($W = 0$) as shown in Fig. 1.31a.

Let *50 J* of heat is transferred from cold reservoir to the hot reservoir without receiving any work thus violating Clausius statement. Further, let *60 J* of heat is transferred from hot reservoir to a heat engine. *50 J* of heat is rejected in the cold reservoir (Fig. 1.31 b) thus (60-40) =*10 J* of work is produced by the heat engine. This heat engine is practically possible and does not violate any law. Now, if the two systems are combined together (Fig. 1.31 c) the net heat transfer in the new system from the hot reservoir is (60-50) = *10 J* while the net heat transfer from the new system to the cold reservoir is (50 -50) = *0* and the net work done by the new system is (60 -50) = *10 J*. This implies that heat pump, heat engine and hot reservoir constitute a device operating in a cycle which produces work without rejecting any heat in the cold reservoir. In other words, the device produces net work while exchanging heat with reservoir at single fixed temperature. This is clearly the violation of Kelvin-Plank statement. Therefore, a system violating Clausius statement also violates Kelvin-Plank statement.

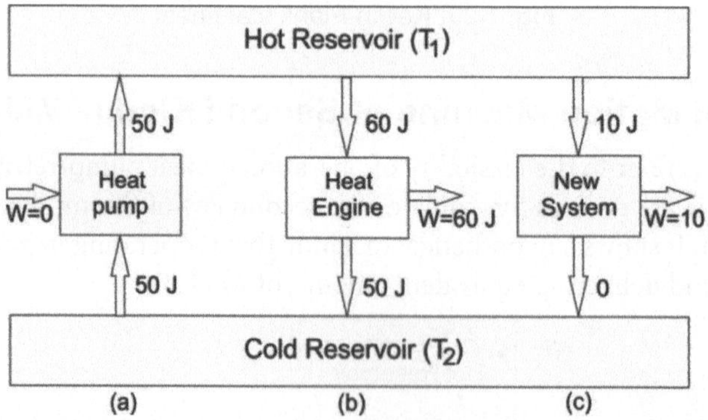

Fig. 1.31: System based on violation of Clausius statement

(ii) System based on violation of Kelvin-Plank statement

Let us assume that there is a heat engine which is operating in a cycle is performing work while exchanging heat from a single reservoir as depicted in Fig. 1.32 a.

Let *50 J* of heat is transferred from the hot reservoir to the heat engine which produces *50 J* of work thus violating the Kelvin-Plank statement. Let *60 J* of heat is transferred from cold reservoir to the heat pump while *50 J* of work is supplied to the heat pump and thus *110 J* of heat is transferred to the hot reservoir (Fig. 1.32 b). This heat pump is practically possible and it does not violate any law. The combination of these two systems will give a new system in which the net work interaction is zero since work done by heat engine is utilized to run heat pump. *60 J* of heat is transferred from cold reservoir to the new system and *60 J* of heat is transferred from new system to the hot reservoir.

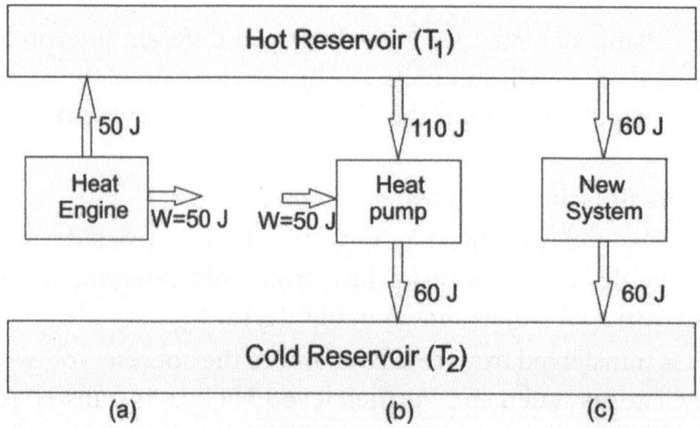

Fig. 1.32: System based on violation of Kelvin-Plank statement

Thus, heat engine, heat pump, hot reservoir and cold reservoir together constitute a system operating in a cycle producing no effect other than transfer of heat from low temperature reservoir to high temperature reservoir without requiring any work from the surroundings (Fig. 1.32 c). This is a clear violation of Clausius statement. Therefore, it may be concluded from the above discussion that the two statements of second law of thermodynamics are equivalent.

1.13 HEAT ENGINE, REFRIGERATION AND HEAT PUMP

(i) Heat Engine

A heat engine is a device which is used to convert heat (thermal energy) into useful work. Generally, heat engines work by generating heat to be used in different systems.

Q_1 is the heat supplied from the reservoir at high temperature T_1 to the working substance of an engine and Q_2 is the heat rejected to the sink at low temperature T_2 as shown in Fig. 1.33 a, then the useful work done by the engine is given as;

$$W = Q_1 - Q_2 \qquad 1.92$$

The performance of the heat engine is expressed by its efficiency.

$$\eta_E \text{or } (COP)_E = \frac{W_E}{Q_E} = \frac{\text{Work done}}{\text{Heat suplied}} = \frac{Q_2 - Q_1}{Q_2} \qquad 1.93$$

(ii) Refrigeration

Refrigeration may be defined in a number of ways, according to the A.S.H.R.A.E. (American Society of Heating, Refrigerating and Air-Conditioning Engineers) refrigeration is "Cooling of a space, substance or system to lower and/or maintain its temperature below the ambient". It may also be defined as the process of eliminating heat, and its practical use is to maintain the temperature lower than the ambient.

The refrigerator's operating substance extracts heat Q_2 from a cold region, it performs external work W on it and discharges heat Q_1 to the hot reservoir at a high-temperature T_1 as shown in Fig. 1.33 b.

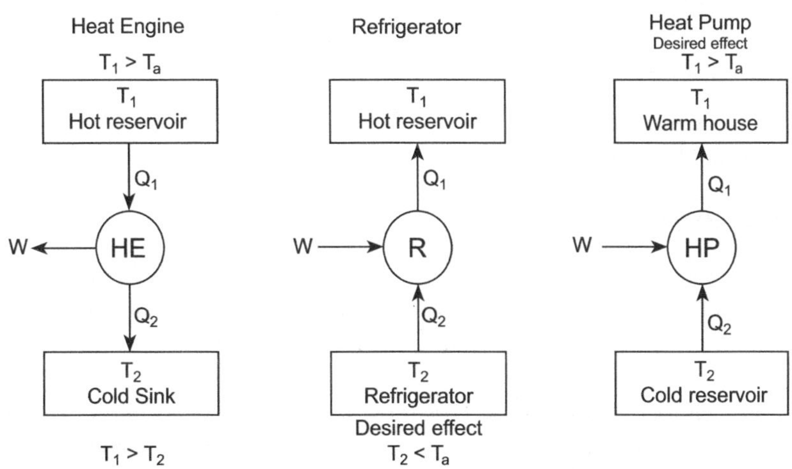

Fig. 1.33: (a) Heat engine, (b) Refrigerator, (c) Heat pump

(iii) Heat Pump

A heat pump is a refrigeration system in which heat is transferred from a cold body and distributes it to a hot body (Fig. 1.33 c). The only thing that separates a heat pump from a refrigerator is the way they operate.

1.13.1 Coefficient of Performance (COP)

The efficiency of a refrigerator or a heat pump is expressed in terms of coefficient of performance and it is defined as the ratio of desired effect to the input required. The desired effect may either be the amount of refrigeration produced in case of a refrigerator or heating of a space in case of a heat pump while the required input is the work done on the refrigerant in a compressor. The value of COP of refrigerator can be greater than unity.

$$\text{COP} = \frac{\text{Desired effect}}{\text{Input required}} = \frac{Q}{W} \tag{1.94}$$

Where Q is the amount of heat extracted to produce refrigeration effect and W is the amount of work done in compressor.

The COP of a refrigerator can be expressed in terms of the amount of heat extracted from a cold body (Q_2) to the amount of work input required (W) as shown in Fig 1.34 (b).

$$\text{COP}_\text{R} = \frac{Q_2}{W} = \frac{Q_2}{Q_1 - Q_2} \tag{1.95}$$

In case of refrigerator the desired effect lies in between cold reservoir and the ambient temperature while in a heat pump, the desired effect lies in between hot reservoir and the ambient temperature as illustrated in Fig. 1.34 (c).

Therefore,

$$\text{COP}_\text{HP} = \frac{Q_1}{W} = \frac{Q_1}{Q_1 - Q_2} \tag{1.96}$$

$$\text{COP}_{HP} = \frac{Q_2}{Q_1 - Q_2} + 1 \tag{1.97}$$

$$\text{COP}_\text{HP} = \text{COP}_\text{R} + 1 \tag{1.98}$$

From Eq. 1.98, it is clear that for fixed values of Q_1 and Q_2 the COP_{HP} is always greater than unity because the COP_R cannot be a negative quantity.

1.14 CARNOT CYCLE

Carnot cycle is a reversible ideal or hypothetical cycle consisting of four processes. This cycle was proposed by Sadi Carnot, a French engineer in 1824, used air as a working fluid for a stationary system of cylinder and a frictionless piston operating with two reversible isothermal and two reversible adiabatic processes.

The sides of the cylinder and piston are assumed to be perfectly insulated and the head of the cylinder is alternatively a perfect conductor or perfect insulator of heat. The engine is considered to operate between a source at a higher temperature and a sink at a lower temperature of infinite capacity. The heat flows into the cylinder during expansion from high temperature source and flows out of the cylinder during compression to the low temperature sink (Fig. 1.34).

Some assumptions made to describe the working of the Carnot engine are as follows:

(i) The piston and cylinder walls are perfect insulators of heat

(ii) The piston is having frictionless movement.

(iii) The head of cylinder should be alternatively a perfect conductor or perfect insulator of heat.

(iv) The compression and expansion processes are reversible.

(v) The transfer of heat does not affect the temperature of source or sink.

(vi) Working medium is a perfect gas (air) and has constant specific heat.

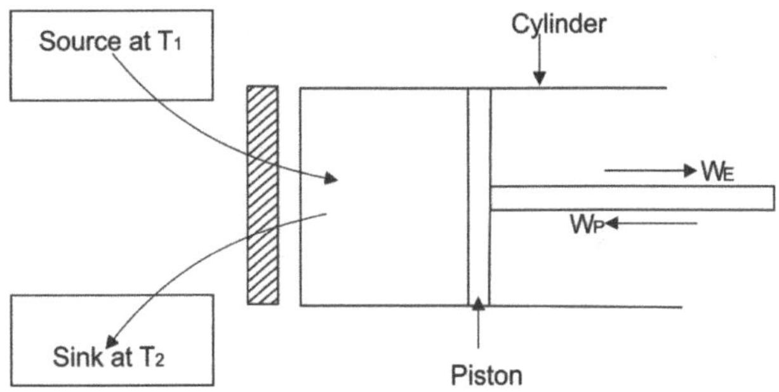

Fig. 1.34: Working of Carnot cycle

Let us assume that the engine cylinder contains m kg of air with p_1, T_1, and V_1 at point 1 on the p-v and T-s diagrams. The p-V and T-s curves are shown in Fig. 1.35.

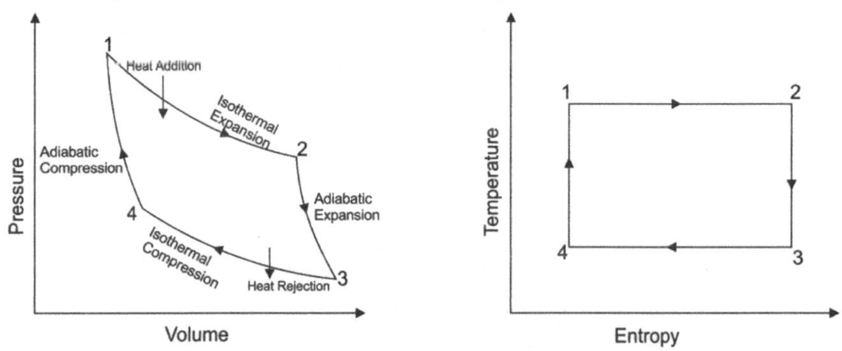

Fig. 1.35: p-V and T-S curve of Carnot cycle

Process 1-2: During this process, the hot energy source at high temperature T_1 is kept in contact with the head of the cylinder. The heat Q_1 is taken in while the air in cylinder expands isothermally from V_1 to V_2 at constant temperature T_1. As the temperature is constant, there is no change in the internal energy of the air. Therefore, the heat supplied during the process is equal to the external work done.

Process 2-3: During the process, the hot source is removed from the head of the cylinder and becomes a perfect insulator so that no heat flow is allowed. The air expands adiabatically and temperature decreases from T_1 to T_2. As no heat transfer takes place by air, the decrease in internal energy is equal to the work done by the air.

Process 3-4: During this process, the cold energy sink at temperature T_2 is kept in contact with the cylinder head. The heat Q_2 is rejected to the sink while the air is compressed isothermally from V_3 to V_4 at constant temperature T_2. As the temperature is constant, there is no change in the internal energy of the air. The heat rejected during the process is equal to work done on the air.

Process 4-1: During this process, the head of the cylinder again becomes a perfect insulator so that no heat flow is allowed. The air is compressed isothermally from V_4 to V_1 and temperature is increased from T_2 to T_1. As no heat transfer takes place by air, the increase in internal energy is equal to the work done on the air.

The work done by the system during the cycle is represented by the area enclosed during the cycle. For a closed cycle, the work done is equal to the difference of heat supplied by the source (Q_1) and the heat rejected to the sink (Q_2).

$$W = Q_1 - Q_2 \tag{1.99}$$

$$\eta = \frac{Work\,done}{Heat\,Supplied} = \frac{Q_1 - Q_2}{Q_1} \tag{1.100}$$

$$= 1 - \frac{Q_2}{Q_1}$$

$$= 1 - \frac{mc_p T_2}{mc_p T_1} \tag{1.101}$$

$$= 1 - \frac{T_2}{T_1} \tag{1.102}$$

Where, m= mass of the air

The engine discussed in the section entirely consists of the reversible processes, which can operate in the reverse direction also. Therefore, it can operate as a heat pump where heat Q_2 is taken from the low temperature reservoir T_2 during the isothermal expansion and heat Q_1 is rejected at high temperature T_1. Work done (W) is required to operate the pump.

This Carnot cycle is an ideal or hypothetical cycle which *cannot be performed in actual due to* following reasons:

1. To transfer the heat, temperature difference is required.
2. In practice, it is impossible to perform a frictionless piston movement.
3. Isothermal process can only be observed when piston moves very slowly and allows the heat transfer to maintain constant temperature. Similarly, adiabatic process can only be observed when piston moves as fast as possible, so that no heat transfer takes place due to very short time. Also, the isothermal and adiabatic process has to take place during the same stroke. Therefore, the piston has to move slowly for a part of the stroke and has to move faster for the remaining part of stroke. Such a motion of the piston during a single stroke is impossible.

1.14.1 Carnot's Theorem

Carnot theorem states that all the heat engines operating between the two same heat reservoirs at a given constant temperature can be more efficient than a reversible heat engine operating between the same two reservoirs.

Let us consider two heat engines one any irreversible heat engine HE_I and other reversible heat engine HE_R operating between heat source at temperature T_1 and sink at temperature T_2 as shown in Fig. 1.36. Now, we have to prove that efficiency of HE_R is more than that of HE_I. If this statement is not true, then; $\eta_A > \eta_B$.

Let the rate of working of the engine be such that $Q_{IA} = Q_{IB} = Q_I$.

Now, consider that engine HE_B be reversed, and as the HE_B is reversible, the magnitude of heat and work transfer will be same, but their directions are reversed as shown in Fig. 1.36. Since $W_I > W_R$, a part of W_I equal to W_R will be used to drive the reversed $(HE_B)_R$. Again, $Q_{IA} = Q_{IB} = Q_I$, then the heat rejected by $(HE_B)_R$ may be supplied to HE_A. Therefore, the source of heat supply may be eliminated. Thus, the net result is that engines HE_A and $(HE_B)_R$ together constitutes a heat engine which generates a network output of W_I - W_R, and exchanges heat with only one reservoir at temperature T_2. This violates the Kelvin-Plank's statement of second law. Hence our assumption of $\eta_A > \eta_B$ is wrong.

Therefore, $\eta_B > \eta_A$

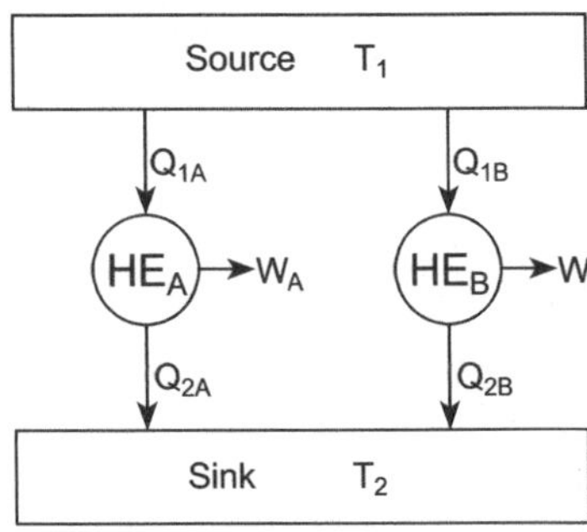

Fig. 1.36: Two cyclic heat engines HE_A and HE_B operating between the same source and sink

1.14.2 Corollary of Carnot Theorem

All the reversible heat engines operating between the same heat source and sink will have the same efficiency.

As discussed above, we have assumed that two engines HE_A and HE_B are reversible engines. Let us consider $\eta_A > \eta_B$, Now if, HE_B is reversed as depicted in Fig. 1.37 and it works as a heat pump which uses a part of work output of HE_A, therefore heat engine and pump together constitutes PMM2 as depicted in Fig. 1.38. Thus, η_A cannot be greater than η_B.

Therefore, $\eta_A = \eta_B$.

So, it can be mentioned that efficiency of any reversible heat engine operating between the same heat source and sink will have the same efficiency. The efficiency will only depend on the temperature limits of source and sink, and is independent of the nature, type or quantity of working substance in the cycle.

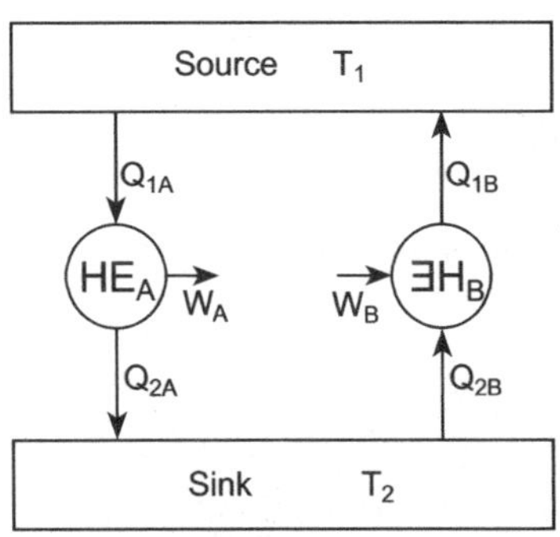

Fig. 1.37: Heat engine HE_B is reversed

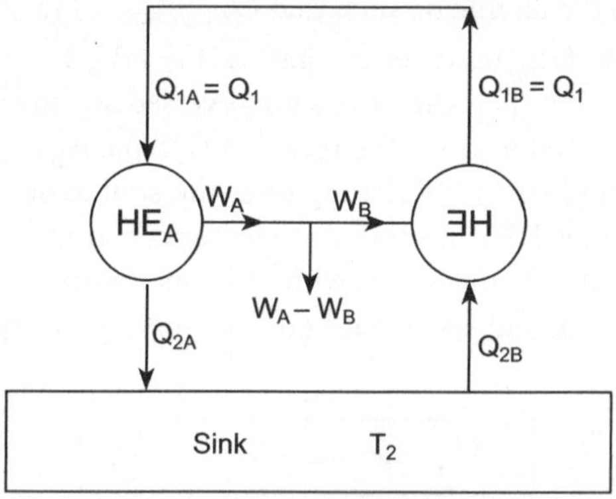

Fig 1.38: Heat engine HE_A and $ƎH_B$ together violate the Kelvin-Planck statement.

Example 1.10: A power cycle operating between two (hot and cold) reservoirs receives heat Q_1 by heat transfer from a hot reservoir at temperature 2000K and rejects heat Q_2 to cold reservoir at 400K. For each of the case find whether the cycle operates reversibly, irreversibly or is impossible.

(a) $Q_1 = 1000$ kJ and $W = 850$ kJ
(b) $Q_1 = 2000$ kJ and $Q_2 = 400$ kJ
(c) $W = 1600$ kJ and $Q_2 = 500$ kJ
(d) $Q_1 = 1000$ kJ and $\eta = 30\%$

Solution:

Given: $T_1 = 2000$ K; $T_2 = 400$ K

(a) $$\eta_{max} = \frac{T_1 - T_2}{T_1} = 1 - \frac{T_2}{T_1}$$

$$\eta_{max} = 1 - \frac{400}{2000} = 0.8 = 80\ \%$$

From the given data

$$\eta = \frac{W}{Q_1} = \frac{850}{1000} = 0.85 = 85\ \%$$

Here $\eta > \eta_{max}$, therefore cycle is impossible Ans

(b) $$\eta = \frac{W}{Q_1} = \frac{Q_1 - Q_2}{Q_1}$$

$$= \frac{2000 - 400}{2000} = 0.80 = 80\ \%$$

Here $\eta = \eta_{max}$, therefore cycle is reversible Ans

(c) $$\eta = \frac{W}{Q_1} = \frac{Q_1 - Q_2}{Q_1}$$

$$W = 1600 = Q_1 - Q_2$$

$$1600 = Q_1 - 500$$

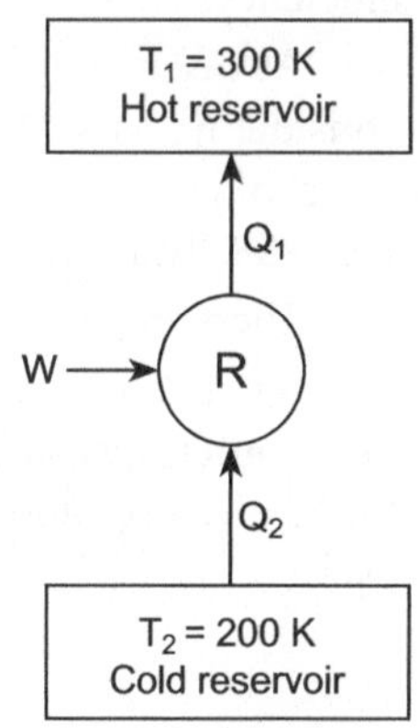

Fig. 1.39: System for example 1.10

$$Q_1 = 2100 \text{ kJ}$$

$$\eta = \frac{1600}{2100} = 0.762 = 76.2\ \%$$

Here $\eta < \eta_{max}$, therefore cycle is irreversible Ans

(d) $\eta = 30\%$ (given)

Here $\eta < \eta_{max}$, therefore cycle is irreversible Ans

Example 1.11: A refrigeration system working between two (hot and cold) reservoirs receives energy Q_2 from cold reservoir at temperature (T_2= 250 K) and rejects energy Q_1 to hot reservoir at temperature (T_1= 300K.) For each of the case find whether the cycle operates reversibly, irreversibly or is impossible.

(a) Q_2 = 1000 kJ and W = 400 kJ
(b) Q_2 = 2000 kJ and Q_1 = 2200 kJ
(c) Q_1 =3000 kJ and W = 500 kJ
(d) W = 400 kJ and COP = 6

Solution:

Given T_1= 300 K; T_2= 250 K

(a) $$COP_{max} = \frac{T_2}{T_1 - T_2}$$

$$= \frac{250}{300 - 250} = 5$$

$$COP = \frac{Q_2}{W} = \frac{1000}{400} = 2.5$$

Here COP < COP_{max}; therefore cycle is irreversible Ans

(b) $$COP = \frac{Q_2}{W} = \frac{2000}{Q_1 - Q_2}$$

$$= \frac{2000}{200} = 10$$

Here COP > COP_{max}; therefore cycle is impossible Ans

(c) $Q_2 = Q_1 - W = 3000 - 500 = 2500$ kJ

$$COP = \frac{2500}{500} = 5$$

Here COP = COP_{max}; therefore cycle is reversible Ans

(d) COP = 6 (given)

Here COP > COP_{max}, therefore cycle is imossible Ans

Example 1.12: During a cycle heat engine receives heat of 600 kJ from high temperature source at 1000 K and converts 150 kJ of heat into work and remaining heat is rejected to low temperature sink at 300 K. Determine whether this engine violates the second law of thermodynamics on the basis of

(a) The Clausis inequality

(b) The Carnot principle

Solution:

Given: Q_1= 600 kJ; W= 150 kJ; T_1= 1000 K; T_2= 300 K

(a) A cycle that violates the Clausius inequality also violates the second law. This is one way of determining whether this cycle violates the second law.

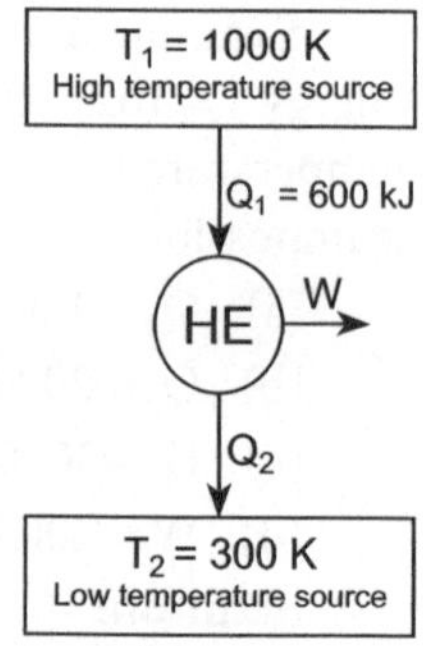

Fig. 1.40: System for example 1.12

$$cyclic\int \frac{\delta Q}{T} = \frac{Q_1}{T_1} - \frac{Q_2}{T_2} \leq 0$$

$$W = Q_1 - Q$$

$$Q_2 = Q_1 - W = 600 - 150 = 45kJ$$

$$cyclic\int \frac{\delta Q}{T} = \frac{600}{1000} - \frac{450}{300}$$

$$= 0.6 - 1.5$$

$$= -0.9kJ / K < 0$$

This cycle satisfies the Clausius inequality and thus the second law of thermodynamics.

(b) The another way of determining the violation of second law is by verifying the Carnot principle.

$$\eta_{th,irr} = 1 - \frac{Q_2}{Q_1} = 1 - \frac{450}{600} = 0.25 = 25\ \%$$

$$\eta_{th,car} = 1 - \frac{T_2}{T_1} = 1 - \frac{300}{1000} = 0.7 = 70\ \%$$

Here, $\eta_{th,irr} < \eta_{th,car}$

This heat engine is in complete compliance with the second law.

A cycle that violates the Clausius inequality will also violates the Carnot principle.

Example1.13: A domestic refrigerator maintains a temperature of -16 °C (source). The ambient air temperature is 32 °C (sink). If heat is continuously removed from the source at the rate of 2.1 kJ/s, what is the minimum power required to pump this heat out continuously?

Solution:

Given: T_1 = 32+273 = 305 K; T_2= -16+273 = 257 K; Q_2= 2.1 kJ/s

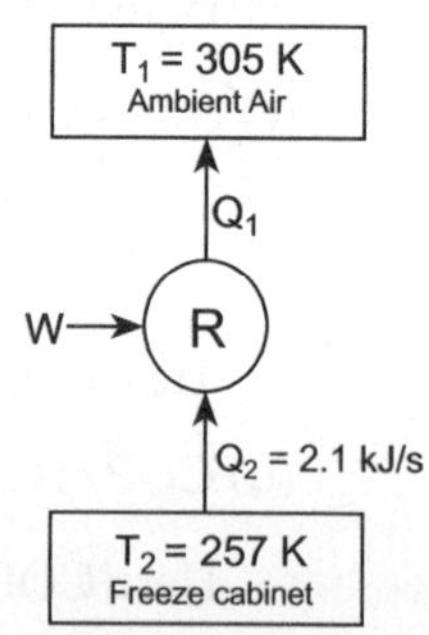

Fig. 1.41: System for example 1.13

$$COP_{carnot,ref} = \frac{T_2}{T_1 - T_2}$$

$$= \frac{257}{305 - 257} = \frac{257}{48} = 5.354$$

COP can also written as $= \frac{Q_{rejected}}{W}$

$$W = \frac{Q_{rejected}}{COP} = \frac{2.1}{5.354} = 0.392 \text{ kW}$$

Example 1.14: An inventor claims that heat engine develops 7.5 kW power. On testing, the engine consumes 1 kg of fuel per hour having calorific value of 45 MJ. The maximum temperature of the cycle is 1600 K and the minimum is 800 K. Find whether the engine is justified in his claim.

Solution:

Given: P = 7.5 kW; mass flow rate = 1 kg/hour;

Calorific value = 45 kJ;

$T_1 = 1600$;

$T_2 = 800$ K

The maximum possible efficiency of the heat engine

$$\eta_{max} = \frac{T_1 - T_2}{T_1} = \frac{1600 - 800}{1600}$$

$$= \frac{800}{1600} = 0.50 \quad \text{or} \quad 50\%$$

$$\eta_{claimed} = \frac{W_{net}}{\text{Heat supplied}}$$

$$= \frac{7.5 \times 60 \times 60}{1 \times 45 \times 10^3} = 0.60 \quad \text{or} \quad 60\%$$

Here the inventor claim is not possible as the efficiency of any engine cannot be greater than the Carnot cycle efficiency.

1.15 CONCEPT OF ENTROPY

We often encounter chaotic situations in our everyday life which is synonymously referred as 'entropy'. The second law leads to the definition of a new property called entropy which was coined by Rudolph Clausius in 1865. In simple words *entropy* may be defined as 'the measure of molecular disorder or molecular randomness'. It is a Greek word which means transformation. As the system becomes more disordered the molecular arrangement becomes unpredictable and hence its entropy increases. As we can see that the molecular position is much ordered in solids so its entropy is less as compared to liquid and gases.

The entropy is a non-conserved quantity and there is nothing like conservation of entropy. The natural processes tend to become more disorganised over time. Entropy deals with the quality of energy which means that one form of energy differs from another in terms of quality. Let us understand this with the example of two libraries containing same number of books. One library contains books well arranged, neatly tagged and placed in well-organized stacks while the other has randomly arranged books in unorganized stacked piles. Although both the libraries contain same number of books but they differ hugely in the quality of service they can provide. Entropy can be thought of as an indicator of the quality of energy, with lower entropy indicating higher quality. Lower entropy energy is stored in an organised manner (the efficient library). Energy stored in a chaotic way (the unorganized library) has high entropy.

1.15.1 The Clausius Inequality

The quantification entropy was established by Clausius inequality which said that the cyclic integral of $\delta Q/T$ is always less than or equal to zero.

$$\oint \frac{\delta Q}{T} \le 0 \qquad 1.103$$

This inequality applies for both reversible an irreversible process. For internally reversible cycle it can be shown that

$$\oint \frac{\delta Q}{T}\bigg|_{\text{int},rev} = 0 \qquad 1.104$$

If the cyclic integral of any quantity is zero then it can be said that the quantity is independent of path and hence a state function and thus it is a *property*. Clausius named this property as entropy and denoted this by S.

Therefore,

$$dS = \frac{dQ}{T}\bigg|_{\text{int},rev} \ (\text{kJ/K}) \qquad 1.105$$

Entropy is an extensive property however if it is considered on per unit mass basis (kJ/kg.K) then it becomes an intensive property. There may be generation of entropy during a process and this is called as *entropy generation*, denoted by S_{gen}.

$$\Delta S = S_2 - S_1 = \int_1^2 \frac{\delta Q}{T} + S_{gen} \qquad 1.106$$

It should be noted that entropy generation S_{gen} is always a positive quantity for irreversible process and zero for reversible process. The value of S_{gen} depends upon the process and thus it is not the *property* of a system. The entropy change may be negative for a system or its surrounding but entropy generation cannot.

Scan to know more about entropy

The entropy of the universe always increases and under the limiting case it may remain constant (reversible process) but will never decrease this is known as increase of entropy principle.

Thus, according to Clausius theorem

$$\oint \frac{\delta Q}{T} < 0 \quad \text{for irreversible process} \qquad 1.107$$

$$\oint \frac{\delta Q}{T} = 0 \quad \text{for reversible process} \qquad 1.108$$

$$\oint \frac{\delta Q}{T} > 0 \quad \text{for impossible process} \qquad 1.109$$

Example 1.15: 0.04 m^3 of nitrogen contained in a cylinder behind a piston is initially at 1 bar and 290 K. The gas is compressed isothermally and reversibly until the pressure is 4.5 bar. Calculate;

(a) The change in entropy

(b) Heat flow and the workdone and sketch the process on p-v and T-s diagram. Assume nitrogen to act as a perfect gas.

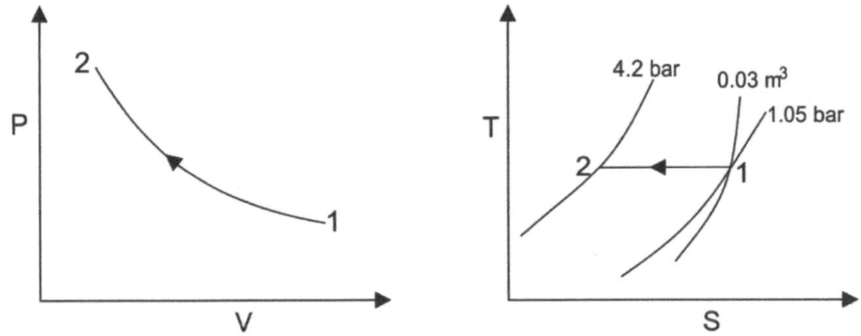

Fig. 1.42: p-V and T-s curve for example 1.15

Solution:

Given: $V_1 = 0.04\ m^3$; $P_1 = 1$ bar; $T_1 = 290$ K; $P_2 = 4.5$ bar

(a) Characterstic gas constant (R) = $\dfrac{\text{Universal gas constant } (R_u)}{\text{Molecular mass of the gas (M)}}$

$$= \frac{8.314}{28} = 0.297 \text{ kJ/kgK}$$

$$p_1\, V_1 = m\, R\, T_1$$

$$1.0 \times 100 \times 0.04 = m \times 0.297 \times 290$$

$$m = \frac{1.0 \times 100 \times 0.04}{0.297 \times 290} = 0.046 \text{ kg}$$

$$S_2 - S_1 = m\, R \ln \frac{p_1}{p_2}$$

$$= 0.046 \times 0.297 \ln \frac{1.0}{4.5}$$

$$= 0.046 \times 0.297 \times (-1.504)$$

$$= -0.013662 \text{ kJ/K} \qquad \text{Ans}$$

(b) $Q = T\,(S_2 - S_1)$

$$= 290\,(-0.013662)$$

$$= 3.9 \text{ kJ} = W \qquad \text{Ans}$$

UNIT SUMMARY

- In this unit we have introduced a thermodynamic system as a control volume, which for a fixed mass is referred as control mass. Such a system can be closed (energy transfer but no mass transfer), open (both energy and mass transfer) or isolated (no mass, momentum, or energy exchange with its surroundings).
- When the properties of a substance change, the state changes and a process occur. A cycle is completed when a substance has gone through several processes and returned to its initial state.

The point function depends on its initial and final state only while a path function depends on the process path.

- A system is in thermodynamic equilibrium when the temperature and pressure are the same at all points and there is no velocity gradient.
- A reversible process is one that can be stopped and reversed at any point so that the system and surroundings are exactly restored to their initial states, whereas an irreversible process is one in which the system and the surroundings cannot be restored to its initial states on reversal of process.
- The *Zeroth law of thermodynamics* establishes the foundation of temperature measurement which states that 'if two thermodynamic sy`stems or bodies are in thermal equilibrium with a third body
- The reversible work for a non-flow system is given as,

$$W_{1-2} = \int_1^2 \delta W = \int_1^2 pdv = p(v_2 - v_1)$$

- The energy accumulated into the system after the heat and work interaction is called as the *internal energy (U)* of the system. Enthalpy simply represents the total heat content in a system and is given as,

$$H = U + PV$$

- The relation between specific heats is given as,

$$c_p - c_v = R; \quad \therefore\ c_v = \frac{R}{\gamma - 1} \text{ where } \gamma = c_p/c_v \text{ and } c_p = \frac{\gamma R}{\gamma - 1}$$

- According to the first law of thermodynamics, heat and work are mutually convertible, but because energy cannot be created or destroyed, the total energy associated with an energy conversion remains constant and it is expressed as,

$$\oint \delta Q = \oint \delta W$$

- Work and heat interaction in non-flow system for:

Constant pressure (isobaric) process

$$W = \int_1^2 pdv = p(v_2 - v_1)$$

$$Q = h_2 - h_1 = c_p(T_2 - T_1)$$

Constant volume (isochoric) process

$$W = 0$$

$$Q = U_2 - U_1 = mc_v(T_1 - T_2)$$

Constant temperature (isothermal) process

$$Q = W = p_1 v_1 \ln\frac{v_2}{v_1} = p_2 v_2 \ln\frac{v_2}{v_1} \qquad Q = p_1 V_1 \ln\frac{p_2}{p_1}$$

Reversible adiabatic (pv^γ = constant) process

$$W = (u_1 - u_2)$$

$$Q = 0$$

Reversible polytropic (pv^n = constant) process

$$W = \frac{R(T_1 - T_2)}{n-1}$$

$$Q = \left(\frac{\gamma - n}{\gamma - 1}\right) W$$

- Relationship between the ratios of p, v and T is given as

$$\frac{p_2}{p_1} = \left(\frac{v_1}{v_2}\right)^{\gamma}$$

$$\frac{T_2}{T_1} = \left(\frac{v_1}{v_2}\right)^{\gamma - 1}$$

$$\frac{T_2}{T_1} = \left(\frac{p_2}{p_1}\right)^{\frac{\gamma - 1}{\gamma}}$$

- In Joule or free expansion, the initial and final internal energy is equal hence, for a perfect gas undergoing free expansion the initial and final temperatures are equal.
- The steady flow energy equation (SFEE) per unit mass is expressed as,

$$q + h_1 + \frac{C_1^{\,2}}{2} + gz_1 = h_2 + \frac{C_2^{\,2}}{2} + gz_2 + w$$

- The work done in a steady flow process is given as, $dW = -vdp$
- The efficiency of a heat engine is expressed as,

$$\eta_E \text{ or } (COP)_E = \frac{W_E}{Q_E} = \frac{\text{Work done}}{\text{Heat suplied}} = \frac{Q_2 - Q_1}{Q_2}$$

- The COP of refrigerator is given as, $COP_R = \frac{Q_2}{W} = \frac{Q_2}{Q_1 - Q_2}$
- The COP of heat pump is given as, $COP_{HP} = \frac{Q_1}{W} = \frac{Q_1}{Q_1 - Q_2}$; also $COP_{HP} = COP_R + 1$
- The efficiency of Carnot cycle is given as $\eta = 1 - \frac{T_2}{T_1}$
- Entropy is a property which is defined as the measure of molecular disorder or molecular randomness. The quantification of entropy is established by Clausius inequality which is given by,

$$\oint \frac{\delta Q}{T} \leq 0$$

- The entropy of the universe always increases called as increase of entropy principle. According to Clausius theorem

$$\oint \frac{\delta Q}{T} < 0 \qquad \text{for irreversible process}$$

$$\oint \frac{\delta Q}{T} = 0 \qquad \text{for reversible process}$$

$$\oint \frac{\delta Q}{T} > 0 \qquad \text{for impossible process}$$

EXERCISE

Multiple Choice Questions

1. A thermodynamic system refers to
   (a) Defined region in space
   (b) Specified mass in fluid flow
   (c) Specified region with constant volume
   (d) Prescribed and identifiable quantity of matter
2. In which of the following systems does mass transfer occur across the system boundary?
   (a) Isolated system
   (b) Closed system
   (c) Open system
   (d) None of the mentioned
3. In a closed thermodynamic system
   (a) No energy or mass transfer across the boundary takes place
   (b) No energy but mass transfer across the boundary takes place
   (c) No mass but energy transfer across the boundary takes place
   (d) Both energy and mass transfer across the boundary takes place
4. For an isolated system executing process
   (a) No heat is transferred
   (b) No mass transfer across the boundary
   (c) No work done
   (d) No chemical reaction within the system
   Which of the above mentioned statements are true?
   (a) i, ii and iii
   (b) i, iii and iv
   (c) ii, iii and iv
   (d) i, ii, iii and iv
5. Which of the following is an extensive property of a thermodynamic system?
   (a) Density (b) Pressure
   (c) Temperature (d) Volume

6. Extensive property of a system is one whose value
   (a) Depends on the mass or extent of the system
   (b) Depend on interaction of the system with its surrounding
   (c) Dependent on the path followed by the system in going from one state to another
   (d) Dependent on nature of boundaries
7. Which of the following is an intensive property of a thermodynamic system?
   (a) Mass (b) Energy
   (c) Temperature (d) Volume
8. Intensive property of a system is one whose value
   (a) Depends on the mass of the system, like volume
   (b) Does not depend on the mass of the system, like temperature, pressure, etc.
   (c) Is not dependent on the path followed, but on the state
   (d) Is dependent on the path followed and not on the state
9. Work done in a free expansion process is
   (a) +ve (b) –ve
   (c) Zero (d) Maximum
10. Heat flow into a system is ____, and heat flow out of the system is ________.
   (a) Positive, positive (b) Negative, negative
   (c) Negative, positive (d) Positive, negative
11. A cyclic machine, as shown below, receives 325 kJ from a 1000 K energy reservoir. It rejects 125 kJ to a 400 K energy reservoir and the cycle produces 200kJ of work as output. Is this cycle reversible, irreversible, or impossible?
   (a) Reversible (b) Irreversible
   (c) Impossible (d) None of the mentioned
12. The value of $n = 1$ in the polytropic process indicates it to be
   (a) Reversible process (b) Isothermal process
   (c) Adiabatic process (d) Irreversible process
13. Heat interaction between system and surrounding
   (a) Represents energy in transit
   (b) Does not depends on the choice of the system.
   (c) Can be identified after the completion of system
   (d) Is a property of the system and its differential is exact
14. For a system to be in thermodynamic equilibrium, the system and surrounding are to be in
   (a) Thermal equilibrium
   (b) Chemical equilibrium
   (c) Mechanical equilibrium
   (d) Thermal, Chemical and Mechanical equilibrium
15. In an isothermal process, the internal energy
   (a) Increases (b) Decreases
   (c) Remains constant (d) First increases, then decreases

16. In an isothermal process,

(a) There is no change in temperature
(b) There is no change in enthalpy
(c) There is no change in internal energy
(d) All of these

17. According to first law of thermodynamics

(a) Work done by a system is equal to heat transferred by the system
(b) Total internal energy of a system during a process remains constant
(c) Internal energy, enthalpy and entropy during a process remain constant
(d) Total energy of a system remains constant

18. Which of the following is the correct statement of the second law of thermodynamics?

(a) There is a definite amount of mechanical energy, which can be obtained from a given quantity of heat energy
(b) It is impossible to transfer heat from a body at a lower temperature to a higher temperature, without the aid of an external source
(c) It is impossible to construct an engine working on a cyclic process, whose sole purpose is to convert heat energy into work
(d) All of the above

19. The work done equals $\int p\, dv$ for

(a) An irreversible process
(b) Unrestricted expansion
(c) Non-flow quasi-static process
(d) The arrangement where work is done on the gas as the paddle wheel turns by a falling weight

20. A quasi-static process is one in which all the states through which a system passes are very close to

(a) Equilibrium state (b) Original state
(c) Same temperature (d) Each other

21. For which condition, the change in the enthalpy of a system equals the heat supplied?

(a) Constant volume
(b) Constant pressure
(c) Constant temperature
(d) Under standard temperature-pressure condition

22. The air is heated in a cylinder fitted with a frictionless piston held by a constant weight. The process is

(a) Isothermal (b) Adiabatic
(c) Isochoric (d) Isobaric

23. The heating of a gas at constant pressure is governed by

(a) Boyle's law (b) Charle's law
(c) Gay-Lussac law (d) Joules law

24. The amount of heat required to raise the temperature of the unit mass of gas through one degree at constant volume, is called

(a) Specific heat at constant volume
(b) Specific heat at constant pressure
(c) Kilo Joule
(d) None of these

25. A machine which can supply mechanical work continuously without consumption of any energy is called as

(a) Perpetual Motion Machine of the First kind (PMM1)
(b) Perpetual Motion Machine of the Second kind (PMM2)
(c) Perpetual Motion Machine of the Third kind (PMM3)
(d) None of the above

26. Zeroth law of thermodynamics helped in the creation of which scale?

(a) Temperature
(b) Heat energy
(c) Pressure
(d) Internal energy

27. In the context of a closed non- flow system, the net energy transferred as heat and work equals the change in

(a) Entropy
(b) Enthalpy
(c) Internal work
(d) Flow work

28. Work output from a system is at the expense of internal energy is a non-flow process carried out at

(a) At constant pressure
(b) At constant volume
(c) Adiabatically
(d) Polytropically

29. Three engines A, B and C are operating on Carnot cycle respectively uses air, steam and helium as the working substance. If all the engines operate within the same high and low temperature limits, then which engine will have the highest efficiency?

(a) Engine A
(b) Engine B
(c) Engine C
(d) All will have the same efficiency

30. Consider the following statements:

The definition of;

(i) Temperature is due to Zeroth law of thermodynamics
(ii) Entropy is due to first law of thermodynamics
(iii) Internal energy is due to second law of thermodynamics
(iv) Reversibility is due to Kelvin- Planks's statement

Identify the correct statements

(a) (i), (ii) and (iii)
(b) (i), (iii) and (iv)
(c) (i) alone
(d) (ii) alone

31. The value of universal gas constant is

(a) 848 mkgf/kg mol/K
(b) 427 mkgf/kg mol/K
(c) 735 mkgf/kg mol/K
(d) 8314 mkgf/kg mol/K

32. Which of the following is true according to Clausius statement?
    (a) It is possible to construct a device that can transfer heat from a cooler body to a hotter body without any effect
    (b) It is impossible to construct a device that can transfer heat from a cooler body to a hotter body without any effect
    (c) It is impossible to construct a device that can transfer heat from a hotter body to a cooler body without any effect
    (d) None of the mentioned
33. The entropy of an isolated system can never ____
    (a) Decrease (b) Be zero
    (c) Increase (d) None of the mentioned
34. What is the relationship between Kelvin-Planck's and Clausius' statements?
    (a) Violation of one doesn't violate the other
    (b) Not connected to each other
    (c) Virtually two parallel statements of the second law
    (d) None of the mentioned
35. Which of the following is known as the inequality of Clausius?
    (a) Cyclic integral of dW/T<=0 (b) Cyclic integral of dW/T>=0
    (c) Cyclic integral of dQ/T>=0 (d) Cyclic integral of dQ/T<=0
36. Carnot cycle consists of........
    (a) Two constant volume and two reversible adibatic processes
    (b) Two isothermal and two reversible adibatic processes
    (c) Two constant pressure and two reversible adibatic processes
    (d) One constant volume, one constant pressure and two reversible adibatic processes
37. The more effective way of increasing efficiency of Carnot engine is to
    (a) Increase higher temperature
    (b) Decrease higher temperature
    (c) Decrease lower temperature
    (d) Increase lower temperature
38. The door of a running refrigerator inside a room was left open. Which of the following statements is correct?
    (a) The room will be cooled to the temperature inside the refrigerator
    (b) The room will be cooled very slightly
    (c) The temperature of the air in room will remain unaffected
    (d) The room will be gradually warmed up
39. 20J of heat energy is extracted from a cold reservoir when 30J of work is done on a refrigerator. What is the coefficient of the refrigerator ?
    (a) 2/3 (b) 5/3
    (c) 3/5 (d) 1/3

40. Efficiency of a heat engine is defined as
(a) Total heat output / net work input
(b) Total heat input / net work output
(c) Net work output / total heat input
(d) Net work input / total heat output

41. The coefficient of performance of a refrigeration working on a reversed Carnot cycle is 4. The ratio of highest absolute temperature the lowest absolute temperature is
(a) 1.2 (b) 4
(c) 1.25 (d) 3.33

Answers for Multiple Choice Questions							
1. (d)	2. (c)	3. (c)	4. (a)	5. (d)	6. (a)	7. (c)	8. (a)
9. (c)	10. (d)	11. (c)	12. (a)	13. (a)	14. (d)	15. (c)	16. (d)
17. (d)	18. (d)	19. (a)	20. (d)	21. (d)	22. (d)	23. (b)	24. (a)
25. (a)	26. (a)	27. (a)	28. (c)	29. (d)	30. (c)	31. (a)	32. (b)
33. (a)	34. (c)	35. (d)	36. (b)	37. (c)	38. (d)	39. (a)	40. (c)
41. (c)							

Short and Long Answer Type Questions

1. What do you mean by system, boundary and surrounding?
2. Explain first law of thermodynamics.
3. Define Zeroth law of thermodynamics.
4. Define the terms; property, state, system and process.
5. Write short notes on steady and non-steady flow process, reversible and irreversible process, open and closed systems, flow and non-flow work.
6. Explain how the heat, work and internal energy are related to each other.
7. What do you understand by thermal and mechanical equilibrium?
8. Explain intensive and extensive properties with examples.
9. Define different gas laws.
10. Define internal energy and enthalpy.
11. Define isothermal, adiabatic, isobaric and isochoric processes.
12. Derive C_p- C_v = R
13. State Kelvin Plank and Clausius statement of second law of thermodynamics.
14. State Kelvin Plank and Clausius statement and show that they are equivalent.
15. Draw and explain Carnot cycle and derive expression of entropy
16. Explain and prove Carnot theorem.
17. Discuss the concept of entropy.
18. Differentiate between heat engine, refrigerator and heat pump.

19. What are PMM-I and PMM-II?
20. Show that al, the reversible engines operating between the same temperatures T_1 and T_2, having T_1 greater than T_2 must have the sane efficiency.
21. Is it possible for a heat engine to operate without any heat rejection? Explain.
22. Explain any process which satisfies the second law of thermodynamics, but violates the first law.

Numerical Problems

1. A perfect gas for which the ratio of specific heats is 1.4 occupies a volume of 0.06 m^3 at 10 bar and 500 K. The gas undergoes expansion to 0.3 m^3. Find the heat absorbed or rejected by the gas for each of the following methods of expansion.
   (a) Constant pressure
   (b) Isothermal
   (c) According to law $PV^{1.125} = C$
   (d) Adiabatic

   For gas consider R=0.25 kJ/kg K

   [840 kJ; 96.56 kJ; 60.14 kJ; 0]

2. A cylinder containing the air comprises the system. Cycle is completed as follows:
   (i) 82000 N-m of work is done by the piston on the air during compression stroke and 45 kJ of heat are rejected to the surroundings.
   (ii) During expansion stroke 100000 N-m of work is done by the air on the piston.

   Calculate the quantity of heat added to the system.

   [Q= 63 kJ]

3. The power developed by a turbine in a certain steam plant is 1200 kW. The heat supplied to the steam in the boiler is 3360 kJ/kg, the heat rejected by the system to cooling water in the condenser is 2520 kJ/kg and the feed pump work required to pump the condensate back into the boiler is 6 kW. Calculate the steam flow round the cycle in kg/s.

   [1.421 kg/s.]

4. A fluid system undergoes a non-flow frictionless process following the pressure-volume relation as $P = \frac{5}{V} + 1.5$ where p is in bar and V is in m^3. During the process the volume changes from 0.15 m^3 to 0.05 m^3 and the system rejects 45 kJ of heat. Determine the:
   (i) Change in internal energy;
   (ii) Change in enthalpy.

   [519 kJ; 504 kJ]

5. 1 kg of gaseous CO_2 contained in a closed system undergoes a reversible process at constant pressure. During this process 42 kJ of internal energy is decreased. Determine the work done during the process.
   Take c_p = 840 J/kg °C and c_v = 600 J/kg °C.

   [- 16.8 kJ]

6. A cylinder contains 0.45 m^3 of a gas at 1 × 105 N/m^2 and 80°C. **The gas is compressed to a** volume of 0.13 m^3, the final pressure being 5 × 105 N/m^2. Determine:
   (i) The mass of gas
   (ii) The value of index 'n' for compression
   (iii) The increase in internal energy of the gas
   (iv) The heat received or rejected by the gas during compression.

   Take γ = 1.4, R = 294.2 J/kg °C.

   [0.433 kg; 1.296; 49.9 kJ; 17.54 kJ]

7. In a gas turbine unit, the gases flow through the turbine is 15 kg/s and the power developed by the turbine is 12000 kW. The enthalpies of gases at the inlet and outlet are 1260 kJ/kg and 400 kJ/kg respectively, and the velocity of gases at the inlet and outlet are 50 m/s and 110 m/s respectively. Calculate the rate at which heat is rejected to the turbine.

   [828 kW]

8. A mass of gas with an internal energy of 1500 kJ is contained in a cylinder which has perfect thermal insulation. The gas is allowed to expand behind a piston until its internal energy is 1400 kJ. Calculate the work done if the expansion follows a law PV^2= Constant, and the initial pressure and volume of gas are 28 bar and 0.06 m^3 respectively. Calculate the final pressure and volume.

   [100 kJ; 4.58 bar; 0.148 m^3]

9. In a 150 mm diameter pipe, air flows at the rate of 2.5 kg/s. it has a pressure of 8 bar and a temperature of 100 °C before it is throttled by a valve to 4 bar. Find the velocity of the air downstream of the restriction, and show that the enthalpy is essentially the same before and after the throttling process.

   [37.87 bar]

10. Air at 100 kPa and 280 K is compressed steadily to 600 kPa and 400 K. The mass flow rate of air is 0.02 kg/s and heat loss of 16 kJ/kg during the process. Assuming the changes in kinetic and potential energies are negligible, calculate the necessary power input to the compressor.

11. A energy source at 1000 K losses 420 kJ heat to a sink at (a) 750 K and (b) 500 K. Determine which heat transfer process is more reversible.

    [Energy sink of 500 K is more irreversible]

12. Air with volume 0.6 m^3, temperature of 500 K and pressure of 6.8 bar according to
    (a) PV = Constant
    (b) $PV^{1.2}$ = Constant
    (c) Adiabatic process to a pressure of 1 bar.

    Find the change in entropy in each case, assuming γ= 1.41; c_v= 0.708 kJ/kg K.

    [-1.447 kJ/K; 0.2309 kJ/K; 0]

13. An engine has power output of 90 kW with thermal efficiency of 28%. Determine the rate of fuel consumption, if the heating value of the fuel is 44000 kJ/kg.

    [26.31 kg/hr]

14. A Carnot heat engine operates between a source at1000 K and a sink at 300 K. If the het engine is supplied with the heat at a rate of 800 kJ/min, determine;

(a) The thermal efficiency
(b) Power output

[70%, 9.33 kW]

15. A house requires 2 × 105 kJ/h for heating in winter. Heat pump is used to absorb heat from cold air outside in winter and send heat to the house. Work required to operate the heat pump is 3 × 104 kJ/h. Determine:
(a) Heat abstracted from outside
(b) Co-efficient of performance.

[170000 kJ/h; 6.66]

16. A Carnot heat engine operating in series between two reservoirs maintained at 327 °C and 27 °C respectively. The energy rejected by the first engine is input to the second engine. If the first engine's efficiency is 20% greater than the second engines efficiency, calculate intermediate temperature.

[135.6 °C]

PRACTICAL

1. To determine the COP and Tonnage capacity of a Mechanical Heat pump.
2. To determine the COP of a vapour compression refrigeration system.

KNOW MORE

- Although the principles of thermodynamics have existed since the beginning of universe, thermodynamics did not emerge as a science until the construction of the first successful atmospheric steam engines in England by Thomas Savery in 1697 and Thomas Newcomen in 1712. These engines were slow and inefficient, but they paved the way for the development of a new science. The study of thermodynamics involve two approaches namely macroscopic approach and microscopic approach. When the knowledge of behaviour of individual particle is not required it is known as macroscopic approach and is covered under classical thermodynamics while the study involving the average behaviour of individual particles is known as microscopic approach and is covered in statistical thermodynamics.
- Dimensions can be used to define any physical quantity. The magnitudes assigned to the dimensions are referred to as units. Some basic dimensions, such as mass m, length L, time t, and temperature T, are designated as primary or fundamental dimensions, while others, such as velocity v, energy E, and volume V, are expressed in terms of the primary dimensions and are referred to as secondary or derived dimensions.
- The thermodynamic property Temperature is defined by the zeroth law of thermodynamics. It is called the zeroth law because it was defined after the first and second laws of thermodynamics. The temperature scale in SI units is Celsius scale (Centigrade scale, °C). The absolute temperature scale in SI unit is Kelvin scale (K). The tripe point of water is 0.01 °C and 0 K corresponds to 273.15 °C
- Pressure is the force exerted by the fluid per unit area. The pressure is only referred for gas or a liquid whereas in solids the term stress is used. The absolute zero pressure corresponds to zero

stress. The unit of pressure is N/m² also called Pascal (Pa).

$$1 \text{ Pa} = 1 \ \frac{\text{N}}{\text{m}^2}$$

$$1 \text{ bar} = 10^5 \text{ Pa}$$

1 bar = (100/9.81) m of water = 10.2 m of water = 750 mm of Hg`

1 standard atmosphere (atm.) = 1.01325 bar = 760 mm of Hg = 10.33 m of water

- It is assumed that the specific heats (c_p and c_v) are constant at all temperatures and pressures but this is not true and it actually varies with temperature and may be approximates as,

  $c_p =$ a + KT and c_v = b + KT

  where, a, b and K are constants
- For solving numerical problems, in calculator we use log10 or $\log_e = \ln$
- The entropy of the universe always increases and this will lead to a stage when the universe will reach the state of maximum entropy and therefore no work can be extracted from the universe. This condition of universe is referred as heat death of the universe.

REFERENCES AND SUGGESTED READINGS

1. Y. A. Cengel, M. A. Boles. Thermodynamics An Engineering Approach, McGraw- Hill, Delhi, India, 2019.
2. M.P. Poonia and S.C. Sharma. Basic Mechanical Engineering – Khanna Publishing House, New Delhi, India, 2005.
3. P. K Nag, Basic and Applied Thermodynamics, Tata McGraw-Hill, Delhi, India, 2006.
4. R. T. Balmer, Modern Engineering Thermodynamics, Academic Press, UK, 2011
5. R. K. Rajput, Engineering Thermodynamics, Laxmi Publications, Delhi, India, 2007.
6. N. D. Hari Dass, The Principles of Thermodynamics, CRC Press, New York, 2014.
7. O. Singh, Applied Thermodynamics, New Age International Publishers, Delhi India, 2009.

2 Heat Transfer & Thermal Power Plant

UNIT SPECIFICS

Through this unit we will discuss the following aspects:

- Various modes of heat transfer;
- Fourier's law of heat conduction;
- General heat conduction equation in Cartesian coordinates, cylindrical coordinates and spherical coordinates;
- Electrical analogy and concept of thermal resistance
- Heat transfer through composite walls;
- Power plant layout and classification various type of boilers;
- Fire-tube and water-tube boilers and their differences;
- Problems related to Fourier's equation, rectangular and cylindrical composite walls.

The topics are discussed in such a way that it covers practical applications along with the amalgamation of basic concepts. This will raise further curiosity and creativity to help improving problem solving capacity of the learners.

Besides giving a large number of multiple-choice questions as well as questions of short and long answer types marked in two categories following lower and higher order of Bloom's taxonomy, assignments through a number of numerical problems, a list of references and suggested readings are given in the unit so that one can go through them for practice. It is important to note that for getting more information on various topics of interest some QR codes have been provided in different sections which can be scanned for relevant supportive knowledge.

After the related practical, based on the content, there is a "Know More" section. This section has been carefully designed so that the supplementary information provided in this part becomes beneficial for the users of the book. This section mainly highlights the initial activity, examples of some interesting facts, analogy, history of the development of the subject focusing the salient observations and finding. The timelines starting from the development of the concerned topics up to the recent time, applications of the subject matter for our day-to-day real life or/and industrial applications on variety of aspects have also been highlighted. Case studies related to environmental, sustainability, social and ethical issues whichever applicable, and finally inquisitiveness and curiosity on the topics of the unit are also the key take away of this section.

RATIONALE

This unit on heat transfer and thermal power plant will help the students to get a primary idea about the various modes of heat transfer and their characteristics behaviour. The unit also deliberates the basic understanding of thermal power plant and its components along with the understanding of various types of boilers.

Conduction of heat transfer is governed by the Fourier's law. The generalized heat conduction equation in Cartesian coordinate system using Fourier's law has been explicitly derived in this unit along with an overview of the same in cylindrical and spherical coordinate system. This unit also explains the electrical analogy and the concept of thermal resistance which will be useful for solving the heat transfer problems. Mathematical formulations for conduction through a plane wall, cylindrical wall, composite wall and cylindrical composite wall have also been discussed to get an idea of various practical situations involving conduction heat transfer. The concept of overall heat transfer coefficient is explained by considering the combined effect of conduction and convection. The numerical problems will further assist the learners to get a deep insight of a number of applications of heat transfer.

Thermal power plant fulfils 60% of the world's electrical energy demand; this necessitates the need for understanding the layout and working of a steam power plant. The unit describes the various components of a steam power plant and their working. The Rankine cycle is the fundamental cycle on which steam power plants operates. This cycle has been described with the help of T-s plot. Steam is the working fluid to run a turbine which is generated in a boiler, therefore various types of boilers and the working of Cochran boiler and Babcock and Wilcox boiler is also discussed in this unit.

PRE-REQUISITES

- Mathematics: Co-ordinate Systems, Differentiation and Integration (Class XII)
- Basic Thermodynamics

UNIT OUTCOMES

List of outcomes of this unit is as follows:

U2-O1: Describe the basic modes of heat transfer

U2-O2: Describe Fourier's law and derive the generalized heat conduction equation in Cartesian coordinate system

U2-O3: Explain the conduction and convection phenomena in plane and composite walls

U2-O4: Explain the working of thermal power plant and steam boilers

U2-O5: Apply the heat conduction equations in solving real life problems

Unit-2 Outcomes	EXPECTED MAPPING WITH COURSE OUTCOMES (1- Weak Correlation; 2- Medium correlation; 3- Strong Correlation)					
	CO-1	CO-2	CO-3	CO-4	CO-5	CO-6
U2-O1	3	2	3	2	-	2
U2-O2	3	1	3	-	-	2
U2-O3	3	2	3	1	-	2
U2-O4	3	2	3	-	-	2
U2-O5	3	1	3	-	-	3

2.1 INTRODUCTION

Heat transfer relates with the generation, transformation, and exchange of heat between physical systems by virtue of temperature difference. Heat transfer answers to how the heat energy is transferred and at what rate the heat exchange will take place. The basic difference between thermodynamics and heat transfer is that the thermodynamics deals with the end states in equilibrium and can tell us the amount of energy required to change a system from one equilibrium state to another but how fast this change may take place in a system is explained by concepts of heat transfer. We can understand this with the help of an example; let us consider a hot steel rod which is placed in a bucket of water. The final equilibrium temperature of steel rod-water system may be predicted from thermodynamics but it will not tell us that how long it will take to reach this equilibrium condition or what will be the temperature of bar after a certain length of time before the attainment of equilibrium condition. Heat transfer may be used to predict the temperature of both steel rod and water as a function of time.

2.2 MODES OF HEAT TRANSFER

The modes of heat transfer are namely conduction, convection and radiation:

2.2.1 Conduction

The transfer of energy that takes place due to molecular agitation in a material without any displacement of the material in totality is known as *conduction*. Alternatively, when the particles having more kinetic energy transfer their energy to the adjacent particles possessing lesser kinetic energy as a result of their mutual interaction the transfer is said to be achieved by conduction. Consider the transfer of heat along a metal rod as shown in Fig. 2.1, one end of the rod is exposed to a heat source. The particles of the rod in the immediate vicinity of the flame gets heated which leads to the vibration of the particles about their mean position.

Consequently, these more active particles collide with the less active particles lying next to them leading to transfer of thermal energy to the less active particles. The process is repeated until the transfer of energy is reached to the other end of the rod. The temperature of the rod decreases as we move from the flame side towards the end of the rod. Heat flows due to the existence of temperature difference along the length of the rod and heat flow rate between the two ends of the rod depends upon the temperature difference between these ends along with the length of the rod and its material.

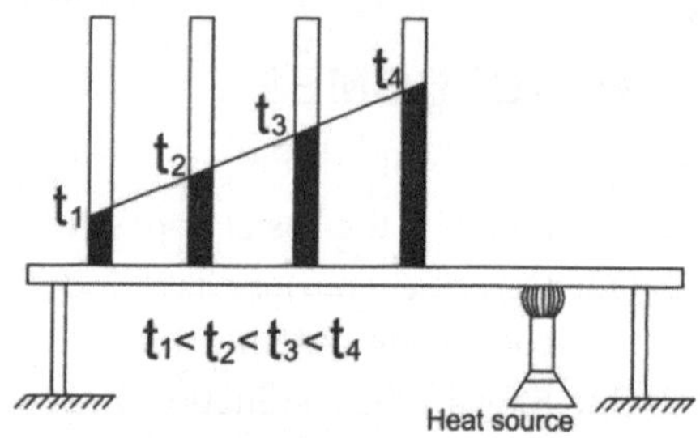

Fig. 2.1: Conduction heat transfer along a metal rod

In case of solids, the conduction occurs by the combined effect of intermolecular lattice vibrations and the energy transfer by the free electrons. Whereas, in liquids and gases, it occurs by virtue of the combined impact of molecular collisions and diffusion during random motion.

2.2.2 Convection

It is a method of transferring heat whereby the molecules of fluids, such as gases and liquids, move about in large quantities. The mode of energy transfer combines the effects of conduction and the motion of fluid and can happen between the solid surface and the moving fluid. Convective heat transfer will increase as fluid motion accelerates.

Consider a pot filled with water which is heated from the bottom (Fig. 2.2). When particles of water at the bottom of the pot receives heat or thermal energy, they move faster and spreads farther apart due to more collisions. This results in the decrease of density of the particles, and the particles starts rising up in the fluid. The particle starts transferring their thermal energy to other particles and gets cooled in this activity. Now, these particles after losing their energy moves slowly and have fewer collisions, and moves closer leading to increased density and settles down, and the same cycle repeats. This results in the formation of loop of moving particles termed as convection current.

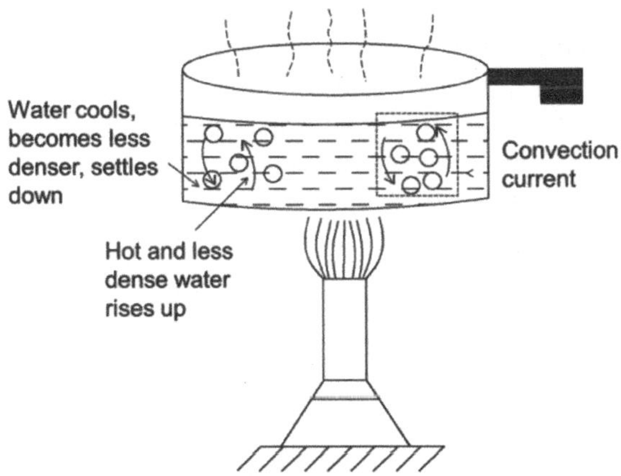

Fig. 2.2: Convection heat transfer in a pot filled with water

Natural or free convection and forced convection are the two types of convection. In natural convection, fluid movement occurs due to buoyancy forces generated by density differences caused by temperature variations in the fluid. Whereas, in forced convection devices like fan, pump, or the wind forces the fluid to flow over the surface.

2.2.3 Radiation

Matter emits energy in the form of electromagnetic waves (or photons) due to changes in the electronic configurations of atoms or molecules. The energy transfer via radiation do not require any intermediate medium and can occur even in a vacuum or a space devoid of substance. All the bodies at finite temperature emit radiation. When we sit next to fire, the heat coming out of fire reaches us by radiation (Fig. 2.3).

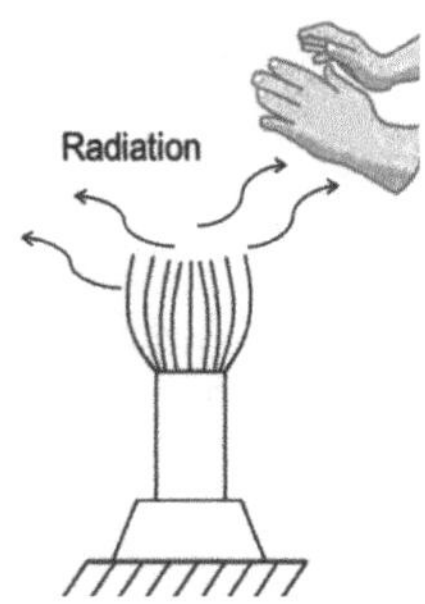

Fig. 2.3: Radiation heat transfer

The propagation of electromagnetic radiations due to the temperature difference is called as *thermal radiation*. Radiation heat transfer is more efficient in vacuum, as an everyday example we can see that the sun transfers its heat to the earth in the form of electromagnetic waves passing through vacuum. But when the sun radiation enters the earth's atmosphere its impact starts reducing due to the presence of air, water, gas and dust particles present in the earth's atmosphere.

2.3 CONDUCTION HEAT TRANSFER (FOURIER EQUATION)

Energy moves from the high to low-temperature region. It may be said that when the energy is transported through conduction heat flux which is heat transfer rate per unit area, is directly proportional to the normal temperature gradient.

$$\frac{Q}{A} \alpha \frac{dT}{dx}$$

When we insert the constant of proportionality

$$Q = -kA\frac{dT}{dx}$$

$$q = \frac{Q}{A} = -k\frac{dT}{dx} \qquad 2.1$$

Where Q is the heat-transfer rate (W), A is the area exposed normal to the of heat flow direction (m^2), q is the heat flux (heat conducted per unit area per unit time, W/m^2), dT is the temperature difference between the two surfaces across which the heat is passing (K or ^oC), dx is the thickness of material along the direction of heat flow (m), dT/dx is the temperature gradient in the direction of the heat flow. The constant k is called the thermal conductivity of the given material (W/mK).

Since, the heat flow takes place in the direction of negative temperature gradient therefore the minus sign is inserted so that the second principle of thermodynamics will be satisfied. Eq. 2.1 is called Fourier's law of heat conduction after the French mathematical physicist Joseph Fourier, who made very significant contributions to the analytical treatment of conduction heat transfer. It is vital to note that this equation is the defining equation for the thermal conductivity.

The following presumptions form the foundation of the Fourier law:

- The heat conduction takes place under steady state which implies that the temperature of the fixed points within a heat conducting body does not change with time.
- Uni-directional heat flow, i.e., to describe the temperature distribution within the heat conducting body, just one space co-ordinate is necessary.
- Bounding surfaces have an isothermal nature, which means that the temperatures at the two faces remain constant and uniform.
- The material is supposed to be homogenous and isotropic, which implies that heat conductivity is constant in all directions.
- The temperature gradient is assumed to be constant with a temperature profile that is linear.
- No generation of heat within.

The following are some key characteristics of the Fourier relation:

- Fourier law predicts how heat is conducted through a medium from a region of higher temperature to a lower temperature.
- The Fourier law applies to all matter, regardless of its state (solid, liquid, or gas).
- The Fourier law is a vector equation which indicates that heat transfer rate is normal to an isotherm and in the direction of temperature decline.
- Fourier law is a generalisation based on experimental evidence; it cannot be derived from first principles.

2.4 HEAT CONDUCTION EQUATION IN CARTESIAN COORDINATES

The validity of Fourier law is primarily for conditions under unidirectional flow of heat and steady state, but in almost all the practical situations the temperature may be the function of space as well as time. Since, heat can flow in all the directions, we need to have a 3-dimensional heat flow equation to cater for both non-uniformity of temperature and irregular boundary surface. To achieve this task relevant energy transfer processes are identified and appropriate energy rate equations are introduced in an elemental volume and solution of the differential equations are sought. The solution of resulting differential equations gives the temperature distribution. After that, the Fourier rate equation is used to calculate the heat transfer rate through the conducting medium.

Consider an infinitesimal volume element in 3-dimensional coordinate system to derive the generalised heat conduction equation in Cartesian coordinate system (Fig. 2.4). The *dx*, *dy*, and *dz* sides run parallel apart to the *x*, *y*, and *z* axes.

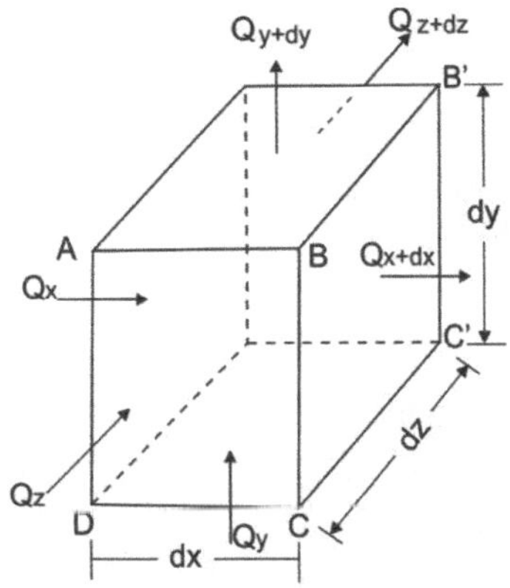

Fig. 2.4: Infinitesimal volume element for conduction analysis in Cartesian coordinates

Let *t* denote the temperature at the differential control volume's left face. Because the area of this face can be made arbitrarily small, the temperature *t* can be assumed to be uniform across the entire surface. The temperature changes along the *x*-axis, and the rate of change is given by $\partial t/\partial x$ (temperature gradient). The temperature change over distance *dx* will then be $(\partial t/\partial x)dx$. This temperature change is depicted graphically in Fig. 2.5. Therefore, the temperature at the right face at a distance *dx* from the left will be given as $[t+(\partial t/\partial x)dx]$.

For setting up a generalized heat conduction equation we apply the Fourier equation in each of the Cartesian coordinate along with the appropriate energy conservation equations. Let k_x be the thermal conductivity on the left face, then heat influx Q_x (quantity of heat flowing into the control volume) through this face in the time interval of $d\tau$ will be:

$$\text{Heat influx, } Q_x = -k_x(dydz)\frac{\partial t}{\partial x}d\tau \qquad 2.2$$

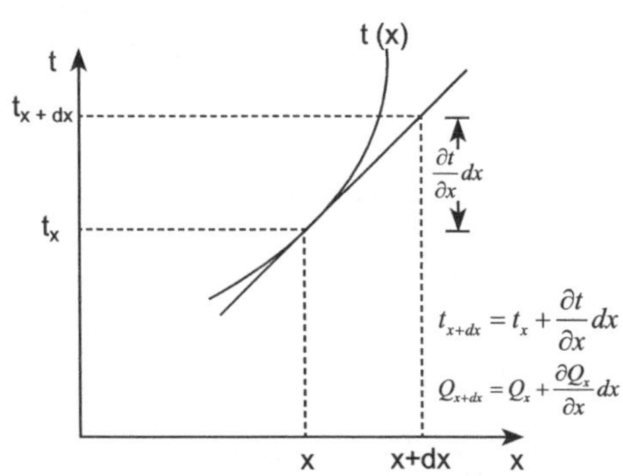

Fig. 2.5: Temperature change with respect to distance

Similarly, heat efflux Q_{x+dx} (heat flowing out from the right-hand face of control volume) is simply the additional amount of heat added in the control volume while passing through a distance of dx and will be given as

Heat efflux, $Q_{x+dx} = Q_x + \frac{\partial}{\partial x}(Q_x)dx$ 2.3

The heat accumulated in the control volume, dQ_x will be equal to:

dQ_x = heat influx-heat efflux = $Q_x - \left[Q_x + \frac{\partial}{\partial x}(Q_x)dx\right]$

$$= \frac{\partial}{\partial x}(Q_x)dx$$

Substituting the value of Q_x from Eq. 2.2

$$= \frac{\partial}{\partial x}\left[-k_x(dydz)\frac{\partial t}{\partial x}d\tau\right]dx \quad 2.4$$

By adopting the same approach, the heat accumulation in y and z-directions may also be as:

$$dQ_y = \frac{\partial}{\partial y}\left[-k_y\frac{\partial t}{\partial y}\right]dxdydzd\tau \quad 2.5$$

$$dQ_z = \frac{\partial}{\partial z}\left[-k_z\frac{\partial t}{\partial z}\right]dxdydzd\tau \quad 2.6$$

The net heat accumulated inside the control volume will be the sum of the heat accumulated in x, y and z directions and is given as,

Total heat accumulated

$$= \left[\frac{\partial}{\partial x}\left(k_x\frac{\partial t}{\partial x}\right) + \frac{\partial}{\partial y}\left(k_y\frac{\partial t}{\partial y}\right) + \frac{\partial}{\partial z}\left(k_z\frac{\partial t}{\partial z}\right)\right]dxdydzd\tau \quad 2.7$$

It is assumed that there is heating arrangement inside the control volume which produces heat q_g in a unit volume and unit time. The net heat produced in the control volume will be:

Net heat generated = $q_g dxdydzd\tau$ 2.8

The net heat accumulated and the net heat generated inside the control volume is given in Eq. 2.7 and 2.8 respectively and it will contribute in increasing the overall thermal energy of the control volume. If ρ and c being the density and specific heat of the material respectively, then the variation in internal energy inside the control volume is represented as,

$$\rho(dxdydz)c\frac{\partial t}{\partial \tau}d\tau \quad 2.9$$

The energy balance equation is written by combining Eq. 2.7, 2.8 and 2.9,

$$\left[\frac{\partial}{\partial x}\left(k_x\frac{\partial t}{\partial x}\right) + \frac{\partial}{\partial y}\left(k_y\frac{\partial t}{\partial y}\right) + \frac{\partial}{\partial z}\left(k_z\frac{\partial t}{\partial z}\right)\right]dxdydzd\tau + q_g dxdydzd\tau = \rho(dxdydz)c\frac{\partial t}{\partial \tau}d\tau$$

Dividing both the sides by $dxdydzd\tau$,

$$\frac{\partial}{\partial x}\left(k_x\frac{\partial t}{\partial x}\right) + \frac{\partial}{\partial y}\left(k_y\frac{\partial t}{\partial y}\right) + \frac{\partial}{\partial z}\left(k_z\frac{\partial t}{\partial z}\right) + q_g = \rho c\frac{\partial t}{\partial \tau} \quad 2.10$$

The Eq. 2.10 is known as *general heat conduction equation.* Thermal conductivity is uniform at all places and in all directions for homogeneous and isotropic materials; thus,

$$k = k_x = k_y = k_z$$

Putting this condition in Eq. 2.10, we get

$$\frac{\partial^2 t}{\partial x^2}+\frac{\partial^2 t}{\partial y^2}+\frac{\partial^2 t}{\partial z^2}+\frac{q_g}{k}=\frac{\rho c}{k}\frac{\partial t}{\partial \tau}=\frac{1}{\alpha}\frac{\partial t}{\partial \tau} \quad 2.11$$

Where $\alpha = k/\rho c$ is known as the thermal diffusivity of the given material. Larger value of α indicates that the heat will diffuse into the given material more quickly. Materials with high thermal conductivity and low heat capacity have better diffusion characteristics. Using conventional calculation methods, Eq. 2.11 can be converted into spherical or cylindrical coordinates. The formulas are as follows.

(a) Cylindrical coordinates:

The coordinate system for cylindrical coordinates is shown in Fig. 2.6

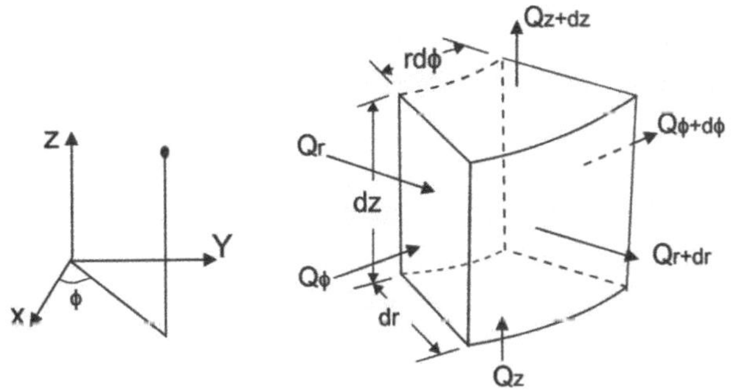

Fig. 2.6: Elemental volume for conduction analysis in cylindrical coordinates

$$\left(\frac{\partial^2 t}{\partial r^2}+\frac{1}{r}\times\frac{\partial t}{\partial r}+\frac{1}{r^2}\times\frac{\partial^2 t}{\partial \phi^2}+\frac{\partial^2 t}{\partial z^2}\right)+\frac{q_g}{k}=\frac{1}{\alpha}\frac{\partial t}{\partial \tau} \quad (2.12)$$

(b) Spherical coordinates:

The coordinate system for spherical coordinates is depicted in Fig. 2.7

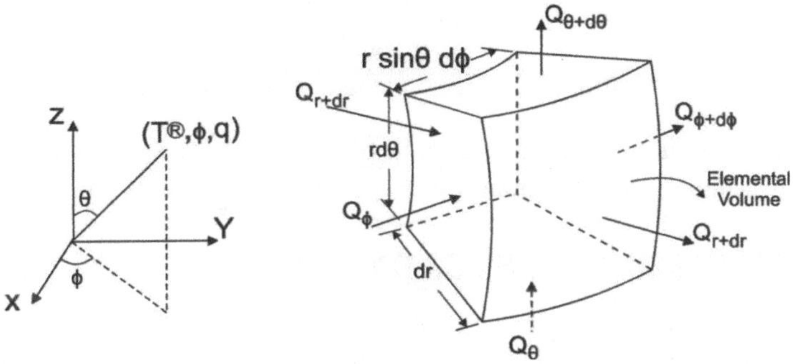

Fig. 2.7: Elemental volume for conduction analysis in spherical coordinates

$$\left[\frac{1}{r^2\sin^2\theta}\frac{\partial^2 t}{\partial \phi^2}+\frac{1}{r^2\sin\theta}\frac{\partial}{\partial \theta}\left(\sin\theta\frac{\partial t}{\partial \theta}\right)+\frac{1}{r^2}\frac{\partial}{\partial r}\left(r^2\frac{\partial t}{\partial r}\right)\right]+\frac{q_g}{k}=\frac{1}{\alpha}\frac{\partial t}{\partial \tau} \quad (2.13)$$

2.5 CONCEPT OF THERMAL RESISTANCE AND ELECTRICAL ANALOGY

In many ways, the conduction of heat through materials resembles that of electricity through conductors. A potential difference in a conductor drives the flow of electricity, just as a temperature differential in a conductor drives the flow of heat. In thermal conduction, heat is transferred from one point of a solid to another by the vibration of the solid's molecules as a result of their increased energy, as opposed to electric conduction, in which the electric charge is transferred from one point in a conductor to another by the motion of the electrons.

From Ohm's Law,

$$\text{Current } (i) = \frac{\text{Potential Difference } (dV)}{\text{Electrical Resistance } (R_e)}$$

and from Fourier's Law,

$$\text{Heat flow Rate } (Q) = \frac{\text{Temperature Difference } (dt)}{\text{Thermal Resistance } (dx/kA)}$$

From the above two relations, an analogy between electric current and heat flow may be established *i.e.*,

- Current is analogous to rate of heat flow.
- Voltage is comparable to temperature difference.
- Resistance is analogous to dx/kA. This quantity is known as *thermal resistance* (R_t).

The concept of thermal resistance is shown in Fig. 2.8.

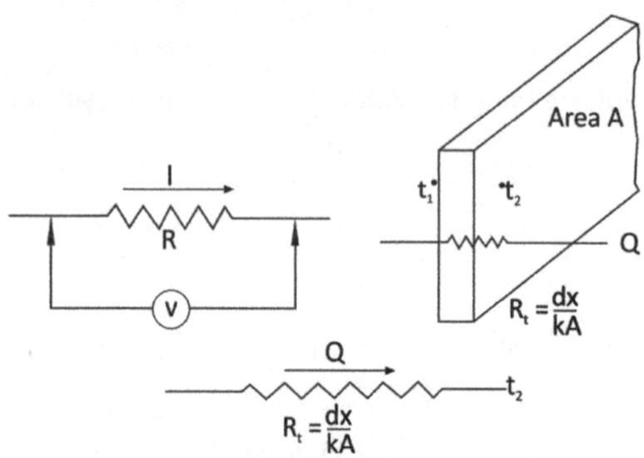

Fig. 2.8: Concept of thermal resistance

The electrical analogy is used to solve complex heat transfer problems with ease. Likewise, the concept of thermal resistance can be advantageously applied while making computations for heat transfer.

2.6 CONDUCTION UNDER STEADY STATE

This section focuses on using diffusion rate equations to calculate steady-state, one-dimensional heat transfer through bodies having simple geometries and no internal heat generation. The condition that exists in a heat conducting medium when temperatures at fixed points do not change with time is referred to as steady state. There is only one prevailing direction in one-dimensional heat flow. The heat flow in the other directions can be safely ignored. Clearly, only single space co-ordinate is needed to define the temperature variation within the body through which heat is conducted. Such situations are uncommon in real-world problems, and the assumption of one-dimensional heat conduction yields reasonably accurate estimates of heat conduction in the following cases.

- Heat flow through a plane wall in areas far from the edges. Edge effects are then ignored, and heat flow is solely determined by the co-ordinates measured normal to the plane of the wall.
- Flow of heat through a long hollow cylinder (such as a pipe) with uniform temperatures on its inside and outside surfaces. Heat conductivity is assumed to be solely dependent on radial distance as the co-ordinate in this case.
- Heat flow from a very thin wire or rod with different temperatures at each end. Temperature can be approximated to be uniform over any section of a sufficiently thin wire.

The appropriate differential heat equation is then suitably modified and solved to observe the temperature distribution using prescribed boundary condition, and the Fourier law is applied to estimate the heat flow rate.

2.6.1 Heat Conduction Through a Plane Wall

Consider heat conduction in one dimension through a homogeneous and isotropic wall of thickness δ having constant thermal conductivity k and uniform cross-sectional area A, as shown in Fig. 2.9. The wall's lateral faces are insulated, and have constant temperature t_1 at one face and t_2 at the other.

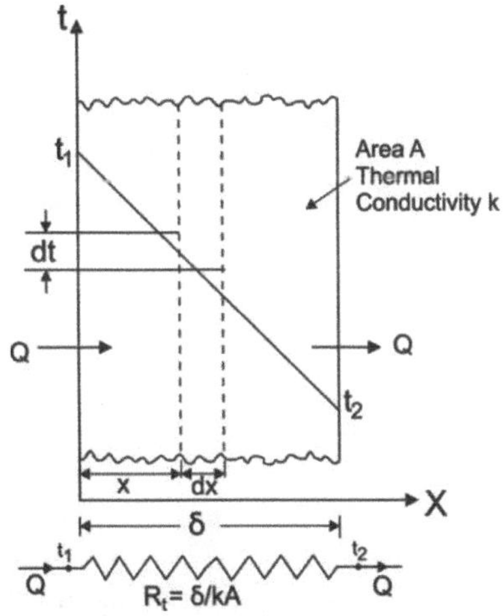

Fig. 2.9: Conduction through a plane wall

It can be noted that temperature varies only in the direction that is normal to the face of the wall. The temperature difference causes heat transfer in the positive x-direction.

Let us solve the general heat conduction equation in cartesian coordinates.

Recalling Eq. 2.11,

$$\frac{\partial^2 t}{\partial x^2}+\frac{\partial^2 t}{\partial y^2}+\frac{\partial^2 t}{\partial z^2}+\frac{q_g}{k}=\frac{\rho c}{k}\frac{\partial t}{\partial \tau}=\frac{1}{\alpha}\frac{\partial t}{\partial \tau}$$

The following conditions may be applied

$$\frac{\partial t}{\partial \tau}=0 \qquad \text{(Steady state condition)}$$

$$\frac{\partial^2 t}{\partial y^2} = \frac{\partial^2 t}{\partial z^2} = 0 \qquad \text{(One-dimensional)}$$

$$\frac{q_g}{k} = 0 \qquad \text{(No internal heat generation)}$$

$$\frac{\partial^2 t}{\partial x^2} = 0 \quad \text{or} \quad \frac{d^2 t}{dx^2} = 0 \qquad 2.14$$

Eq. 2.13 is the differential equation of second order which can be solved by double integration with respect to x.

$$\frac{dt}{dx} = C_1 \text{, and}$$

$$t = C_1 x + C_2 \qquad 2.15$$

The constants of integration C_1 and C_2 are assessed in light of the boundary conditions pertinent to the flow scenario. The boundary conditions in this case involve known temperatures which are as follows:

$$t = t_1 \quad \text{at} \quad x=0 \quad \text{and} \quad t = t_2 \quad \text{at} \quad x = \delta$$

Applying these boundary conditions in Eq. 2.15

$$t_1 = 0 + C_2 \quad \text{and} \quad t_1 = C_1\delta + C_2$$

The above two identities will give the constants of integrations as

$$C_2 = t_1 \quad \text{and} \quad C_1 = \frac{t_2 - t_1}{\delta}$$

Substituting the values of C_1 and C_2 in Eq. 2.15, we get

$$t = \left(\frac{t_2 - t_1}{\delta}\right)x + t_1$$

The above equation can be written in dimensionless form as,

$$\frac{t - t_1}{t_2 - t_1} = \frac{x}{\delta} \qquad 2.16$$

It is evident that the temperature distribution is linear across the wall and as there is no involvement of thermal conductivity. Thus, it may be concluded that temperature distribution is independent of the type of material.

Now, when we have obtained the expression of temperature gradient, we can proceed for heat flow computations by substituting the values in the Fourier equation.

$$Q = -kA\frac{dT}{dx}$$

$$\frac{dt}{dx} = \frac{d}{dx}\left[\left(\frac{t_2 - t_1}{\delta}\right)x + t_1\right] = \frac{t_2 - t_1}{\delta}$$

$$\therefore \quad Q = -kA\frac{t_2 - t_1}{\delta} = kA\frac{t_1 - t_2}{\delta} \qquad 2.17$$

It is clear from the above expression that heat flow is independent of x. Eq. 2.17 can also be written in the form of thermal resistance as,

$$Q = \frac{t_1 - t_2}{\delta / kA} = \frac{t_1 - t_2}{R_t} \qquad 2.18$$

Where R_t is the thermal resistance to heat flow.

Example 2.1: A slab of thickness L = 0.25 m has a temperature of 100 °C at one surface and 0 °C at the other. Find the heat flux across the slab if the material of the slab is copper (k = 388 W/mK).

Solution:

Given: L = 0.25 m; t_1 = 100 °C; t_2 = 0 °C and k = 388 W/mK.

Applying Fourier's Law,

$$Q = -kA\frac{dt}{dx}$$

Net heat flux,

$$q = \frac{Q}{A} = -kA\frac{(t_2 - t_1)}{L}$$

$$= -388 \times \frac{(0-100)}{0.25}$$

$$= 1.55 \times 10^5 W / m^2 \qquad \text{Ans.}$$

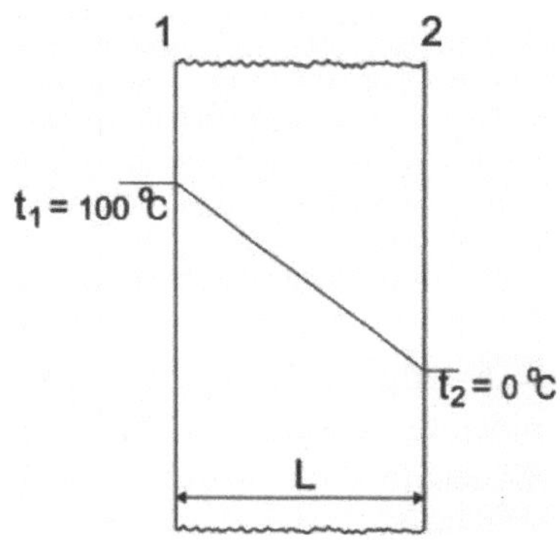

Fig. 2.10: Slab for example 2.1

Example 2.2: A 10 cm thick plane wall of area 3 m² is made with a material having a conductivity of 8.5 W/mK. The wall temperatures are 100 °C and 30 °C respectively. Find the temperature gradient and the heat flow across the wall.

Solution:

Given: t_2 = 30 °C, t_1 = 100 °C, δ = 10 cm, k = 8.5 W/mK

We know that the temperature gradient in the direction of heat flow is given by

$$\frac{dt}{dx} = \frac{(t_2 - t_1)}{\delta} = \frac{(30-100)}{0.1} = -700\ °C/m \qquad \text{(Ans.)}$$

The heat flow across the wall is calculated by Fourier's equation

$$Q = -kA\frac{dt}{dx}$$

$$= -8.5 \times 3 \times (-700)$$

$$= 17850\ W \text{ or } 17.85\ kW \qquad \text{Ans.}$$

Example 2.3: A cylindrical rod of length 20 cm and diameter of 3 cm is maintained at a temperature of 100 °C and 10 °C at its two ends. The cylindrical surface of the rod is fully insulated and the heat is flowing through the rod at a rate of 6 W. Find the thermal conductivity of the rod material.

Solution:

Given: δ = 20 cm or 0.2 m; D_c = 3 cm or 0.03m; Q = 6 W; t_1 = 100 °C; t_2 = 10 °C

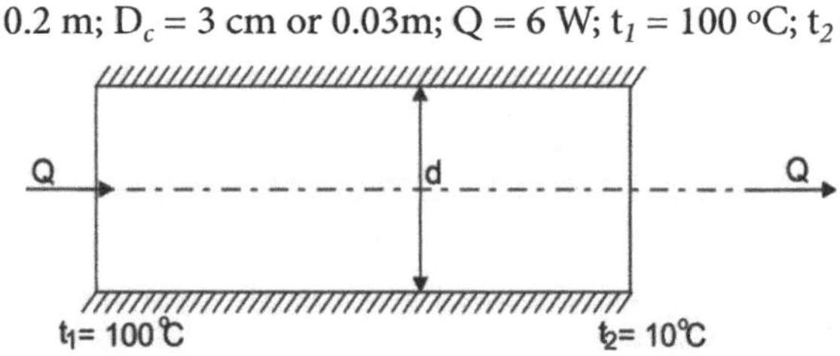

Fig. 2.11: Cylindrical rod for example 2.3

The heat flow through the rod is given by

$$Q = -kA_c \frac{dt}{dx} = \frac{kA_c(t_1 - t_2)}{\delta}$$

The area of cylindrical rod is $A_c = \pi/4 \times D_c^2$

$$\therefore \qquad Q = \frac{k\left(\pi/4 \times D_c^2\right)(t_1 - t_2)}{\delta}$$

$$6 = \frac{k\left(\pi/4 \times 0.03^2\right)(100 - 10)}{0.2}$$

$\therefore$ Thermal conductivity of the rod material is $k = \dfrac{6}{0.318} = 18.87\ W/mK$ Ans.

Example 2.4: The heat flux through a wood slab 50 mm thick, whose inner and outer surface temperatures are 40 and 20°C respectively, has been determined to be 40 W/m. What is the thermal conductivity of the wood?

Solution:

Given: $t_2 = 20$ °C, $t_1 = 40$ °C, L = 50 mm, q = 8.5 W/K

Recalling the Fourier conduction equation, the heat flux q is given

$$q = \frac{Q}{A} = -k\frac{dt}{dx} = k\frac{(t_1 - t_2)}{L}$$

Rearranging the above equation

$$k = q \times \frac{L}{(t_1 - t_2)}$$

$$= 40 \times \frac{0.05}{(40 - 20)}$$

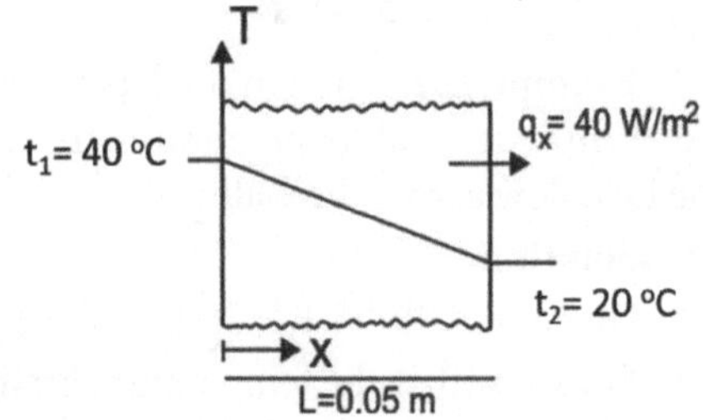

Fig. 2.12: Wooden slab for example 2.4

$\therefore$ Thermal conductivity of the rod material is $k = 0.1\ W/mK$ Ans.

Example 2.5: A plane slab having a thickness of 60 cm and thermal conductivity of 17.5 W/mK is exposed to a heat source from left side. The heat flux of the slab is given to be 530 W/m^{2}and the right side of the slab is maintained at a constant temperature of 38 °C. Find the maximum temperature of the slab and also the temperature at the mid-span of the slab. Assume steady state conduction with no heat generation.

Solution:

Given: thickness (δ) = 60 cm = 0.6 m; k = 17.5 W/mK; q = 530 W/m^2; t_2 = 38 °C (temperature on the right side)

Since, the slab is exposed to the heat source from the left side therefore, the rate of heat conducted from the slab

$$\text{Heat flux } q = k\frac{(t_1 - t_2)}{\delta}$$

$$530 = 17.5\frac{(t_1 - 38)}{0.6}$$

∴ Temperature at the left side of the slab

$$t_1 = \frac{530 \times 0.6}{17.5} + 38 = 56.17\ ^\circ C \qquad \text{Ans.}$$

This temperature is also the maximum temperature within the slab.

Recalling the expression for temperature distribution

$$t = t_1 + \frac{t_2 - t_1}{\delta} x$$

The temperature at the mid-span of the slab $x = 30$ cm

$$t = 56.17 + \frac{38 - 56.17}{0.6} \times 0.3$$

$$= 47.08\ ^\circ C$$

2.6.2 Heat Conduction Through a Composite Wall

A composite wall is one that is made up of several layers of different materials with varying thickness and thermal conductivity. The layers of varying conductive resistance are arranged in such a way that the composite wall achieves the requisite heat resistance as well as other characteristics such as lightweight, increased strength, durability, and so on. Each layer is made up of various materials such as wood, metals, refractory bricks, foam, glass wool, and more. As a result, the composite wall serves multiple functions in addition to thermal insulation.

Fig. 2.13 depicts a composite wall with three layers of various materials tightly bonded together. The layer thicknesses are δ_1, δ_2, δ_3, with thermal conductivities k_1, k_2, k_3 corresponding to average temperature conditions. The surface temperatures of the wall are t_1 and t_2, and the interface temperatures are t_2 and t_3.

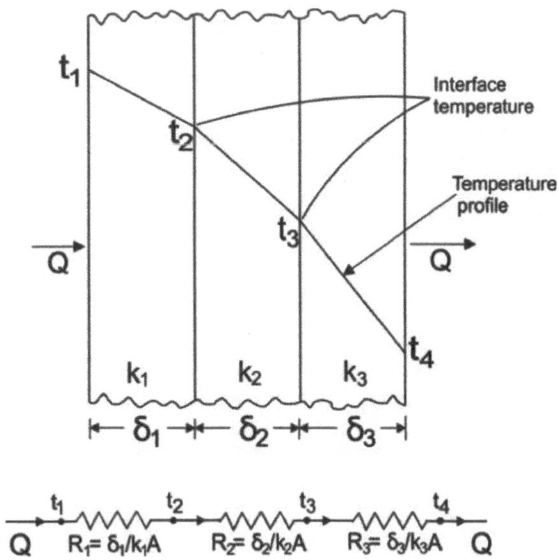

Fig. 2.13: Conduction through a composite wall

The heat flow is uniform on the entire wall under steady-state conditions, i.e., it is the same for each layer. Thus

$$Q=\frac{k_1A}{\delta_1}(t_1-t_2)=\frac{k_2A}{\delta_2}(t_2-t_3)=\frac{k_3A}{\delta_3}(t_3-t_4)$$

The above equation can be rewritten in terms of temperature drop across each layer,

$$(t_1-t_2)=\frac{Q\delta_1}{k_1A};\quad (t_2-t_3)=\frac{Q\delta_2}{k_2A}\quad \text{and}\quad (t_3-t_4)=\frac{Q\delta_3}{k_3A}$$

The overall temperature difference across the wall can be obtained by summing the above three equations,

$$(t_1-t_4)=Q\left(\frac{\delta_1}{k_1A}+\frac{\delta_2}{k_2A}+\frac{\delta_3}{k_3A}\right)$$

Therefore,

$$Q=\frac{(t_1-t_4)}{\left(\frac{\delta_1}{k_1A}+\frac{\delta_2}{k_2A}+\frac{\delta_3}{k_3A}\right)}$$

or, $$Q=\frac{(t_1-t_4)}{\left(R_{t_1}+R_{t_2}+R_{t_3}\right)} \qquad 2.19$$

It is to be noted that analysis of the composite wall makes an assumption that all layers are perfectly in contact with one another and that there is no temperature drop at the material interface.

Example 2.6: The wall of a reactor 32 cm thick, is having an inner layer of fire brick (k = 0.84 W/mK) and an insulating layer (k = 0.16 W/mK) next to it. The operating temperature of the reactor is 1325 °C and the ambient temperature is 25 °C. Find the thickness of the fire brick and the insulation. Also, find the heat loss presuming that the insulating material has a maximum temperature of 1200 °C.

Solution:

Given: t_1 = 1325 °C; t_2 = 1200 °C; t_3 = 25 °C; k_A = 0.84 W/mK; k_B = 0.16 W/mK;

$\delta_A + \delta_B = \delta$ =32 cm =0.32 m; $\delta_B = (0.32 - \delta_A)$

The heat flux is constant throughout the wall under steady state conditions,

$$q=\frac{t_1-t_3}{\delta_A/k_A+\delta_B/k_B}=\frac{t_1-t_2}{\delta_A/k_A}=\frac{t_2-t_3}{\delta_B/k_B}$$

Taking the first two quantities

$$\frac{1325-25}{\delta_A/0.84+\delta_B/0.16}=\frac{1325-1200}{\delta_A/0.84}$$

$$\frac{1300}{1.19\delta_A+6.25(0.32-\delta_A)}=\frac{105}{\delta_A}$$

$$\frac{1300}{2-5.06\delta_A}=\frac{105}{\delta_A}$$

$$1300\delta_A=105(2-5.06\delta_A)$$

$$1300\delta_A=210-531.3\delta_A$$

$$\delta_A = \frac{210}{1300+531.3} = 0.1146 \quad m \text{ or } 114.6 \; mm$$

$\therefore$ Thickness of insulation $\delta_B = 320 - 114.6 = 205.4 \quad mm$ Ans.

Heat loss per unit area (heat flux),

$$q = \frac{t_1 - t_2}{\delta_A / k_A} = \frac{1325-1200}{0.1146/0.84} = 916.23 \; W/m^2$$ Ans.

Example 2.7: Determine the rate of heat flow through a composite wall as shown in the Fig. 2.14 assuming that there is one-dimensional heat flow.

The parameters are as follows:

k_A = 150 W/mK

k_B = 30 W/mK

k_C = 65 W/mK

k_D = 50 W/mK

Solution:

Given: k_A = 150 W/mK; k_B = 30 W/mK; k_C = 65 W/mK; k_D = 50 W/mK; δ_A − 3 cm = 0.03 m; $\delta_B = \delta_C$ = 8 cm =0.08 m; δ_D = 5 cm = 0.05 m; t_1 = 400 °C; t_2 = 60 °C

The corresponding thermal circuit for the given composite wall is depicted in Fig. 2.14

The areas of the corresponding sections can be computed as follows:

$$A_A = 0.1 \times 0.1 = 0.01 \; m^2$$

$$A_B = 0.1 \times 0.03 = 0.003 \; m^2$$

$$A_C = 0.1 \times 0.07 = 0.007 \; m^2$$

$$A_D = 0.1 \times 0.1 = 0.01 \; m^2$$

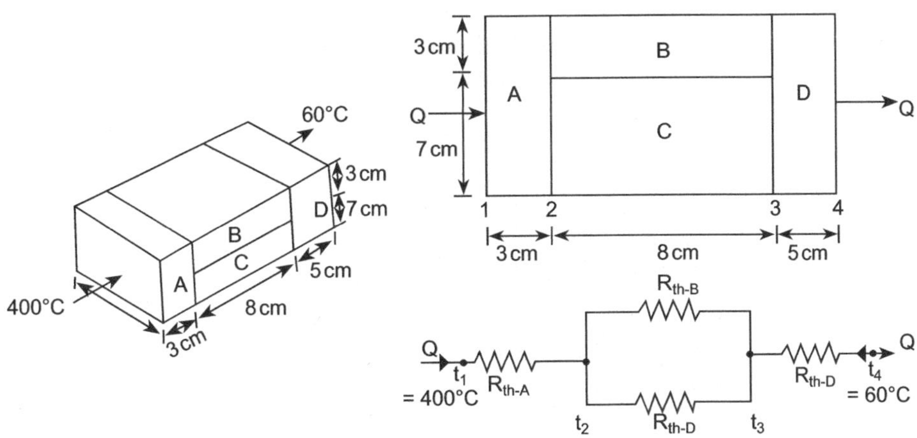

Fig. 2.14: Conduction through a composite wall for example 2.7

The thermal resistances of the various sections are given by

$$R_A = \frac{\delta_A}{k_A A_A} = \frac{0.03}{150 \times 0.01} = 0.02 \; K/W$$

$$R_B = \frac{\delta_B}{k_B A_B} = \frac{0.08}{30 \times 0.003} = 0.89\ K/W$$

$$R_C = \frac{\delta_C}{k_C A_C} = \frac{0.08}{65 \times 0.007} = 0.176\ K/W$$

$$R_D = \frac{\delta_D}{k_D A_D} = \frac{0.05}{50 \times 0.01} = 0.1\ K/W$$

The resistances R_B and R_C are in parallel and their equivalent resistance R_{eq} is given by

$$\frac{1}{R_{eq}} = \frac{1}{R_B} + \frac{1}{R_C} = \frac{R_B + R_C}{R_B \times R_C}$$

$$R_{eq} = \frac{R_B \times R_C}{R_B + R_C} = \frac{0.89 \times 0.176}{0.89 + 0.176} = 0.147\ K/W$$

The equivalent resistance is now in series with R_A and R_D. The total resistance for the entire circuit is given by

$$\sum R_t = R_A + R_{eq} + R_D$$

$$= 0.02 + 0.147 + 0.1$$

$$= 0.267\ K/W$$

Therefore, the heat transfer rate through the given system is

$$Q = \frac{\Delta t}{\sum R_t} = \frac{400 - 60}{0.267} = 1273.4W \qquad Ans.$$

2.6.3 Heat Conduction Through a Cylindrical Wall

The majority of process industries, oil refineries, and power plants all depend on cylindrical metal tubes. The heat exchangers, condensers, and boilers all have tubes inside of them, and these units are all connected via tubes. Therefore, it follows that the rate of radial heat transmission via a tube or any possible insulation that surrounds it is crucial.

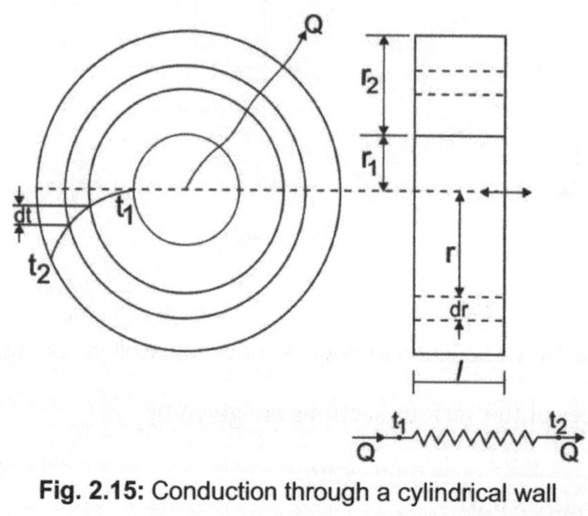

Fig. 2.15: Conduction through a cylindrical wall

Consider the transfer of heat via a cylindrical tube as shown in Fig. 2.15 with an inner radius of r_1 and outer radius of r_2, and a length of l. The temperature of the tube's interior and exterior surfaces, t_1 and t_2 are both constant, and the material's thermal conductivity k is constant over the specified temperature range. If the insulation at both ends is perfect, only radial heat flow is possible. Additionally, heat moves outward in a radial pattern if temperature t_1 at the inner surface is higher than temperature t_2 at the outer surface.To find the heat transfer through a cylindrical wall let us consider Eq. 2.12,

$$\left(\frac{\partial^2 t}{\partial r^2}+\frac{1}{r}\frac{\partial t}{\partial r}+\frac{1}{r^2}\frac{\partial^2 t}{\partial \phi^2}+\frac{\partial^2 t}{\partial z^2}\right)+\frac{q_g}{k}=\frac{1}{\alpha}\frac{\partial t}{\partial \tau}$$

Under the condition of steady state i.e., $\partial t/\partial \tau = 0$ and unidirectional heat flow i.e., $\{t \neq f(\varphi, x)\}$ and with no heat generation i.e., $q_g = 0$, the above equation will reduce to,

$$\frac{d^2t}{dr^2}+\frac{1}{r}\frac{dt}{dr}=0, \text{ or}$$

$$\frac{1}{r}\frac{d}{dr}\left(r\frac{dt}{dr}\right)=0$$

We know that $\frac{1}{r} \neq 0$

Therefore, the only choice is that

$$\frac{d}{dr}\left(r\frac{dt}{dr}\right)=0,$$

or $$r\frac{dt}{dr} = \text{constant (say } C_1) \qquad 2.20$$

On integration of Eq. 2.20 we get,

$$t = C_1 \log_e r + C_2 \qquad 2.21$$

In order to solve the above equation, we need to apply the relevant boundary conditions i.e.,

$t = t_1$ at $r = r_1$ and $t = t_2$ at $r = r_2$

By applying these boundary conditions, we can obtain the values of C_1 and C_2 as,

$$C_1 = \frac{t_1 - t_2}{\log_e \frac{r_2}{r_1}} \quad \text{and} \quad C_2 = t_1 + \frac{t_1 - t_2}{\log_e \frac{r_2}{r_1}}\log_e r_1 \qquad 2.22$$

Substituting the values of the constants from Eq. 2.22 into Eq. 2.21 we get,

$$t = t_1 + \frac{t_1 - t_2}{\log_e \frac{r_2}{r_1}}\log_e r_1 - \frac{t_1 - t_2}{\log_e \frac{r_2}{r_1}}\log_e r$$

or, $$(t - t_1)\log_e \frac{r_2}{r_1} = (t_1 - t_2)\log_e r_1 - (t_1 - t_2)\log_e r$$

On rearranging the above equation, we get

$$= (t_2 - t_1)\log_e r - (t_2 - t_1)\log_e r_1 \text{ or}$$

$$= (t_2 - t_1)\log_e \frac{r}{r_1}$$

Therefore, the above equation can be obtained in the dimensionless form as,

$$\frac{t-t_1}{t_2-t_1}=\frac{\log_e \frac{r}{r_1}}{\log_e \frac{r_2}{r_1}} \qquad 2.23$$

Apparently, radial conduction through a cylinder results in a logarithmic temperature distribution rather than a linear one like a plane wall. Further, the temperature in the cylinder can be stated solely as a function of radius at any point. Using the temperature distribution and the Fourier law, the conduction heat transfer rate is calculated between the inner and outer cylinder boundaries along isotherms or lines of constant temperature.

$$Q=-kA\frac{dt}{dr}$$

$$=\frac{d}{dr}\left[t_1+\frac{(t_1-t_2)}{\log_e \frac{r_2}{r_1}}\log_e r_1-\frac{(t_1-t_2)}{\log_e \frac{r_2}{r_1}}\log_e r\right]$$

$$=k(2\pi rl)\left(\frac{-(t_1-t_2)}{r\log_e \frac{r_2}{r_1}}\right)$$

$$Q=2\pi kl\frac{(t_1-t_2)}{\log_e \frac{r_2}{r_1}} \quad \text{or} \quad =\frac{(t_1-t_2)}{R_t} \qquad 2.24$$

It can be seen from Eq. 2.24 that for the conduction through a hollow cylinder, the thermal resistance is in the following form

$$R_t=\frac{\log_e \frac{r_2}{r_1}}{2\pi kl}$$

2.6.4 Heat Conduction Through a Composite Cylindrical Wall

Metal pipes designed to handle a hot fluid usually have multilayer cylinder walls to prevent heat losses. The pipe is typically covered with one or more layers of heat insulation. For instance, a steam pipe used in a steam power plant to transport high pressure steam may have a cylindrical metal wall, a layer of insulating material, and then a layer of protective plaster. The configuration is also known as a *pipe lagging system.*

Consider a composite cylindrical wall with three layers of different materials as depicted in Fig. 2.16. The temperature at the interface of any two adjoining layer is assumed to be same. We can obtain the heat transfer equation under steady state condition by applying the Fourier equation for cylindrical coordinates.

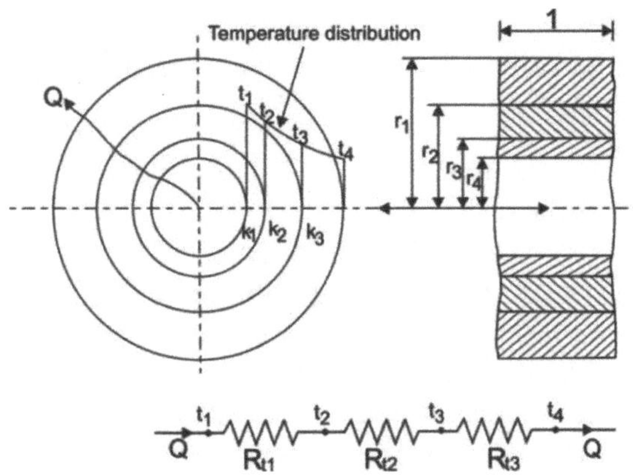

Fig. 2.16: Conduction through a cylindrical composite wall

$$Q = 2\pi k_1 l \frac{(t_1 - t_2)}{\log_e \frac{r_2}{r_1}} = 2\pi k_2 l \frac{(t_2 - t_3)}{\log_e \frac{r_3}{r_2}} = 2\pi k_3 l \frac{(t_3 - t_4)}{\log_e \frac{r_4}{r_3}} \quad 2.25$$

Rearranging Eq. 2.25 in the terms of temperature difference at each layer, we get,

$$t_1 - t_2 = \frac{Q}{2\pi k_1 l} \log_e \frac{r_2}{r_1}$$

$$t_2 - t_3 = \frac{Q}{2\pi k_2 l} \log_e \frac{r_3}{r_2}$$

$$t_3 - t_4 = \frac{Q}{2\pi k_3 l} \log_e \frac{r_4}{r_3}$$

On summation of the above equations we obtain,

$$t_1 - t_4 = Q\left[\frac{1}{2\pi k_1 l} \log_e \frac{r_2}{r_1} + \frac{1}{2\pi k_2 l} \log_e \frac{r_3}{r_2} + \frac{1}{2\pi k_3 l} \log_e \frac{r_4}{r_3}\right] \quad 2.26$$

Thus, the heat flow with three layers is,

$$Q = \frac{(t_1 - t_4)}{\left[\frac{1}{2\pi k_1 l} \log_e \frac{r_2}{r_1} + \frac{1}{2\pi k_2 l} \log_e \frac{r_3}{r_2} + \frac{1}{2\pi k_3 l} \log_e \frac{r_4}{r_3}\right]} \quad 2.27$$

In terms of thermal resistance, $Q = \dfrac{(t_1 - t_4)}{R_{t_1} + R_{t_2} + R_{t_3}}$ 2.28

2.6.5 Combined Conduction and Convection

We have seen many situations in our daily life which requires heating and cooling of fluids where heat exchange process occurs between surface and the surroundings. A thin boundary layer forms next to the wall when a flowing fluid interacts with a stationary surface, and in this layer, there is no relative velocity at the surface. This layer is known as the *stationary film* in a heat exchange process, and in this layer conduction and convection processes both contributes to the heat transfer phenomena. In case of flowing fluids, the heat exchange process mainly occurs due to convection. Therefore, the convective heat transfer rate between the fluid and the solid surface can be estimated by Newton's law of cooling given as,

$$Q = hA(t_s - t_f) \quad 2.29$$

Where h is the convective heat transfer coefficient or film coefficient, A is the surface area of the wall exposed to the heat transfer, t_s is the surface temperature of the wall and t_f is the temperature of the hot fluid. The unit of h is W/m²K and the *factor 1/hA* is known as *thermal resistance stationary film.*

Consider a hot fluid A and a cold fluid B having temperature t_a and t_b respectively being separated by a wall of thickness δ and thermal conductivity k as shown in Fig. 2.17. The film coefficients of the hot and cold fluids are h_a and h_b respectively and the temperatures at the interfaces of the hot and cold fluids are t_1 and t_2 respectively. The transfer of heat will take place from the hot fluid to the wall through convention then the heat conduction will take place through the wall and finally the heat will be transported from the wall to the cold fluid through convection.

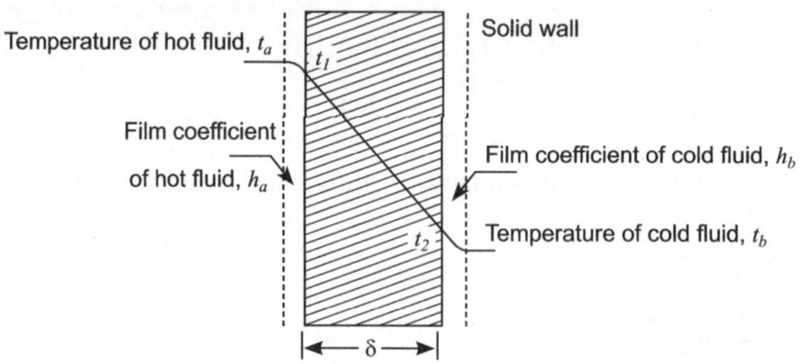

Fig. 2.17: Combined conduction and convection through a wall separating two fluids

The heat flow equations following the above-mentioned conditions can be expressed as,

$$Q = h_a A(t_a - t_1) = \frac{kA}{\delta}(t_1 - t_2) = h_b A(t_2 - t_b) \quad 2.30$$

The above expressions can be represented in the form of temperature difference as,

$$t_a - t_1 = \frac{Q}{h_a A};\ t_1 - t_2 = \frac{Q\delta}{kA} \text{ and } t_2 - t_b = \frac{Q}{h_b A}$$

The summation of above three equations will yield

$$t_a - t_b = Q\left[\frac{1}{h_a A} + \frac{\delta}{kA} + \frac{1}{h_b A}\right]$$

$$\text{Therefore, } Q = \frac{(t_a - t_b)}{\frac{1}{h_a A} + \frac{\delta}{kA} + \frac{1}{h_b A}} = \frac{(t_a - t_b)}{R_{t_1} + R_{t_2} + R_{t_3}} \quad 2.31$$

As seen in Eq. 2.31, the denominator is the sum of thermal resistances of the different sections through which heat is flowing. In general practice the heat flow equation through a composite section is expressed as;

$$Q = UA(t_a - t_b) \quad 2.32$$

Where U is referred as *overall heat transfer coefficient* and its unit is W/m²K.

On comparing Eq. 2.31 and Eq. 2.32, we get

$$UA(t_a - t_b) = \frac{(t_a - t_b)}{\frac{1}{h_a A} + \frac{\delta}{kA} + \frac{1}{h_b A}}$$

or $$U = \left[\frac{1}{h_a} + \frac{\delta}{k} + \frac{1}{h_b}\right]^{-1} \qquad 2.33$$

The overall heat transfer coefficient is computed to solve problems related to composite walls such as heat exchangers, walls of boilers, refrigerators, building structures etc. and it depends on wall geometry, film coefficients of the fluids at different surfaces and its thermal properties.

Example 2.8: A steel tube of thick wall has inner and outer diameters as 2 cm and 4 cm respectively. The periphery of the tube is covered with a 3 cm asbestos layer (k = 0.2 W/mK). The temperature outside the insulation layer is 1000 °C and the inside wall temperature of the tube is maintained at 300 °C. Determine the heat loss per metre length of the tube. Ignore the thermal conductivity of steel.

Solution:

Given: r_1 = 2/2 cm = 1 cm = 0.01 m; r_2 = 4/2 cm =2 cm = 0.02 m; r_3 = 2+3 = 5 cm = 0.05 m; t_1 = 600 °C; t_3 = 1000 °C; k_B = 0.2 W/mK

The heat transfer per metre length of the tube, *Q/L*:

$$Q = \frac{2\pi L(t_1 - t_3)}{\frac{\ln(r_2/r_1)}{k_A} + \frac{\ln(r_3/r_2)}{k_B}}$$

$$\therefore \frac{Q}{L} = \frac{2\pi(t_1 - t_3)}{\frac{\ln(r_3/r_2)}{k_B}}$$

$$= \frac{2\pi(600 - 1000)}{\frac{\ln(0.05/0.02)}{0.2}} = -548.6\ W/m \qquad Ans$$

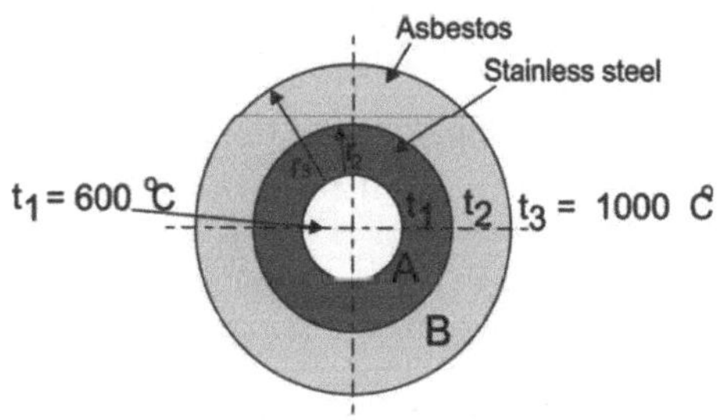

Fig. 2.18: Steel tube for example 2.8

Example 2.9: A metal tube is employed to carry steam having an outer diameter of 10 cm. The tube is insulated with two layers each having a thickness of 2.5 cm. The thermal conductivity of one material is 3 times that of the other while the temperatures at the junctions of the tube are constant. In order to minimize the heat loss, investigate the position of better insulation layer. Also, find the percentage saving in the heat dissipation with the given arrangement.

Solution:

Given: r_1 = 5 cm = 0.05 m; r_2 = 5 + 2.5= 7.5cm = 0.075 m; r_2 = 5 + 2.5= 7.5cm = 0.075 m; r_3 = 7.5 + 2.5= 10 cm = 0.1 m; k_1 = k; k_2 = 3k

Assuming the thermal conductivities of the two insulating materials be *k* and *3k*. It is known that the material with low thermal conductivity is a better insulator (superior).

Case I: Better insulating (superior) material placed next to the tube.

The thermal resistance of the given arrangement

$$R_{t1} = \frac{1}{2\pi k_1 L}\ln\frac{r_2}{r_1} + \frac{1}{2\pi k_2 L}\ln\frac{r_3}{r_2}$$

$$= \frac{1}{2\pi(k)L}\ln\frac{r_2}{r_1} + \frac{1}{2\pi(3k)L}\ln\frac{r_3}{r_2}$$

$$= \frac{1}{2\pi kL}\left[\ln\frac{0.075}{0.05} + \frac{1}{3}\ln\frac{0.1}{0.075}\right] = \frac{0.251}{\pi kL}$$

Heat dissipated with this arrangement

$$Q_1 = \frac{\Delta t}{R_{t1}} = \frac{\Delta t \times \pi kL}{0.251} = 3.99\pi kL\Delta t$$

Case II: Inferior insulating material placed near to the tube.

The thermal resistance of the given arrangement

$$R_{t2} = \frac{1}{2\pi k_2 L}\ln\frac{r_2}{r_1} + \frac{1}{2\pi k_1 L}\ln\frac{r_3}{r_2}$$

$$= \frac{1}{2\pi(3k)L}\ln\frac{r_2}{r_1} + \frac{1}{2\pi(k)L}\ln\frac{r_3}{r_2}$$

$$= \frac{1}{2\pi kL}\left[\frac{1}{3}\ln\frac{0.075}{0.05} + \ln\frac{0.1}{0.075}\right] = \frac{0.212}{\pi kL}$$

Heat dissipated with this arrangement

$$Q_2 = \frac{\Delta t}{R_{t2}} = \frac{\Delta t \times \pi kL}{0.212} = 4.73\pi kL\Delta t$$

The ratio of heat dissipation with superior material first to that of inferior material first is:

$$\frac{Q_1}{Q_2} = \frac{3.99}{4.73} = 0.843$$

Evidently, $Q_1 < Q_2$ which implies that heat dissipation is small when we place the material with low thermal conductivity (superior material) next to the steam tube and inferior insulating material afterwards.

Percentage saving in heat dissipation

$$= \frac{4.73 - 3.99}{3.99} \times 100 = 18.55\% \qquad \text{Ans}$$

Example 2.10: The inside diameter of a steel pipe is 20 mm ant it is 2 mm thick. The pipe is wrapped with 20 mm thick insulating material of thermal conductivity 0.05 W/mK. The inside and outside convective heat transfer coefficients are 10 and 5 W/m^{2}K respectively. Determine the overall heat transfer coefficient based on inside diameter of the pipe.

Solution:

Given: r_1 = 10 mm; r_2 = 10 + 2 = 12 mm; r_3 = 12 + 20 = 32 mm; k_1 = 0.05 W/mK; h_i = 10 W/m^{2}K; h_o = 5 W/m^{2}K

To compute the heat transfer through the insulated pipe let us start with the thermal resistance offered at each section of the pipe

$$\sum R_t = R_{t1} + R_{t2} + R_{t3} + R_{t4}$$

$$=\frac{1}{h_iA_i}+\frac{\ln\frac{r_2}{r_1}}{2\pi k_1 L}+\frac{\ln\frac{r_3}{r_2}}{2\pi k_2 L}+\frac{1}{h_oA_o}$$

$$=\frac{1}{h_i\times 2\pi r_1 L}+\frac{\ln\frac{r_2}{r_1}}{2\pi k_1 L}+\frac{\ln\frac{r_3}{r_2}}{2\pi k_2 L}+\frac{1}{h_o\times 2\pi r_3 L}$$

Therefore, heat transfer through the insulted pipe is given by

$$Q=\frac{\Delta t}{\sum R_t}=\frac{2\pi L(t_i-t_o)}{\frac{1}{h_ir_1}+\frac{1}{k_1}\ln\frac{r_2}{r_1}+\frac{1}{k_2}\ln\frac{r_3}{r_2}+\frac{1}{h_or_3}}$$

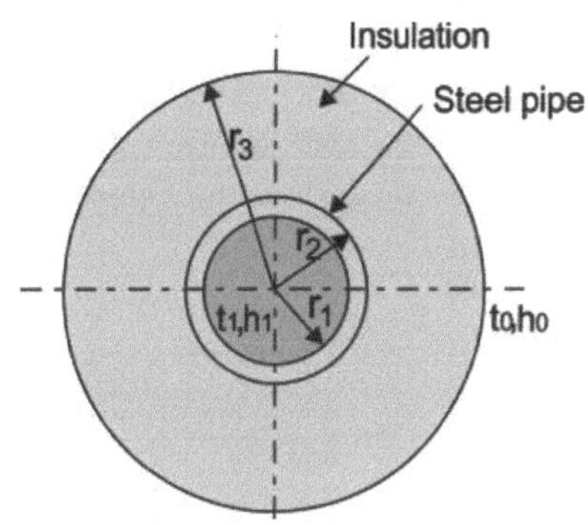

Fig. 2.19: Steel pipe for example 2.10

The thermal conductivity of steel pipe is not given, it can be readily ignored because the thermal resistance offered by steel is much less due to high thermal conductivity as compared to insulating material.

$$\therefore Q=\frac{2\pi L(t_i-t_o)}{\frac{1}{h_ir_1}+\frac{1}{k_2}\ln\frac{r_3}{r_2}+\frac{1}{h_or_3}}$$

Let U_i be the overall heat transfer coefficient based on inside area of the steel pipe, the heat rate can be written as

$$Q=U_iA_i(t_i-t_o)=U_i\times 2\pi r_1L(t_i-t_o)$$

Comparing above equations we get;

$$U_i\times 2\pi r_1L(t_i-t_o)=\frac{2\pi L(t_i-t_o)}{\frac{1}{h_ir_1}+\frac{1}{k_2}\ln\frac{r_3}{r_2}+\frac{1}{h_or_3}}$$

or $$\frac{1}{U_ir_1}=\frac{1}{h_ir_1}+\frac{1}{k_2}\ln\frac{r_3}{r_2}+\frac{1}{h_or_3}$$

or $$\frac{1}{U_i}=\frac{1}{h_i}+\frac{r_1}{k_2}\ln\frac{r_3}{r_2}+\frac{r_1}{h_or_3}$$

$$\frac{1}{U_i}=\frac{1}{10}+\frac{0.01}{0.05}\ln\frac{0.32}{0.12}+\frac{0.01}{5\times 0.32}$$

$$=0.1+0.196+0.063=0.359$$

Therefore ,

$$U_i=\frac{1}{0.359}=2.79\ W/m^2K$$ Ans

Example 2.11: The three layers A, B and C that make up a furnace wall are 250 mm thick, 100 mm thick, and 150 mm thick, respectively, with thermal conductivities of 1.65 W/mK, k, and 9.2 W/mK. The inside surface is exposed to hot gases at 1250°C with a convection coefficient of 25 W/m²K which heats the inside surface to a temperature of 1100°C. The outside surface faces air at 25°C with a convection coefficient of 12 W/m²K. Determine: (i) The thermal conductivity k (ii) The total heat transfer coefficient; (iii) All surface temperatures.

Solution:

Given: L_A = 0.25 m; L_B = 0.1 m; L_C = 0.15 m; k_A = 1.65 W/mK; k_B = k W/mK; k_C = 9.2 W/mK; t_h = 1250 °C (temperature of hot gases); t_1 = 1100 °C (inside wall surface temperature); t_c = 25 °C (temperature of cold air); h_h = 25 W/m²K; h_c = 12 W/m²K

(i) The thermal conductivity k or k_B:

The heat flux from the hot gases to the furnace wall is computed as

$$q = h_h(t_h - t_1) = 25(1250 - 1100) = 3750 \ W/m^2$$

We know that, $q = \dfrac{(t_h - t_c)}{\sum R_t}$

$$3750 = \frac{(1250-1100)}{\frac{1}{h_h}+\frac{L_A}{k_A}+\frac{L_B}{k_B}+\frac{L_C}{k_C}+\frac{1}{h_c}} = \frac{1225}{\frac{1}{25}+\frac{0.25}{1.65}+\frac{0.1}{k_B}+\frac{0.15}{9.2}+\frac{1}{12}}$$

$$= \frac{1225}{0.291+\frac{0.1}{k_B}}$$

$$3750\left(0.291+\frac{0.1}{k_B}\right) = 1225$$

$$\frac{0.1}{k_B} = \frac{1225}{3750} - 0.291 = 0.036$$

$$k_B = k = \frac{0.1}{0.036}$$

$$k = 2.82 \ W/mK$$ Ans.

(ii) The overall heat transfer coefficient, U:

$$U = \frac{1}{\sum R_t} = \frac{1}{\frac{1}{h_h}+\frac{L_A}{k_A}+\frac{L_B}{k_B}+\frac{L_C}{k_C}+\frac{1}{h_c}} = \frac{1}{\frac{1}{25}+\frac{0.25}{1.65}+\frac{0.1}{2.82}+\frac{0.15}{9.2}+\frac{1}{12}}$$

$$U = \frac{1}{0.327} = 3.06 \ W/m^2K$$ Ans.

(ii) All surface temperatures: t_1, t_2, t_3, t_4:

The heat flux will be equal for each section, therefore

$$q = q_A = q_B = q_C = 3750 \ W/mK$$

For the wall section A:

$$3750 = \frac{k_A(t_1 - t_2)}{L_A} = \frac{1.65(1100 - t_2)}{0.25}$$

$t_2 = 531.8\,^{\circ}C$ Ans

Similarly for wall section *B*:

$$3750 = \frac{k_B(t_2 - t_3)}{L_B} = \frac{2.82(531.8 - t_3)}{0.1}$$

$t_3 = 398.6\,^{\circ}C$ Ans

For wall section *C*:

$$3750 = \frac{k_C(t_3 - t_4)}{L_C} = \frac{9.2(398.6 - t_4)}{0.15}$$

$t_3 = 337.5\,^{\circ}C$ Ans

2.7 THERMAL POWER PLANT LAYOUT

Thermal power plant (or thermal power station) is a type of station where the heat energy obtained from the fossil fuel (e.g., coal, natural gas) is utilized to generate electricity. The heat from the burning of fuel is used to produce high pressure steam to run a steam turbine or can be directly used in gas turbines. Thermal power plants fulfil almost two third of the world's electrical energy demand.

The steam power plant makes use of heat to boil the water in a boiler drum to generate high-pressure superheated steam that drives a steam turbine coupled to an electric generator. The low-pressure exhaust steam from the turbine exit is allowed to enter a steam condenser where it is cooled to produce condensate. This condensate is recirculated to the boiler drum by passing it through the feed water pump and the cycle repeats. This is referred to as a *Rankine cycle.* A typical layout of coal based thermal power plant is shown in Fig. 2.20.

We will now discuss the basic working principle and components of a steam power plant:

- **Coal:** The requirement of coal is fulfilled by coal supplied from the coal mines which is stored in large storage yards. The coal is first cleaned in magnetic separator to filter out magnetic particles. The coal is washed to remove mud and other impurities, then the coal clinkers are crushed in ball mills to convert them into fine powder (to a consistency of talcum powder), this fine powder is called as *pulverized coal.* Pulverization is done for efficient and complete combustion of coal thereby improving the efficiency of boiler.
- **Boiler:** It is the component where water is converted into high pressure superheated steam. The pulverized coal is supplied to the furnace (cyclone furnace) where it is burnt and the heat produced is utilized to generate steam. Steel tubes run all along the boiler walls called as water walls. The flue gases from the boiler pass through the superheater, economizer, and air preheater before being discharged to the atmosphere through the chimney.
- **Superheater:** The hottest area of the boiler contains the superheater tubes. The superheater tube superheats the saturated steam generated in the boiler tubes to a temperature of roughly 540 °C. After that, the steam turbine is supplied with the superheated high-pressure steam.
- **Economizer:** As the flue gas leaves the superheater tubes, it still has a considerably high temperature which can be utilized to heat the feed water. An economizer is employed to heat the feed water before supplying to the boiler.
- **Alternator:** It is coupled with the steam turbine to produce electricity. This generated electrical energy is then amplified in a transformer and transported to the place where it has to be used.

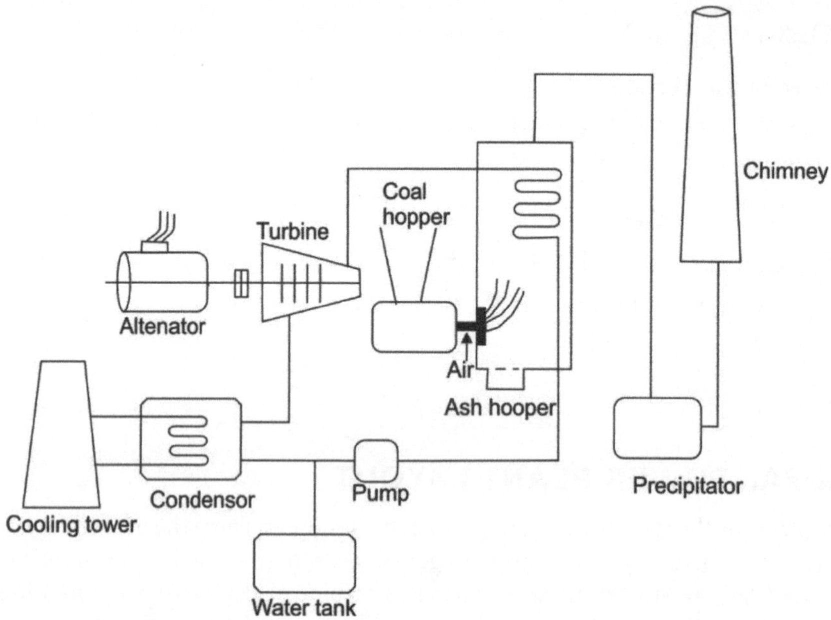

Fig. 2.20: Typical layout of a coal based thermal power plant

- **Air pre-heater**: The air pre-heater subsequently heats the air, which the primary air fan draws from the atmosphere. Coal is then injected into the boiler with previously heated air. Preheating the air has the benefit of improving coal combustion.
- **Steam Turbine:** It is fed with high pressure superheated steam, which makes the turbine blades to rotate. The steam turbine, which works as the primary mover, converts the energy of the steam into mechanical energy. As the steam goes through the turbine, its pressure and temperature decrease, and its volume increases. The condenser exhausts the low pressure expanded steam.
- **Condenser:** The exhaust steam is condensed in the condenser via cold water circulation. Here, both pressure and temperatures are reduced and steam is converted back into water. Condensing is necessary because compressing a gaseous fluid demands a large amount of energy in comparison to compressing a liquid. As a result, condensing improves cycle efficiency.
- **Feed Water Pump:** It pumps and returns the condensed water back to the boiler drum. Some water may be lost during the cycle, which can be replaced by water from an external source.

2.8 RANKINE CYCLE

The Rankine cycle is a thermodynamic cycle that turns heat into mechanical energy. It is also known as the *Rankine vapour cycle.* Heat energy is supplied to the system through a boiler where the working fluid (typically water) is converted to a high-pressure gaseous state (steam) in order to turn a turbine. After passing over the turbine the fluid is allowed to condense back into a liquid state as waste heat energy is rejected before being returned to boiler completing the cycle. Friction losses throughout the cycle are often neglected for the purpose of simplifying calculation, as such losses are usually much less significant than thermodynamic losses, especially in larger systems.

A simple steam power plant cycle and the related Rankine cycle on T-s plot is shown in Fig. 2.21 and Fig. 2.22 respectively. The four processes in a Rankine cycle are discussed below:

- **Process 1-2**: During this process the working fluid (saturated liquid) is pumped from low pressure to high pressure. Because the fluid is in the form of liquid at this point, the pump consumes less energy to operate. This is also referred as isentropic compression.
- **Process 2-3:** In this process, liquid under high-pressure enters the boiler drum and is converted into superheated steam under constant pressure by an external heat source. The required input energy can be determined graphically (using an enthalpy-entropy chart, or Mollier diagram) or numerically (using steam tables or software). Process 2-3 is a constant pressure heat addition into the boiler.
- **Process 3-4:** Power is generated when superheated steam expands through a turbine. As a result, the vapour's temperature and pressure are reduced. The condition of steam after expansion in turbine is wet vapour. Isentropic expansion occurs in process 3-4.
- **Process 4-1:** In this process, the wet vapour is allowed to enter into a condenser, where condensation process takes place and wet vapour is condensed into a saturated liquid at a constant pressure. Constant pressure heat rejection in a condenser is process 4-1.

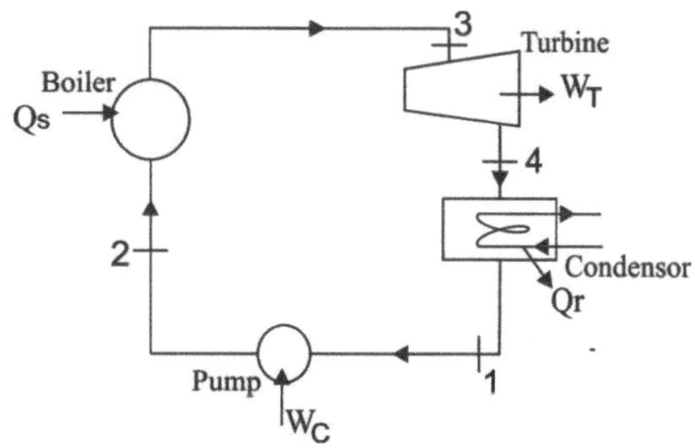

Fig. 2.21: A simple steam power plant cycle

The efficiency of Rankine cycle is given as

$$\eta_{rankine} = \frac{W_{net}}{Q_s} = \frac{W_T - W_P}{Q_s}$$

$$= \frac{(h_3 - h_4) - (h_2 - h_1)}{h_3 - h_2} \qquad 2.34$$

Usually, the pumping work is very small as compared to turbine work hence the pump work may be neglected. Therefore Eq. 2.34 reduces to,

$$\eta_{rankine} = \frac{h_3 - h_4}{h_3 - h_2} \qquad 2.35$$

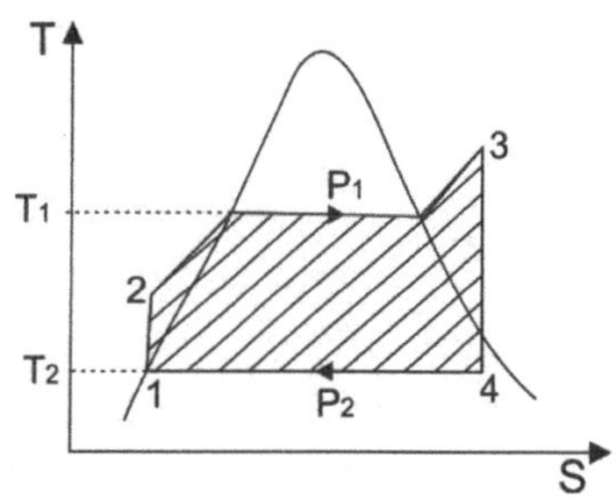

Fig. 2.22: T-S plot of Rankine cycle

Where h_1, h_2, h_3 and h_4 are the specific enthalpies at the respective points.

2.9 STEAM GENERATORS (BOILERS)

A boiler is a device that generates steam. Steam is produced at the desired temperature and pressure using thermal energy released during burning of fuel. The steam produced from the boiler is used for various purposes like;

(i) Production of mechanical work by expanding it in a steam turbine.

(ii) Execution of specific processes in the sugar mill, chemical, and textile sectors.

(iii) Heating residential and industrial structures.

There are certain requirements that a boiler need to fulfil are;

(i) The boiler should be safe under any operating condition and should confirm to the safety regulations laid down in the Indian Boiler Act.
(ii) It should be capable to supply steam as per the requirements.
(iii) In order to have maximum efficiency, the boiler should be designed to absorb maximum amount of heat.
(iv) Its initial cost should be low.
(v) The construction should be simple and free of any maintenance.
(vi) The various parts of the boiler should be easily accessible for repair and maintenance.
(vii) It should have no joints exposed to the flame.
(viii) It should be quick to start and load.
(ix) It should be sufficiently strong against corrosion and wear.

2.9.1 Classification of Boilers

The boilers can be classified on the following basis:

(1) On the basis of flow of water and hot flue gases
  (a) Fire tube boilers
  (b) Water tube boilers
(2) On the basis of water circulation
  (a) Natural circulation boilers
  (b) Forced circulation boilers
(3) On the basis of steam generation rate
  (a) Low-capacity boilers (< 20,000 kg/h)
  (b) Medium-capacity boilers (20,000 to 75,000 kg/h)
  (c) High-capacity boilers (> 75,000 kg/h)
(4) On the basis of steam pressure
  (a) Low pressure boilers (< 30 bar)
  (b) Medium pressure boilers (30 to 70 bar)
  (c) High pressure boilers (>150 bar)
  (d) Super critical boilers (>225 bar)
(5) On the basis of application
  (a) Stationary boilers
  (b) Mobile boilers (Marine, Locomotive)
(6) On the basis of orientation
  (a) Vertical
  (b) Horizontal
  (c) Inclined

2.9.2 Fire Tube Boiler

Fire tube boiler makes use of hot flue gases to heat water in the boiler drum. In this type of boiler, the hot flue gases are allowed to pass through the tubes, whereas the water is passed through the shell. Steam is produced as the water begins to boil. This steam is routed away from the boiler and into other areas of the process, while makeup water is injected to compensate for fluid loss. The water level within the shell must continually be maintained in this type of system so that the tubes are covered. Otherwise, the tubes may overheat and damage.

The various types of fire tube boiler are:

- Cochran Boiler
- Cornish Boiler
- Locomotive Boiler
- Velcon Boiler
- Simple Vertical and
- Scotch Marine Boiler.

2.9.2.1 *Cochran Boiler*

The Cochran fire tube boiler is a vertical, multi-tube boiler used for the generation of steam in small capacity. Cochran boilers are made in different sizes with steam generation capacities varying from 150 to 3000 kg/h with a working pressure up to 15 bar. The Cochran Boiler has an external cylinder shell and a firebox as depicted in Fig. 2.23. The hemispherical shape of the shell and firebox, as well as the boiler's hemispherical head provides maximal room and strength for withstanding the pressure generated up in Cochran boiler. In addition, the firebox's hemispherical crown is also advantageous for resisting heat losses.

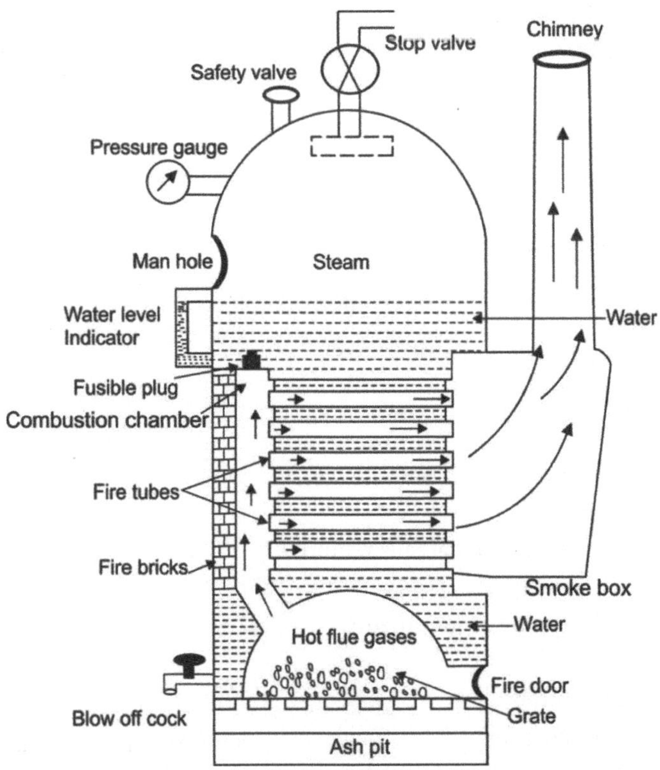

Fig. 2.23: Cochran boiler

A short pipe connects the firebox and combustion chamber. A series of tubes transport flue gases from the combustion chamber to the smoke box. These gases escape from the smoke box via a chimney. The combustion chamber is equipped with firebricks to resist high temperatures. On the shell, there is a cleaning manhole towards the top of the crown. It is located at the bottom of the firebox and is used to feed the grate and coal through a fire hole that is kept closed by a fire door.

2.9.2.2 Working of Cochran Boiler

Initially, the coal is put into the grate through a fire hole for burning. The ash produced during combustion is collected in an ash-pit beneath the grate and removed manually. Hot gases from the grate move through the combustion chamber towards horizontal fire tubes, where they transfer heat to water through convection. Exhaust gases from the fire tubes pass through the smoke box and are expelled into the atmosphere via a chimney. The smoke box has a door for cleaning the fire pipe and smoke box. The Cochran boiler can be used in variety of applications like paper and pulp industries, chemical processing units, refining units and power generation.

The main parts of Cochran boiler are as follows:

(i) Shell
(ii) Combustion chamber
(iii) Grate and ash pit
(iv) Fire hole
(v) Furnace
(vi) Flue gas pipes
(vii) Fire brick lining
(viii) Smoke box
(ix) Hemispherical dome
(x) Manhole
(xi) Chimney

The mounting and accessories are:

(i) Water level indicator
(ii) Pressure gauge
(iii) Stop valve and safety valve
(iv) Feed check valve
(v) Blow off cock
(vi) Fusible plug

Advantages of Cochran Boiler

(i) It is very compact so less floor area is required.
(ii) All types of fuel can be used.
(iii) Suitable for small capacity requirements.
(iv) Easy to transport.

Disadvantages of Cochran Boiler

(i) Steam generation rate is low.
(ii) Inspection and maintenance are difficult.
(iii) The pressure range of this boiler is limited.

2.9.3 Water Tube Boiler

The tubes in a water tube type boiler are filled with water rather than combustion gases, as opposed to combustion gases in a fire tube boiler. The combustion gases travels over the exterior surfaces of the tubes and transmit heat to the water within the tubes. The boilers of this class are used in modern high-capacity boilers. Because this boiler produces more steam, the boiler's overall efficiency is good. Water tube boilers can be designed without the need of extremely large and thick-walled pressure vessels makes them particularly appealing in applications that requires:

- A high output of steam (up to 500 kg/s)
- Steam at high pressure (up to 160 bar)
- Superheated steam (up to 550°C)

Some examples of water tube boilers are,

- Babcock and Wilcox Boiler
- Lamont Boiler
- Benson Boiler
- Loeffler boiler
- Yarrow boiler

2.9.3.1 *Babcock and Wilcox Boiler*

Babcock and Wilcox boiler is a stationary type water tube boiler comprising a steam-water drum and other components as illustrated in Fig. 2.24. A small tube links the steam water drum to the uptake header and down header. Each down space header is equipped with a mud box, and the mud accumulated is removed. The coal is fed from the hopper by a slow-moving chain. The firebrick baffle wall diverts the hot gases, allowing them to exit from the chimney. The dampers, which are controlled by a chain and a pulley are used to regulate the draught.

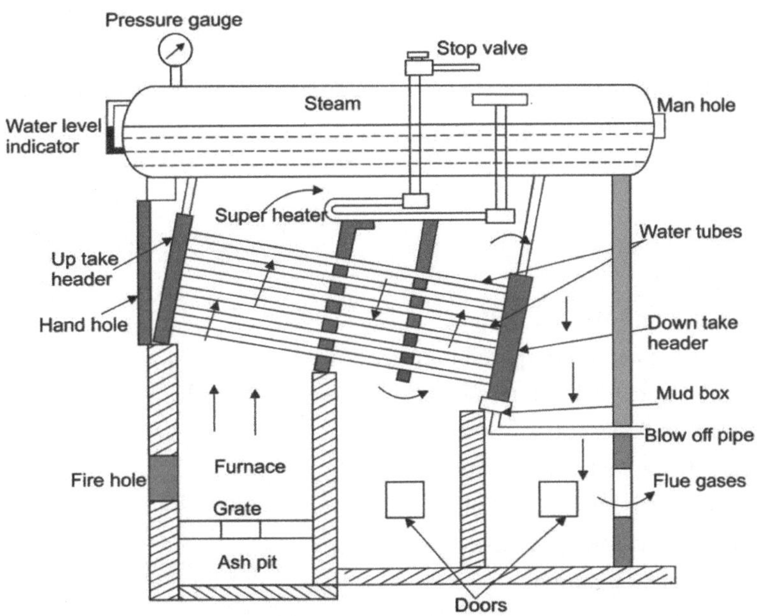

Fig. 2.24: Babcock and Wilcox boiler

2.9.3.2 Working of Babcock and Wilcox Boiler

Coal is allowed to be fed into the travelling fire-grate via fire door where it is burnt. The high temperature exhaust gases expand upward and travel through the bank of the water tubes. The baffles deflect the flue gases, causing them to move in a zig-zag pattern across the water tubes and alongside the superheater. The exhaust gases exit the atmosphere through the chimney. A blow-off cock is provided to remove mud and sediments from the mud box on a regular basis.

The section of the water tubes directly above the furnace is heated to a higher temperature than the rest of tubes. Water enters the drum through the uptake header. The steam and water are circulated in the drum at this point. Steam, which is lighter, is collected in the upper half of the drum. The water from the drum enters the water tubes through the down header. As a result, the constant circulation of water from the drum to the water tubes is maintained. Water circulation is maintained by convective currents and is referred to as *natural circulation*. Steam is transported from the steam space to the superheater tubes. In the superheater tubes, the steam gets superheated which is passed to the steam turbines to generate power.

The important components of Babcock and Wilcox boiler constitute:

1. Water drum
2. Uptake header
3. Down take header
4. Water tubes
5. Baffle plates
6. Furnace
7. Travelling grate
8. Fire door
9. Mud box
10. Feed check valve
11. Dampers

The various boiler mounting and accessories utilised in this type of boiler include the following:

1. Water level indicator
2. Pressure gauge
3. Safety valve
4. Superheater
5. Steam stop valve

Advantages of Babcock and Wilcox Boiler

1. It requires less space as compared to other boilers.
2. It has a high steam generation rate up to 2000 to 40000 kg/hr.
3. Replacement of boiler tubes is easier.
4. It can use solid as well as liquid fuel.
5. The draught loss is low in this boiler.
6. Its repair and cleaning are simple.
7. Its overall efficiency is high.

Limitations of Babcock and Wilcox Boiler

1. High maintenance cost.
2. It is not appropriate for water that is polluted or sedimentary. Scale may build up in the tubes when using impure or sedimentary water, which can lead to the tubes overheating and bursting. Before feeding into a boiler, the water treatment is required.
3. To function properly, a constant supply of feed water is required. The boiler overheats if water is not supplied continuously even for a short period of time. Therefore, water level must be strictly regulated.

The difference between fire tube and water tube boilers are mentioned in Table 2.1.

Table 2.1: Comparison between fire tube boiler and water tube boiler

Fire tube boiler	Water tube boiler
Hot flue gases travel inside the tubes, while water flows outside.	Water flows inside the tube, while hot flue gases pass through the tube.
Steam generation rate is low.	Steam generation rate is high.
Feed water treatment is not necessary since formation of scale may not lead to bursting of tube.	Feed water treatment is compulsory since formation of scale leads to bursting of tube due to the establishment of thermal stress.
The construction is difficult and complicated	The construction is simple.
Due to large shell, its transportation is difficult.	Due to smaller size of boiler shell, the transportation is easier.
Large floor space is required for a given power	It occupies the less floor space for a given power.
It is used for low pressure applications.	It is used for high pressure applications.

UNIT SUMMARY

- In this unit we have discussed about the three basic modes of heat transfer namely conduction, convection and radiation. In conduction the heat transfer occurs due to the lattice vibrations where more energetic particles transfer their energy to the less energetic particles. In convection the heat transfer occurs due to bulk motion and the distance between the particles is more than that in conduction (e.g., liquid and gases). In radiation there is no need of any medium and heat transfer takes place due to the electromagnetic waves, therefore it can travel in vacuum also.
- The governing equation for conduction heat transfer is expressed by Fourier's law given as

$$Q = -kA\frac{dt}{dx}$$

- The negative sign is inserted to satisfy the second principle of thermodynamics because heat flow takes place in the direction of negative temperature gradient. The generalized heat conduction equation for cartesian, cylindrical and spherical coordinate systems for the case of constant thermal conductivities are expressed as,

$$\frac{\partial^2 t}{\partial x^2}+\frac{\partial^2 t}{\partial y^2}+\frac{\partial^2 t}{\partial z^2}+\frac{q_g}{k}=\frac{\rho c}{k}\frac{\partial t}{\partial \tau}=\frac{1}{\alpha}\frac{\partial t}{\partial \tau}$$

$$\left(\frac{\partial^2 t}{\partial r^2}+\frac{1}{r}\times\frac{\partial t}{\partial r}+\frac{1}{r^2}\times\frac{\partial^2 t}{\partial \phi^2}+\frac{\partial^2 t}{\partial z^2}\right)+\frac{q_g}{k}=\frac{1}{\alpha}\frac{\partial t}{\partial \tau}$$

$$\left[\frac{1}{r^2\sin^2\theta}\frac{\partial^2 t}{\partial \phi^2}+\frac{1}{r^2\sin\theta}\frac{\partial}{\partial\theta}\left(\sin\theta\frac{\partial t}{\partial\theta}\right)+\frac{1}{r^2}\frac{\partial}{\partial r}\left(r^2\frac{\partial t}{\partial r}\right)\right]+\frac{q_g}{k}=\frac{1}{\alpha}\frac{\partial t}{\partial \tau}$$

Where, the property $\alpha = k/\rho c$ is known as the thermal diffusivity of the material.

- The flow of electricity through conductors has a resemblance with the heat conduction through materials. Therefore, we can apply electrical analogy to the heat conduction in a way that current and voltage are analogous to heat flow rate and temperature difference respectively and the thermal resistance R_t can be expressed as dx/kA.

- The heat conduction through a plane wall and a cylindrical wall under one-dimensional, steady state condition without heat generation is expressed as;

$$Q=\frac{t_1-t_2}{\delta/kA}=\frac{t_1-t_2}{R_t}$$

$$Q=2\pi kl\frac{(t_1-t_2)}{\log_e\frac{r_2}{r_1}} \quad \text{or} \quad Q=\frac{(t_1-t_2)}{R_t}$$

- The heat conduction through a composite wall and a composite cylindrical wall under one-dimensional, steady state condition without heat generation is expressed as,

$$Q=\frac{(t_1-t_4)}{\left(\frac{\delta_1}{k_1A}+\frac{\delta_2}{k_2A}+\frac{\delta_3}{k_3A}\right)} \quad \text{or} \quad Q=\frac{(t_1-t_4)}{(R_{t1}+R_{t2}+R_{t3})}$$

$$Q=\frac{(t_1-t_4)}{\left[\frac{1}{2\pi k_1 l}\log_e\frac{r_2}{r_1}+\frac{1}{2\pi k_2 l}\log_e\frac{r_3}{r_2}+\frac{1}{2\pi k_3 l}\log_e\frac{r_4}{r_3}\right]} \quad \text{or} \quad Q=\frac{(t_1-t_4)}{(R_{t1}+R_{t2}+R_{t3})}$$

- The combined effect of conduction and convection through a wall separating hot fluids A and cold fluid B can be expressed as,

$$Q=\frac{(t_a-t_b)}{\frac{1}{h_aA}+\frac{\delta}{kA}+\frac{1}{h_bA}}=\frac{(t_a-t_b)}{R_{t_1}+R_{t_2}+R_{t_3}}$$

- The overall heat transfer coefficient, U for the above case can be expressed as;

$$U=\left[\frac{1}{h_a}+\frac{\delta}{k}+\frac{1}{h_b}\right]^{-1}$$

- The thermal power plant utilizes heat energy from the fossil fuels to generate electricity. The steam power plant works on Rankine cycle and the major components of this cycle are boiler, steam turbine, condenser and a feed water pump.
- The efficiency of a Rankine cycle neglecting pump work is expressed as;

$$\eta_{rankine}=\frac{h_3-h_4}{h_3-h_2}$$

- The high pressure and high temperature steam is generated in a large pressure vessel called as a *steam generators or boilers.* Although, there are a number of ways we can classify boilers but the major one is based on the flow of water and hot flue gases i.e., fire tube boilers and water tube boilers. The fire tube boilers are used in industrial and domestic applications while water tube boilers are used for power generation. The components and working of Cochran (fire tube) boiler and Babcock and Wilcox (water tube) boiler is discussed in this unit. The advantages and limitations of both type of boilers is also explained along with the differences between fire tube and water tube boilers.

EXERCISE

Multiple Choice Questions

1. The necessary condition for heat transfer from one body to another is
   (a) One body's heat content must be greater than that of the other.
   (b) Both bodies must be in physical contact with one another.
   (c) One of the bodies must have a high thermal conductivity value.
   (d) A temperature difference between the bodies must exist.
2. Heat transfer in liquid and gases occurs mainly due to
   (a) Conduction (b) Convection
   (b) Radiation (d) All of the above
3. Which of the heat transfer process is unaffected by the material medium between the heat source and receiver
   (a) Conduction (b) Convection
   (b) Radiation (d) Both conduction and convection
4. What is meant by steady state heat flow
   (a) No temperature difference between the two bodies
   (b) Negligible heat flow
   (c) Uniform rate in temperature rise of a body
   (d) Constant heat flow rate i.e. heat flow rate independent of time
5. Which of the statement is wrong
   (a) The heat transfer process is an irreversible process
   (b) A temperature gradient must exist for heat exchange to take place
   (c) A material medium is always necessary for heat transfer
   (d) Heat flow always occurs from higher temperature lower temperature
6. Identify the wrong statement with respect to the Fourier's heat conduction equation
   (a) Fourier law applies to all situations, regardless of their state
   (b) From first principles, Fourier law can be deduced
   (c) Fourier law is a vector expression that indicates heat flows in the direction of decreasing temperature
   (d) Fourier law helps to define thermal conductivity of the heat conducting medium

7. Thermal conductivity is the rate of heat transfer
   (a) Per unit area per unit temperature difference
   (b) Per unit area per unit thickness
   (c) Per unit area per unit temperature difference and per unit wall thickness
   (d) None of the above
8. Variation of temperature in steady state heat conduction through a wall is
   (a) Linear (b) Logarithmic
   (c) Hyperbolic (d) Parabolic
9. A composite wall has two layers of different materials with thermal conductivities k_1 and k_2; if each layer is the same thickness, the wall's equivalent thermal conductivity is
   (a) $\frac{k_1 k_2}{k_1 + k_2}$ (b) $\frac{2k_1 k_2}{k_1 + k_2}$
   (c) $\frac{2k_1}{k_1 + k_2}$ (d) $\frac{k_1 k_2}{2(k_1 + k_2)}$
10. Thermal resistance in case of combined conduction and convection is
   (a) 1/UA (a) L/UA
   (b) U/kA (d) None of the above
11. Heat is transferred from a hot fluid to a cold fluid through a plane wall of thickness δ, surface area A, and thermal conductivity k. The heat transfer coefficients for the given setup are h_1 and h_2. What is the thermal resistance for the given setup
   (a) $A\left(h_1 + \frac{k}{\delta} + h_2\right)$ (b) $A\left(\frac{1}{h_1} + \frac{\delta}{k} + \frac{1}{h_2}\right)$
   (c) $\frac{1}{A}\left(h_1 + \frac{k}{\delta} + h_2\right)$ (d) $\frac{1}{A}\left(\frac{1}{h_1} + \frac{\delta}{k} + \frac{1}{h_2}\right)$
12. The temperature distribution associated with radial conduction through a cylinder for steady state and constant value of thermal conductivity is
   (a) Linear (b) Parabolic
   (c) Exponential (d) Logarithmic
13. A composite wall is composed of two layers of thickness δ_1 and δ_2 with thermal conductivities k and $2k$ and equal surface area normal to the direction of heat flow. The exterior surface of the composite wall is 100 °C and 200 °C. The minimum surface temperature at the intersection is 150 degrees Celsius. What is the wall thickness ratio?
   (a) 1:1 (b) 2:1
   (c) 1:2 (d) 2:3
14. A composite slab is made up of two layers with thermal conductivities that are in the ratio of 1:2. If the thickness of each layer is the same, the slab's equivalent thermal conductivity is
   (a) 2 (b) 1/3
   (c) 2/3 (d) 4/3

15. The rate of heat conduction in a cylindrical tube is commonly expressed as
(a) Length only
(b) Area only
(c) Per unit area
(d) Per unit length

16. A hollow cylinder with inner radius r_1 and outer radius r_2 experiences steady-state heat transfer, resulting in constant surface temperatures t_1 and t_2 at radii r_1 and r_2, respectively. The radial heat flow per unit length of cylinder for constant thermal conductivity k is given by
(a) $\dfrac{(t_1 - t_2)}{2\pi k \log_e \dfrac{r_2}{r_1}}$
(b) $\dfrac{2\pi(t_1 - t_2)}{k \log_e \dfrac{r_2}{r_1}}$
(c) $\dfrac{k(t_1 - t_2)}{2\pi \log_e \dfrac{r_2}{r_1}}$
(d) $\dfrac{2\pi k(t_1 - t_2)}{\log_e \dfrac{r_2}{r_1}}$

17. Identify the correct sequence in the context of thermal power plant layout
(a) Turbine → Pump → Condenser → Boiler
(b) Pump → Condenser → Turbine → Boiler
(c) Boiler → Turbine → Condenser → Pump
(d) Turbine → Condenser → Pump → Boiler

18. A Condenser is employed in a steam power plant to
(a) Condense the low pressure steam coming out of the turbine exit
(b) Condense high pressure steam coming out of the boiler drum
(c) Pump the water to the boiler drum
(d) None of the above

19. Water tube boilers are those in which.....
(a) Flue gases pass through tubes and water around it
(b) Water passes through the tubes and flue gases around it
(c) Water and flue gases are independent to move anywhere
(d) There is no steam drum

20. For a given power fire tube boilers require
(a) Less floor area
(b) Large floor area
(c) Moderate floor area
(d) Nothing can be said

21. For high pressure applications
(a) Water tube boilers are used
(b) Fire tube boilers are used
(c) Neither can be used
(d) Both can be used

22. The steam generation capacity of water tube boiler as compared to fire tube boiler is
(a) Equal
(b) Low
(c) High
(d) None of the above

23. Which one is not an advantage of of a fire tube boiler
(a) Low initial cost
(b) Quick response to change in load
(c) Reliability in operation
(d) More draught is required

24. Cochran boiler is a......
   (a) Horizontal fire tube boiler
   (b) Horizontal water tube boiler
   (c) Vertical water tube boiler
   (d) Vertical fire tube boiler
25. The orientaiton of water tubes in a Babcock and Wilcox boiler are
   (a) Horizontal
   (b) Vertical
   (c) Inclined
   (d) None of the above
26. Babcock and Wilcox boiler is a forced circulation boiler
   (a) True
   (b) False
27. The function of baffle plates in Babcock and Wilcox boiler is
   (a) To direct the flow of water
   (b) To direct the flow of hot flue gases
   (c) To direct the flow of steam
   (d) To direct the flow of air in combustion chamber
28. Inspection and maintenance of Cochran boiler is
   (a) Easy
   (b) Difficult
   (c) Not possible
   (d) None of the above
29. Babcock and Wilcox boiler is an externally fired boiler
   (a) True
   (b) False
30. In case of Cochran boiler the feed water treatment is compulsory
   (a) True
   (b) False

Answers for Multiple Choice Questions							
1. (d)	2. (b)	3. (c)	4. (d)	5. (c)	6. (b)	7. (c)	8. (a)
9. (b)	10. (a)	11. (d)	12. (d)	13. (c)	14. (d)	15. (d)	16. (d)
17. (c)	18. (a)	19. (b)	20. (b)	21. (a)	22. (c)	23. (d)	24. (d)
25. (c)	26. (b)	27. (b)	28. (b)	29. (a)	30. (b)		

Short and Long Answer Type Questions

1. Explain the phenomena of: (i) conduction (ii) convection (iii) radiation
2. Write the Fourier heat conduction equation. What is the physical significance of each term?
3. Explain the reason for why there is a negative sign in the Fourier rate equation for heat transfer by conduction?
4. Describe the concept of electrical analogy by considering a plane composite wall of a building.
5. By using general conduction equation with suitable assumptions show that the temperature distribution through a plane wall is linear?
6. Demonstrate that the temperature change for heat conduction through a cylindrical wall with constant thermal conductivity is logarithmic?
7. What do you understand by overall heat transfer coefficient? In terms of film coefficients and pipe radii, obtain an expression for the overall heat transfer coefficient based on outer area for the transfer of heat from a hot flowing fluid in a pipe to a cold fluid to which the pipe is exposed?

8. Develop an expression for the thermal resistance for heat transfer through the wall of a hollow cylinder with inner and outer radii of r_i and r_o. The temperature of the cylinder's inside and exterior surfaces is kept at T_i and T_o, respectively, with T_i being higher than T_o.
9. Derive the expression for heat transfer from a composite cylindrical wall having 3 layers?
10. Draw a schematic diagram of a steam power plant and explain its working?
11. What are the essential requirements that are to be fulfilled by a boiler?
12. Classify steam boilers. Explain with the help of examples?
13. Briefly describe the essential components of a steam power plant and their functions?
14. What do you understand by fire tube and water tube boiler and water tube boiler? List the names of boilers of each type?
15. Explain Rankine cycle with superheat with the help of neat sketch. Also draw the T-s plot for the same?
16. Explain the working of Cochran boiler with the help of neat sketch. Also explain its advantages and disadvantages?
17. What are the differences between fire tube and water tube boiler?
18. Describe the working of Babcock and Wilcox boiler with the help of neat sketch. List the advantages and disadvantages of this boiler?

Numerical Problems

1. Calculate the heat flux through a homogeneous slab that is 3.9 cm thick and whose two faces are kept at constant temperatures of 35°C and 25°C. Take the thermal conductivity of the wall material as 2.0×10^{-4} kW/mK.

   [0.075 kW/m²]

2. The inside walls of an oven is maintained at a temperature of 850 °C with the help of a thermostat. The walls of oven are 50 cm thick (k = 0.3 W/mK), compute the thermal resistance when the outside wall temperature is 250 °C and the heat flow per m² of the wall surface. Also find the temperature at a distance of 20 cm from interior side.

   [1.67 K/W, 610 °C]

3. The top and bottom surfaces of a solid cylindrical rod are maintained at constant temperatures of 20°C and 95°C while the side surface is perfectly insulated. The rate of heat transfer through the rod is to be determined for the cases of copper, steel, and granite rod. The thermal conductivities are given to be $k = 380$ W/mK for copper, $k = 18$ W/mK for steel, and $k = 1.2$ W/mK for granite.

   [373.1 W, 17.7 W, 1.2 W]

4. A furnace's wall is constructed of an inner layer of silica brick that is 120 mm thick, covered by an outer layer of 240 mm thick magnesite brick. The temperatures at the silica brick wall's interior surface and the magnesite brick wall's exterior surface are 725°C and 110°C, respectively. If silica and magnesite bricks have thermal conductivities of 1.7 W/mK and 5.8 W/mK, respectively, determine the rate of heat loss per unit area of walls.

   [5324.7 W/m²]

5. Calculate the heat flux through a 10 cm thick brick wall (k = 0.69 W/mK) with 105°C and 10°C maintained on its two faces. How does this heat flux change when 3 cm thick layers of magnesia insulation are applied to both the inner and outside faces of the wall? Determine the interface temperature for the composite wall as well.

[972 W/m^2, 140.5 W/m^2, 93.1 °C, 72.8 °C]

6. A plane wall 2.5 m long and 2 m high must be insulated so that the amount of heat flowing per hour through this wall does not exceed 8500 kJ. The temperature of the wall beneath the insulation is 400°C, while the exterior surface of the insulation is 50°C. Calculate the following: (a) thickness of the insulation layer, which is made of asbestos material with a thermal conductivity of k = 0.537 W/mK (b) temperature in the insulation layer at 0.05 m from the inner surface.

[0.114 m, 247 °C]

7. Water at 90 °C is stored in a mild steel tank (k = 50 W/mK) of thickness 10 mm. The inside and outside surface heat transfer coefficients are 2800 and 11 W/m^{2}K, respectively. The atmospheric temperature is 20 °C, determine the heat flux and the outside surface temperature of the tank.

[820 W/m^2, 90 °C]

8. 5 cm of high temperature insulation (k = 0.85 W/mK) and 4 cm of low temperature insulation (k = 0.72 W/mK) are wrapped around a 15 cm outside diameter steam pipe. The steam temperature is 500°C, and the ambient air temperature is 40°C. Calculate the heat loss from 1000 m of pipe length while ignoring thermal resistances on the steam and air sides and the metal wall.

[2928 kW]

9. A furnace wall is made up of two layers: 8 cm of fire clay (k =1.2 W/mK) close to the fire box and 0.6 cm of mild steel (k = 35 W/mK) on the outside. The inside surface of the brick is 900 K, while the steel is surrounded by air at 300 K, with an outer surface coefficient of 5 W/m^{2}K. Compute the rate of heat through the furnace wall and the temperature of steel at the outside surface.

[6.28 kW, 397 °C]

10. A furnace's flat wall is composed of fire brick (k = 1.4 W/mK), insulating brick (k = 0.2 W/mK), and building brick (k = 0.8 W/mK) of 25 cm, 12.5 cm, and 25 cm thicknesses, respectively. The temperature inside the wall is 600°C, while the outside temperature is 20°C. If the exterior surface's heat transfer coefficient is 10 W/m^{2}K. Estimate the heat loss per sq. m of the wall area and outside wall surface temperature of the furnace.

[460 W/m^2, 67 °C]

11. A household refrigerator door has an area of 0.7 m^2 and is made up of a thin metal sheet with a 25 mm thick layer of insulation on the inside. This insulation has a thermal conductivity of 0.25 W/mK and a heat transfer of 10 W/m^{2}K on each side of the door. The temperature in the chilled compartment and the room is 0°C and 20°C, respectively. Ignore the thermal resistance by sheet metal and determine the heat flow rate via the door and also the sheet metal temperature.

[46.6 W, 13.3 °C]

12. Hot air at 60°C is flowing through a steel pipe with a diameter of 100 mm. The pipe is insulated with two layers of different insulating materials with thicknesses of 50 mm and 30 mm respectively, and thermal conductivities of 0.23 and 0.37 W/mK. Inside and outside heat transfer coefficients are 58 and 12 W/m^{2}K respectively. The ambient temperature is 25°C. Determine the rate of heat loss from a 50 m long pipe. Ignore the resistance of the steel pipe.

[2.33 kW]

13. A metallic pipe of internal and external diameters of 5 cm and 6.5 cm respectively (k = 45 W/mK), is carrying steam at 200 °C and the atmospheric temperature is 25°C. The pipe is covered with a high temperature insulation of 2.75 cm thickness (k = 1.12 W/mK). If the inside and outside heat transfer coefficients are 4651 and 11.6 W/m^{2}K respectively, compute the heat loss per meter of the pipe and the temperatures at the interfaces.

[545 W, 200 °C, 198.7 °C, 150.5 °C]

14. A steam pipe of thickness 5 mm and diameter 16 cm (k = 59 W/mK) is wrapped with two layers of insulating materials. The first layer is 3 cm thick (k = 0.18 W/mK) and the second one is 5 cm thick (k = 0.094 W/mK). The steam passing through the pipe has a temperature of 300 °C and the ambient temperature is 30 °C. Considering the inside and outside heat transfer coefficients as 31 and 6 W/m^{2}K respectively, calculate the heat loss per metre length of the pipe.

[217 W/m]

15. A heater element has a rating of 2000 W and its area is 0.04 m^2 is lagged at its bottom with an insulation of 5 cm thickness (k = 1.5 W/mK) and on the front side by a 10 cm thick plate (k = 45 W/mK). The back side is exposed to air at 5 °C with a convection coefficient of 10 W/m^{2}K and the front is exposed to air at 15 °C with a convection coefficient of 250 W/m^{2}K. Find the temperature of heater element and heat flow rate in the room.

[1937 W, 219 °C]

PRACTICAL

1. To determine the thermal conductivity of metal rod.
2. To determine the thermal conductivity of a liquid.
3. To find out the overall thermal heat transfer coefficient and plot the temperature distribution in case of a composite wall.
4. To find out the Heat Transfer Coefficient of vertical cylinder in natural convection.
5. Study of various types of boilers with the help of models.
6. Study of various types of mountings and accessories

KNOW MORE

- There are many interesting applications and facts related to heat transfer and power generation and we are going to discuss some of them in this section. When we talk of heat transfer from an insulating body it is always anticipated that addition of insulation on a material decreases the heat transfer, but there are instances when addition of insulation outside a cylindrical or spherical wall increases the heat loss. In fact, under certain conditions, it actually increases the heat transfer up to a certain thickness of insulation. The radius of insulation at which heat transfer is maximum is called as critical radius or critical thickness of insulation (r_c). The critical radius of insulation only depends on the thermal quantities k and ho.
- Thus, for a cylindrical body $r_c = \frac{k}{h_o}$ and for a spherical body $r_c = \frac{2k}{h_o}$

- In electrical wires the insulation is provided to enhance the heat transfer so that the wires are not overheated while in power plants the insulation is provided to prevent the heat loss so that maximum efficiency is achieved.
- In case of power plant boilers, it is important to note that modern power plants are employing supercritical boilers also called as once-through boilers. These are drumless boilers and operate above supercritical pressure (221 bar, 374 °C). These boilers are able to produce steam at higher pressure and temperature as compared to conventional water-tube boilers. Therefore, more work is obtained with less fuel consumption which means higher efficiency of the plant with reduced CO_2 emission. With the further advancements in material technology, newer materials are coming into the market which can withstand very high temperature and pressure. This has facilitated the engineers and scientists to work on Advanced Ultra-Supercritical (AUSC) boilers

REFERENCES AND SUGGESTED READINGS

1. J. P. Holman. Heat Transfer, McGraw-Hill, New York, 2002
2. M.P. Poonia and S.C. Sharma. Basic Mechanical Engineering – Khanna Publishing House, New Delhi, India, 2005.
3. F. P. Incropera and D. P. DeWitt. Introduction to Heat Transfer, John Wiley & Sons, New York, 2002.
4. D. S. Kumar. Heat & Mass Transfer, S. K. Kataria & Sons, New Delhi, India, 2014.
5. R. K. Rajput. Heat and Mass Trnasfer, S. Chand & Company, New Delhi, India, 2016.
6. Yunus A. Cengel and Afshin J. Ghajar. Heat and Mass Transfer Fundamentals and Applications, McGraw-Hill, New York, 2011.
7. John H. Lienhard IV and John H. Lienhard V. A Heat Transfer Textbook, Phlogiston Press, Cambridge, Massachusetts, 2006.

3 Steam Turbines, IC Engines and Refrigeration

UNIT SPECIFICS

Through this unit we will discuss the following aspects:

- Principle and working of steam turbines, their classification, and compounding of turbines; `
- Types, key components of condensers and cooling towers;
- Fundaments of IC Engines, their nomenclature and air standard cycles;
- Working of 2 stroke and 4 stoke CI and SI engines and their differences;
- Various refrigeration systems, their units, COP;
- Working principle of air refrigeration, vapour compression and vapour absorption refrigeration systems.

The topics discussed in this unit come under the domain of thermal engineering, which is a vast platform for a variety of practical applications such a power generation, internal combustion engines, refrigeration and air conditioning systems and heat transfer etc. The subject provides creativity as well as technical knowledge to develop problem solving capabilities.

This unit provides basic knowledge and insight into various aspects of steam turbines, IC engines and refrigeration with the help of practical applications in our daily life. The language is simple so that the subject matter may be easily grasped by a student of average standard. The large number of multiple-choice questions as well as questions of short and long answer types are provided to apply the concepts and to test the students understanding towards the subject.

A large number of solved and unsolved numerical problems along with a list of references and suggested readings are given in the unit so that one can go through them for practice. QR codes have been provided in different sections which can be scanned for relevant supportive knowledge.

A practical section related to theoretical concepts is included to provide a better understanding of the subject. Based on the content, there is a "Know More" section, which has been carefully designed so that the essential and supplementary information for getting more exposure on various topics of interest becomes beneficial for the users of the book. This section mainly highlights the initial activity, examples of some interesting facts, analogy, and history of the development of the subject focusing on the salient observations and findings.

Timelines starting from the development of the concerned topics up to the recent time, applications of the subject matter for our day-to-day real life or/and industrial applications on a variety of aspects, case study related to environmental, sustainability, social and ethical issues whichever applicable, and finally inquisitiveness and curiosity topics of the unit.

RATIONALE

The steam turbine is a prime mover to run the generator and hence, the student will get the idea of working of various types of steam turbines. The speed of steam turbines is very high which needs to be reduced to practical limits; hence, compounding is explained in simple language. Similar to a turbine, the condenser is the key component of the thermal power plant used to cool the exhaust steam which is discussed in details. In continuation the cooling tower is employed to cool the water used to condense steam in the condenser, and the different cooling tower technologies are discussed.

The basic gas power cycles such as Otto, Diesel and Dual power cycles which explain the working of internal combustion engines are also discussed. This fundamental unit helps the students to get a primary idea about the two-stroke and four-stroke spark ignition (petrol) and compression ignition (diesel) engines. All the basic aspects such as their working, differences and their performance parameters, are elaborated with the help of numerical problems.

Refrigeration is an important topic and having domestic and commercial applications is also discussed in length to develop a basic understanding of the subject. The operating cycles related to vapour compression and vapour absorption refrigeration system are explained in detail. The coefficient of performance (COP) which defines the performance of a refrigerator is also discussed with the help of numerical problems.

PRE-REQUISITES

- Basic Mathematics: Class XII
- Physics: Class XII

UNIT OUTCOMES

List of outcomes of this unit is as follows:

U3-O1: Understand steam turbines, their classification and working principles.

U3-O2: Understand the working of condensers and cooling towers, their types and applications.

U3-O3: To learn basic concepts of gas power cycles used to explain the working of internal combustion engines.

U3-O4: Develop the understanding of two stroke and four stroke petrol and diesel engines.

U3-O5: Understand the refrigeration system and cycles with vapour and air as fluids.

Unit-3 Outcomes	EXPECTED MAPPING WITH COURSE OUTCOMES (1- Weak Correlation; 2- Medium correlation; 3- Strong Correlation)					
	CO-1	CO-2	CO-3	CO-4	CO-5	CO-6
U3-O1	3	2	1	3	-	1
U3-O2	3	2	2	3	-	2
U3-O3	3	2	2	3	-	1
U3-O4	3	2	1	3	-	1
U3-O5	3	2	1	3	-	1

3.1 INTRODUCTION

The steam turbine is a rotary prime mover employed to convert the energy of high-pressure and high-temperature (preferably superheated) steam coming from the boiler into useful mechanical work. Steam turbines are most widely used for generating electricity and fall under the category of work-producing turbo machines. The high pressure and temperature of the steam are utilized in the nozzles fixed on the casing of the turbine, where the pressure drops inside the nozzle passage leading to a significant increase in velocity. This high-velocity jet of steam is allowed to pass through a series of moving blades mounted on a shaft supported by bearings.

The passage of high-velocity steam through one rotor blade section is shown in Fig. 3.1. Between each row of moving blades; there is a ring of fixed blades. Each pair of fixed blades and moving blades is referred to as a *stage* of a turbine; all modern power plants have several *stages* of extracting maximum energy of the steam before it exits the turbine. The basic purpose of fixed blades is to guide the steam to the next ring of moving blades in such a way that the steam glides over the moving blades without shock. As the steam flows over the moving blade, its curved surface leads to a change in the direction of steam, giving rise to a change in momentum, forcing the blades to move.

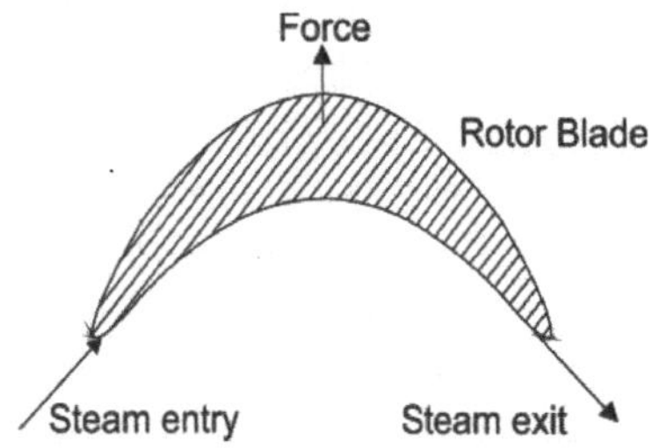

Fig. 3.1: Passage of steam through rotor blade section

3.2 TYPES OF TURBINES

There are various methods to describe steam turbines. The most essential and popular distinction is based on the action of the steam, as:

(a) Impulse turbine
(b) Reaction turbine or impulse-reaction turbine

There are other ways in which the steam turbines may be described, like:

1. **On the basis of the number of cylinders**
   (i) Single-cylinder turbines
   (ii) Multi-cylinder turbines
2. **On the basis of steam flow direction**
   (i) Axial flow turbines: steam flow takes place in the direction parallel to the axis of the turbine shaft. Modern high-capacity power plants use this type of turbine.
   (ii) Radial flow turbines: steam flow takes place in the direction perpendicular to the axis of the turbine shaft.
3. **On the basis of the number of pressure stages**
   (i) Single-stage turbines: in these turbines, the steam is expanded in a single stage to obtain work, i.e., the kinetic energy of the steam is utilized in one stage of the rotor only.
   (ii) Multi-stage turbines: in these types of turbines, the steam is expanded in a number of stages to obtain work. The steam is passed through multiple stages to reduce the speed of the turbine and to obtain maximum work which would otherwise go wasted if used in a single stage.

4. **On the basis of the pressure of steam**
   (i) Low-pressure turbines: the steam pressure varies from 1 to 2 bar.
   (ii) Medium pressure turbines: these turbines use steam from 40 to 50 bar.
   (iii) High-pressure turbines: these turbines use steam above 50 bar.
   (iv) Supercritical pressure turbines: these turbines are operated above a steam pressure of 225 bar.
5. **On the basis of heat drop:**
   (i) Condensing turbines: in these turbines, the steam is passed through a condenser having a pressure less than atmospheric, where the low-pressure steam from the exit of the turbine is converted into liquid.
   (ii) Back pressure turbines: the exhaust steam coming from the turbine is used for heating purposes in process industries. The steam is supplied to the customers at various pressures and temperatures according to their requirements.
   (iii) Topping turbines: these turbines are also a type of back-pressure turbine, but the difference is that the exhaust steam of these turbines is further utilized in running low and medium-pressure condensing turbines.
6. **On the basis of their industrial usage**
   (i) Stationary turbines: these turbines come with constant speed and variable speed. A constant-speed turbine is used for running alternators, while variable-speed turbines are meant for driving pumps, blowers, and air circulators.
   (ii) Non-stationary turbines: these turbines come with variable speeds of operation and serve the purpose of running steamers, ships, and locomotive steam engines.

3.2.1 Common Types of Turbines

The common types of steam turbines are:

1. **Impulse turbine.**
2. **Reaction turbine**

The impulse turbine can be (a) simple (b) velocity stage (c) pressure stage, or (d) a combination of pressure and velocity stages. The reaction turbine may work as (a) 50% reaction (Parson's) turbine and (b) an impulse-reaction type turbine

3.2.1.1 Simple Impulse Turbine

These turbines work on the principle of impulse, where the steam is expanded in the nozzle only, and there is no change of pressure in the moving blade. It is composed primarily of a nozzle or set of nozzles, a rotor positioned on a shaft, one set of moving blades connected to the rotor, and a case. The pressure drops from steam pressure to condenser pressure in the nozzle only, resulting in a significantly high velocity (supersonic) of the steam which is utilized to move the rotor. It can be observed that the steam enters and leaves the moving blades at condenser pressure. It can be stated that the transformation of energy takes place only in the nozzle, while the moving blades (rotor) are only responsible for the transfer of energy (Fig. 3.2).

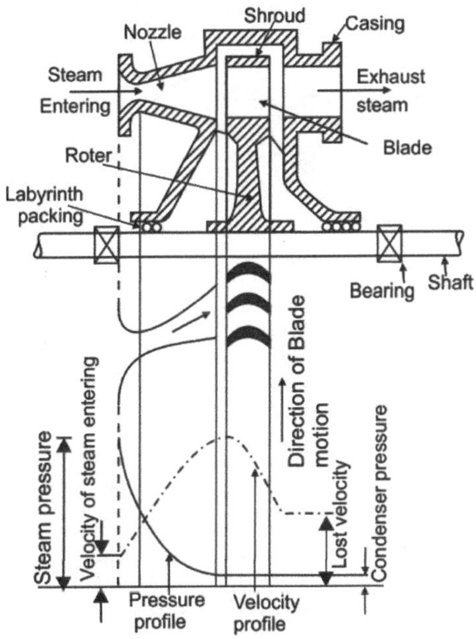

Fig. 3.2: Simple impulse turbine (de-Laval turbine)

A very large portion of steam velocity remains unutilized which implies that its leaving velocity is appreciably high, resulting in loss of energy referred as *'carry over loss'* or *'leaving velocity loss'*. The turbine of this type is well known as the *'de-Laval turbine'*. The passage of the rotor blade is of constant cross-section area so that the pressure remains constant at the inlet and outlet of moving blades. Therefore, the rotor blades do not accelerate the fluid leading to greater chances of flow separation resulting in lower stage efficiency.

The lost velocity in the case of this turbine may amount to 10 to 11 percent of the maximum velocity at the nozzle exit. Also, since all the kinetic energy is to be absorbed in only one rotor (set of moving blades), the rotational speed of the wheel is very high (25000 to 30000 rpm) which is of no use in the case of power generation. Therefore, different measures of speed reduction may be employed so that it may be utilized for the purpose of power generation. This kind of turbine is often used in applications with low power requirements and small rotor diameters.

3.2.1.2 Reaction Turbine (Impulse-Reaction Turbine)

In this type of turbine, the drop in pressure of flowing steam takes place in the fixed blades as well as moving blade which means that both impulse and reaction principle is utilized. A number of rows of moving blades are attached to the rotor, and an equal number of stationary blades are attached to the casing. In this form of the turbine, the fixed blades, which are arranged in the opposite direction as the moving blades, correspond to the nozzles mentioned in relation to the impulse turbine. Steam is admitted for the entire circumference due to the row of fixed blades at the entrance rather than the nozzles, resulting in all-round or total admittance.

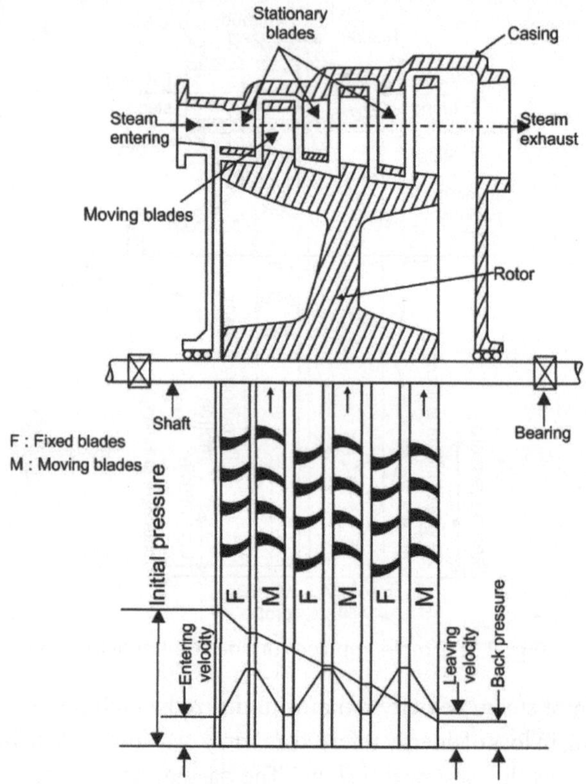

Fig. 3.3: Impulse-reaction turbine

The function of fixed blades in this type of turbine is the same as that of the nozzle implying that they allow pressure drop of steam through their passage to increases the kinetic energy of the steam. Since the pressure of steam drops in the moving blades also, therefore, the moving blades are made of converging cross-sections. The pressure drop experienced by steam while passing through the moving blades generates additional kinetic energy inside the moving blades, causing a reaction and adding to the propulsive force provided to the turbine shaft via the rotor. As we can see that both the impulse and reaction principle applies to this type of turbine; therefore, it is commonly called as impulse-reaction turbine instead of a reaction turbine (Fig. 3.3).

The energy transformation takes place in both fixed and moving blades. As the specific volume of steam is higher at lower pressure therefore, the area of the turbine (size of the blades) must increase after each stage to accommodate for the increased volume to extract maximum energy from the steam. Also, in this type of turbine, the pressure drop per stage is low therefore, a large number of stages are required as compared to an impulse turbine of the same capacity. The power plants with a capacity greater than 30 MW are all impulse-reaction turbines.

3.2.2 Compounding of Impulse Turbine

As we have seen in the previous section that in case of simple impulse turbine the whole of the kinetic energy of the steam has to be absorbed by a single rotor leading to a very high rotational speed of about 30000 rpm. This speed is too high for practical use, and a great amount of energy is lost in the form of leaving loss. Therefore, it becomes imperative to assimilate some improvements in simple impulse turbines to bring their rotational speed to practical limits and achieve high performance. To fix the above problem a method called *compounding* is employed to reduce the rotational speed of the impulse turbine. In this method, arrangements are made in such a way that the energy of steam is utilized in more than one set of nozzles and moving blades keyed together to a common shaft. The speed reduction can be achieved by either dropping the pressure in the set of nozzles through several stages till it reaches the condenser pressure or by dropping the entire pressure in the first set of nozzles and then guiding the high-velocity steam jet through a series of moving blades till maximum energy is absorbed. The leaving loss is significantly reduced by incorporating the above methods. Compounding of impulse turbines can be done in three ways.

(1) Pressure compounding
(2) Velocity compounding and
(3) Pressure-velocity compounding

3.2.2.1 Pressure Compounding

In this type of compounding, the main focus is kept on the pressure of steam only; therefore, the speed reduction is achieved by expanding the steam from steam chest pressure to condenser pressure in several stages. This is done by attaching a number of simple impulse stages (nozzle and moving blades) on a common shaft, as shown in Fig. 3.4.

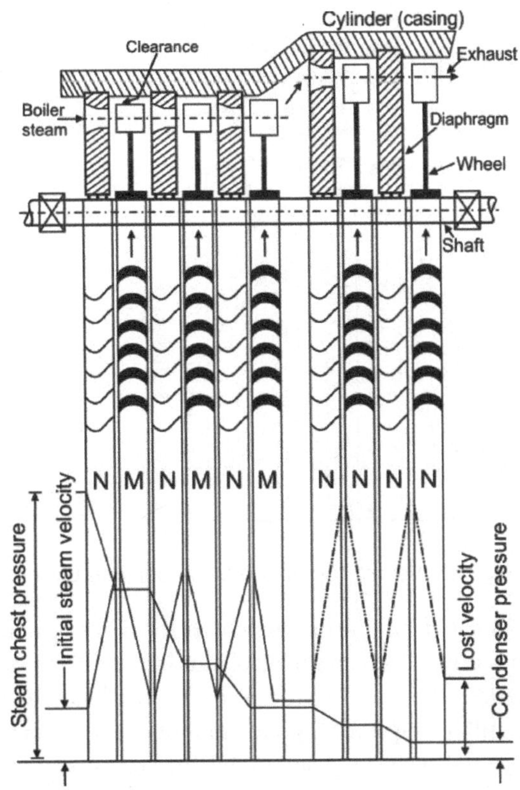

Fig. 3.4: Pressure compounded impulse turbine

3.2.2.2 Velocity Compounding

In this type of compounding, the main focus is kept on the velocity of steam only, which means that speed reduction is achieved by splitting the velocity gained from the exit of the nozzles into many small drops through several sets of moving blades mounted on a common shaft. This type of setup consists of a set of nozzles and rows of moving blades attached to a rotor. Each row of moving blades is followed by a row of fixed blades mounted on the casing, as shown in Fig. 3.5. The function of fixed blades is to guide the steam to the succeeding row of moving blades at a correct angle such that the steam glides smoothly over the moving blades to prevent shock. The whole expansion of steam from the steam chest pressure to the condenser pressure takes place in the first set of nozzles; this means that the pressure remains constant in both moving blades and fixed blades.

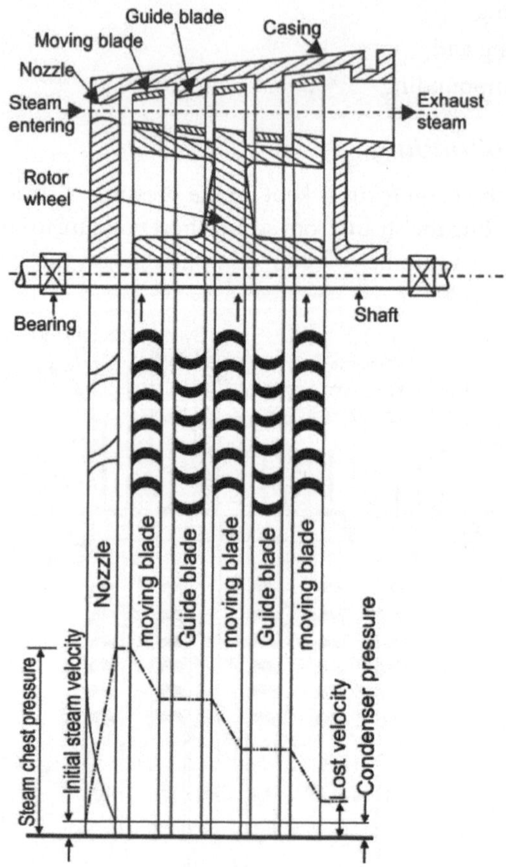

Fig. 3.5: Velocity compounded impulse turbine

The steam at a very high velocity from the nozzles enters the first row of moving blades, where some part of the kinetic of the steam is utilized to run the blades and the rest is reissued to the first row of fixed blades (guide blades) which guides the steam to the second row of moving blades where some part of velocity is reduced. The steam, still having a fairly good amount of energy in itself is again reissued to the second row of fixed blades which redirects the steam to the third row of moving blades. Finally, the steam leaves the rotor at a certain speed which is very less as compared to a simple impulse turbine. It accounts for about 1 to 2 percent of the total initial kinetic energy of steam. It is to be noted that the efficiency of this type of turbine is low; therefore, the stages are limited to 2 to 3 only. The velocity compounded impulse turbines are also called as *Curtis Turbine.*

3.2.2.3 Pressure-Velocity Compounding

This type of compounding utilizes principles of both pressure and velocity compounding, the diagrammatic arrangement of which is shown in Fig. 3.6. It is evident from the figure that there are two sets of velocity compounded impulse turbines attached to a common shaft with an arrangement such that the pressure is allowed to partially drop in the first set of nozzles and the high velocity attained is utilized to run two rows of moving blades. The steam with reduced pressure and velocity is again allowed to drop in the second set of nozzles, and the velocity attained by this pressure drop is again utilized to run another two rows of moving blades. Hence, it can be inferred that velocity drop is utilized to run the moving blades, which can be correlated to velocity compounding, and at the same time, pressure is partially dropped in two sets of nozzles, therefore, correlated with pressure compounding. This type of compounding is more popular than pressure compounding due to its simplicity in construction but is rarely used due to low efficiency.

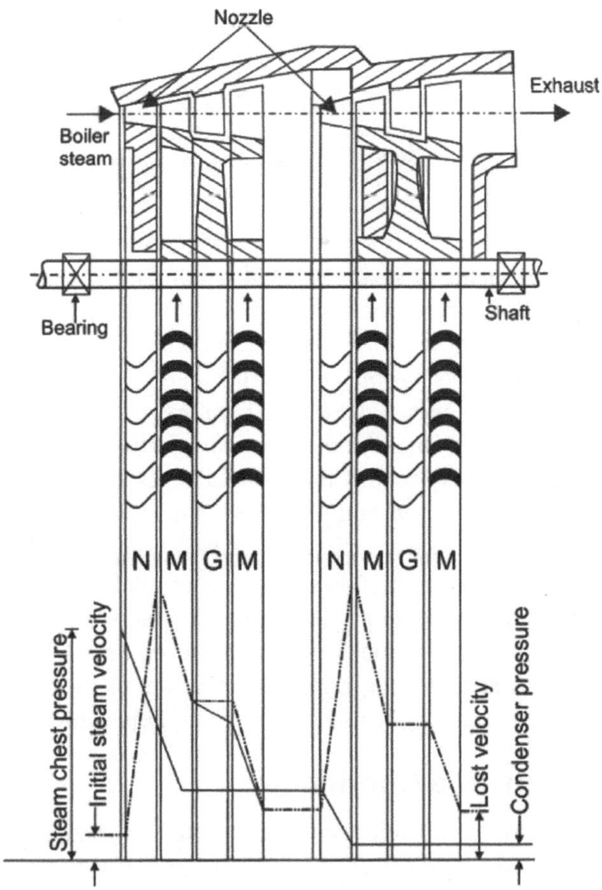

Fig. 3.6: Pressure-velocity compounded impulse turbine

3.2.3 Difference Between Impulse and Reaction Turbines

The comparison between impulse and reaction turbine on the basis of parameters are shown in table 3.1.

Table 3.1: Comparison between impulse and reaction turbine

Parameters	Impulse turbine	Reaction turbine
Pressure drop	Only in nozzles and not in moving blade ring.	In fixed blades (nozzles) as well as moving blades
Blade channel area	Constant.	Varying (converging type).
Type of blade	Profile type.	Aerofoil type.
Steam admission	Not from all round.	All round or complete.
Position of nozzles/fixed blades	Nozzles are contained in the diaphragm.	Fixed blades, similar to moving blades serve as nozzles and are secured on the turbine casing.
Space	Occupies less space for the same power.	Occupies more space for the same power.
Efficiency	Low	High
Steam velocity	High	Low
Power developed	Less	High
Blade manufacturing	Easy and cheap	Difficult and costly

3.3 CONDENSERS

Condenser is one of the key elements of steam-based thermal power plants. *It is defined as a closed vessel in which the exhaust steam from the exit of the turbine is condensed with the help of cooling water while maintaining a vacuum, resulting in an increase in work output and efficiency of the plant and the recirculation of condensate as feed water to the boiler.* The energy loss associated with the condenser accounts for 50 to 60 percent of the energy of steam; therefore, the condenser offers a great energy-saving opportunity in a steam power plant. Another attractive feature of a condenser is that it absorbs only the latent heat of the steam and hence only the phase changes form vapour to liquid at a constant temperature. This significantly reduces the pumping work since handling a liquid is far easier than vapour and the size of the pump and the work requirement is also significantly reduced.

As we have discussed above that about 50 to 60 percent of total heat supplied in a power plant is rejected in a condenser therefore, the cooling system plays a vital role in condensing the steam. The cooling medium taken in power plants is mostly water; however, air may also be taken as a cooling medium, but it is seldom used. When selecting a site thermal power plant, availability of water is the key factor where the cooling water system may be used in the form of open and closed type arrangement.

3.3.1 Key Components of a Condenser

The water-cooled condensing unit has the following key components, as depicted in Fig. 3.7.

(i) **Condenser drum:** It is a closed vessel where the steam exhausted from the low-pressure (LP) turbine condenses. It is a type of heat exchanger where the latent heat of steam is absorbed by the cooling water.

(ii) **Air extraction pump (AEP):** The dry air which is leaked into the condenser through joints and other non-condensable gases are removed by this pump in order to maintain a constant vacuum.

(iii) **Condensate extraction pump (CEP):** The purpose of CEP is to extract condensed steam collected in a hot-well attached to a condenser and pump the condensate to the feed water line.

(iv) **Boiler feed pump (BFP):** It pumps the condensate to the boiler. Since the boiler pressure and temperature is very high thus, the BFP is designed to handle very high pressure and temperature.

(v) **Circulating cooling water pump (CCWP):** It is used to circulate the cooling water in the condenser.

(vi) **Cooling tower or spray pond:** It cools the hot water coming out of the condenser. A cooling tower is employed where there is a scarcity of water, and this arrangement is called as closed cooling water system. In an open cooling water system, the hot and cold water is allowed to flow in an open channel like a river, lake, or pond.

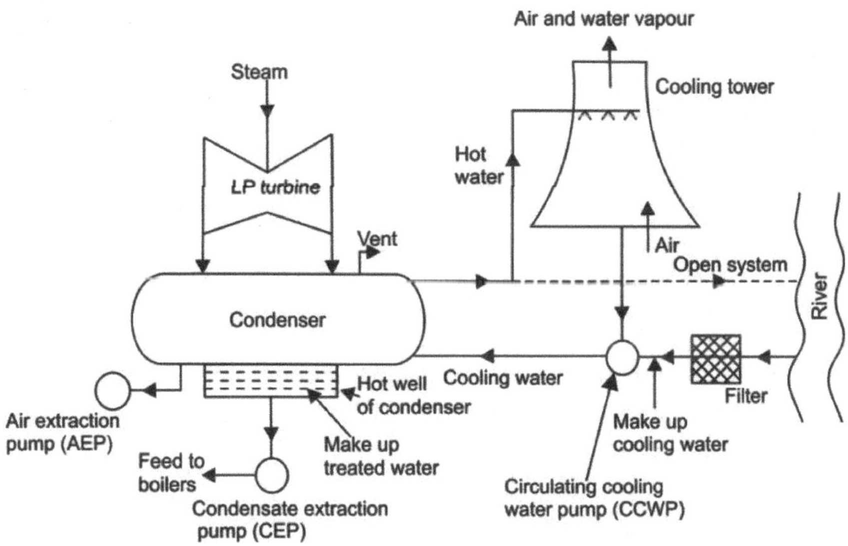

Fig. 3.7: Key components of a condenser

3.3.2 Condenser Types

There are mainly two types of condensers:

3.3.2.1 Direct Contact Type Condensers

In these types of condensers, the exhaust steam from the turbine and cooling water is allowed to mix directly, and they come out as a single stream. There are mainly three kinds of direct contact type condensers.

(a) Spray condenser, (b) Barometric condenser and (c) Jet condenser

1. **Spray condenser**

   The mechanism of the spray condenser is depicted in Fig. 3.8. Steam is passed through a chamber where it mixes with a fine spray of water, due to which steam gets condensed. The condensate so obtained is partly pumped to the feed water line and partly dry cooling tower. The condensate that is cooled in the dry cooling tower is sprayed on the steam in the chamber and this process repeats continuously. The non-condensable gases are ejected through the *steam jet air ejector (SJAE)*. The major limitation of this condenser is that it requires water of a very high level of purity, as impure water is harmful to the boiler operation.

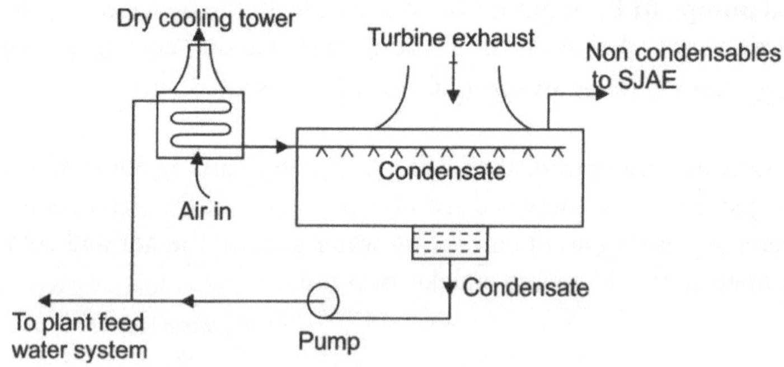

Fig. 3.8: Mechanism of spray condenser

2. **Barometric condenser**

In this type of condenser, the cooling water is allowed to baffle down in a series of baffles such that it forms a water curtain or a sheet. Steam is fed from below in a large volume which comes in direct contact with water. In this way the steam condenses and the mixture is collected in the hot well through a tail pipe as shown in Fig. 3.9. There is no need for a pump because the static head compresses the mixture to atmospheric pressure in the tail pipe. It can be noted that a larger height, *H*, of the tail pipe assists in the flow of mixture from the tail pipe to the hot well. Similar to the spray condenser the barometric condenser also requires highly pure water to prevent fouling and scaling of the pipe.

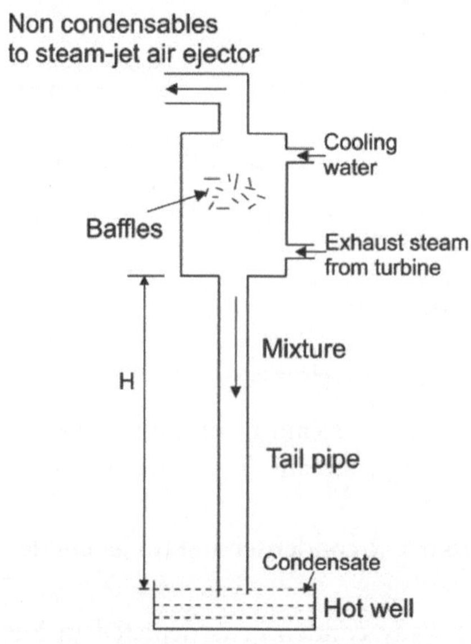

Fig. 3.9: Barometric condenser

3. **Jet condenser**

   A jet condenser employs cascades instead of baffles as in a barometric condenser as shown in Fig. 3.10. Just below the cascaded structure, a diffuser is provided which works on the same principle as of the diverging section of a converging-diverging nozzle for subsonic flow. The advantage of having a diffuser is that the height of the tail pipe is significantly reduced and the mixture is separated and cooled, similar to the spray condenser.

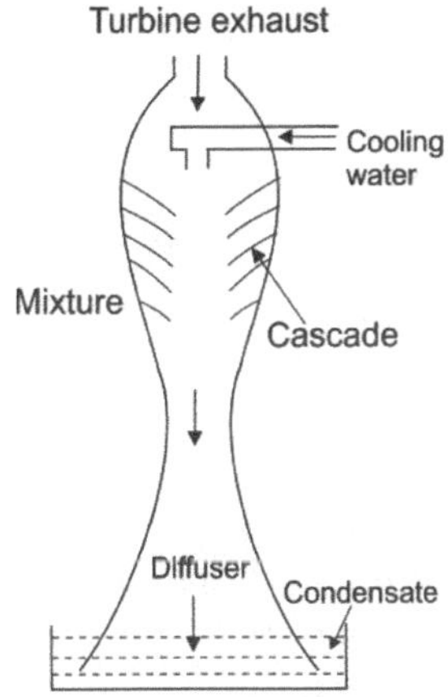

Fig. 3.10: Jet condenser

3.3.2.2 *Indirect Contact Type or Surface Condensers*

In these types of condensers, there is no mixing of cooling water and steam. These are tubular heat exchangers where the steam in the shell surrounds the tubes and gives its heat to the circulating cooling water flowing inside the tubes (shell and tube heat exchanger). The design depends upon the amount of heat exchange required. Surface condensers are often configured with one, two, or four cooling water passes. The number of passes influences the size and efficacy of a condenser. In a single pass condenser, the cooling water flows in the condenser tubes only once from one end to another as shown in Fig. 3.11.

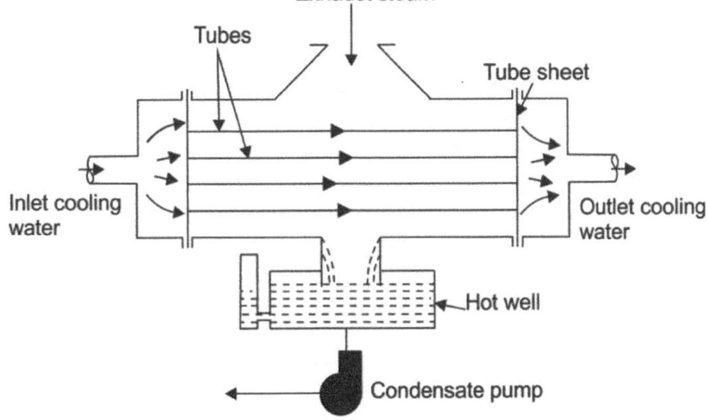

Fig. 3.11: Single pass surface condenser

The shell is made of steel and a tube bank secured in a tubes sheet for the flow of cooling water is installed inside the shell having an inlet and outlet water box for an even flow of water in the tube bank. Exhaust steam enters the shell normally to the tube bank and transfers its heat to the cooling water flowing in the tubes; as a result, the steam condenses. The condensed steam is collected in the reservoir called a hot well provided at the bottom of the condenser. The condensate is then pumped with the help of feed water pumps into the feed water line. Invariably, surface condensers are used in practically all steam power stations.

3.4 COOLING TOWERS

A cooling tower cools the hot water coming from the condenser and then recirculates it back to the condenser. Cooling water requirement is greatly reduced by using a cooling tower, and this facilitates the running of a power plant in the areas where availability of water is limited. The cooling towers are mainly of two types namely *wet* and *dry* type.

3.4.1 Wet-Type Cooling Towers

A wet cooling tower structure consists of a distribution system for spraying hot water evenly over a packing of closely fitted horizontal bars called as *fill*. The cooling phenomena takes place by the dissipation of heat to the environment with the help of air and evaporation of a fraction of recirculated water itself. As the water is splashed from the top, it travels down from one fill level to other by gravity and during this time, the air is thoroughly mixed with the falling water. The outside air enters the tower from the bottom through louvers arranged in the form of horizontal bars along the sides of the tower (Fig. 3.12).

The bars are angled in such a way that water remains inside the tower, which is stored in the reservoir or basin at the bottom of the tower. The thorough mixing of air and water results in an enhanced heat and mass transfer responsible for the cooling of water. The hot and moist air leaves the tower from the top, and the cold water from the basin is pumped back to the condenser in case of a closed system or to the water body in case of an open system. The wet cooling towers may be classified into the following.

(i) Natural draught cooling tower.

(ii) Mechanical draught cooling tower.

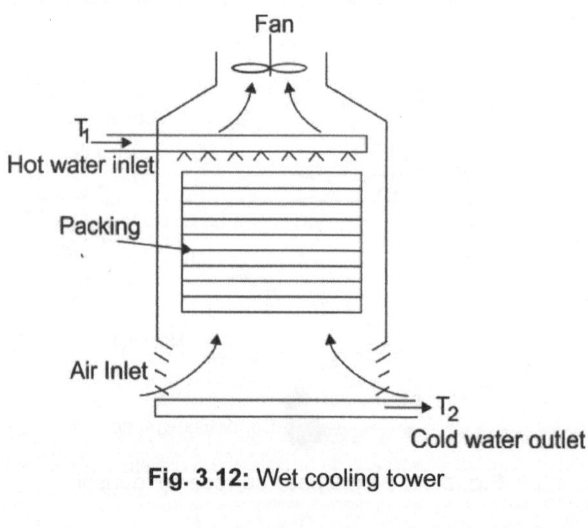

Fig. 3.12: Wet cooling tower

3.4.2 Dry-Type Cooling Towers

Dry cooling towers are a closed type of towers and are very suitable where the availability of cooling water is very scarce. Since there is no contact between the ambient air and the working fluid, there is no loss of water. The dry cooling tower uses finned tubes over which the cooling air circulates. The provision of fins increases the surface area for heat transfer and therefore cooling becomes more effective. The air in a dry cooling tower can be introduced either through a mechanical draught or a natural draught. The dry cooling towers can be of two types: *direct* or *indirect*.

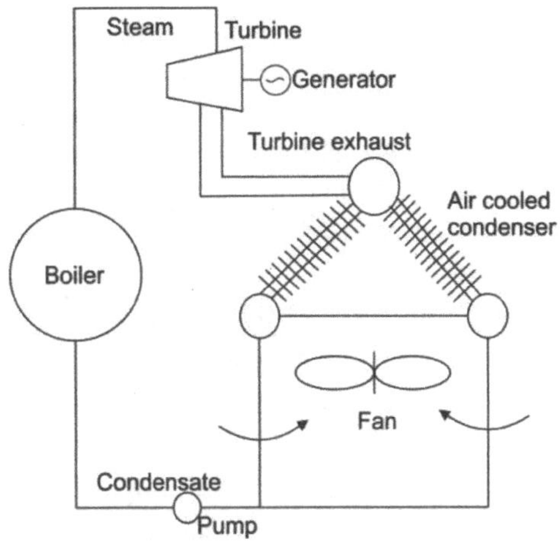

Fig. 3.13: Direct dry cooling tower

A direct dry cooling tower condenses turbine exhaust steam directly through a large number of finned tubes or coils arranged parallel to each other using ambient air *flowing in natural or forced mode*. As illustrated in Fig. 3.13, the turbine exhaust enters the finned tubes directly. At the turbine output, a large-diameter pipe is employed to handle the comparatively low steam densities (compared to water) and minimize the pressure drop across the route. The exhaust steam runs within the finned tubes and is condensed by transferring heat to the circulating air outside the heat exchanger's surface. The condensate gets collected in a receiver from where it is pumped back to the feed water line.

As depicted in Fig. 3.14, an indirect dry cooling tower condenses turbine exhaust steam in a condenser which is a surface heat exchanger, where the heat from the steam is transferred to the cold water pumped from an internal air-cooled heat exchanger.

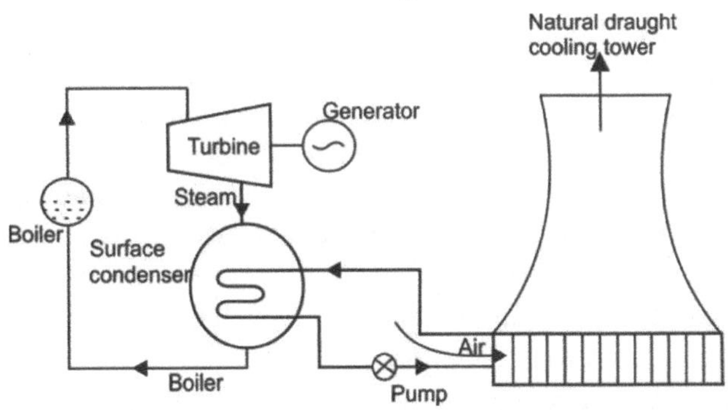

Fig. 3.14: Indirect dry cooling tower

3.5 AIR STANDARD CYCLES

Most of the engines producing power make use of gas or air as a working fluid and operate on air standard cycles. It has constant specific heats C_p and C_v. An air standard cycle is an ideal cycle working within closed boundaries, and the working fluid undergoes chemical reactions. This results in a change in the pressure and volume of the fluid, resulting in the movement of the piston which is connected to a crank shaft and other power transmission mechanisms. An internal combustion (IC) engine, which runs automobiles are of two types; spark-ignition (SI) engines and compression-ignition (CI) engines.

Internal combustion engines are a majorly reciprocating type of engine, where the working fluid is fuel and air and combustion takes place inside the cylinder. The chemical energy of the fluid is transformed into thermal energy. This thermal energy changes pressure and temperature allowing the gas at high pressure to expand and move the piston, which in turn rotates the output shaft. These internal combustion engines are open-cycle heat engines, where the high-temperature fluid after combustion is thrown out during the exhaust cycle through the exhaust valve and the fresh air or mixture of air and fuel is allowed to enter the cylinder. Therefore, it does not undergo a complete thermodynamic cycle. In order to make a simplified analysis of IC engines, the following assumptions are made, assuming air is a fluid.

- Air is the working fluid.
- It has constant specific heat and follows ideal gas equations.
- All the processes in a cycle are considered reversible.
- No heat generation in the system.
- The mass of the working fluid remains constant.
- The heat is added and rejected with the external reservoir.
- The cycle is considered to be closed, i.e., the same fuel is used again and the cycle has non-flow processes.
- There is no frictional loss.

There are various air standard cycles on which the working of petrol, diesel and dual-fuel engine depends. These cycles are discussed in the subsequent sections. The air standard cycles are the ideal cycles and they give the maximum efficiency. The actual cycles have lower efficiency than ideal cycles therefore, a better engine will be the one in which the efficiency is close to the ideal one.

3.5.1 Otto Cycle

The Otto cycle is an ideal thermodynamic cycle used for gas and petrol engines. The cycle has a sequence of processes that describes the functioning of spark ignition (petrol) engines, and it was first given by Nikolaus Otto in 1876. Nowadays, the majority of automobiles work on the Otto cycle to run an engine. It consists of four internally reversible processes; two isentropic processes and two constant volume processes. The p-V curve of the Otto cycle is shown in Fig. 3.15. The air-fuel mixture is sucked into the cylinder, which is then compressed. The combustion of fluid takes place to convert chemical energy into thermal energy, and then expansion takes place. Finally, the combustions products are exhausted.

The Otto cycle is also called as *constant volume* cycle, which has the following processes.

- Process (1-2) is the isentropic compression of air fuel mixture drawn into a cylinder during suction stroke.
- Process (2-3) is constant volume heat addition, through the hot body and combustion of compressed mixture leads to rising in pressure.

$$Q_{2-3} = C_v(T_3 - T_2) \qquad 3.1$$

- Process (3-4) is an isentropic expansion of the combustion products, and work is obtained from the system.
- Process (4-1) is the constant volume heat rejection.

$$Q_{4-1} = C_v(T_4 - T_1) \qquad 3.2$$

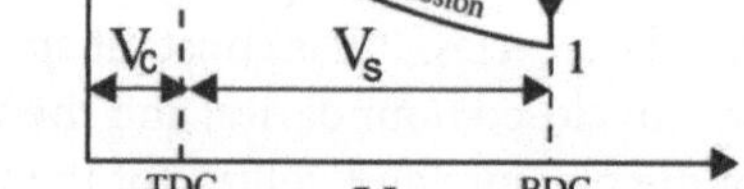

Fig. 3.15: p-V curve of Otto cycle

Work done during the cycle = Heat supplied – Heat rejected

$$= C_v(T_3 - T_2) - C_v(T_4 - T_1) \quad 3.3$$

$$\text{The efficiency of the cycle } (\eta) = \frac{\text{Output}}{\text{Input}}$$

$$= \frac{\text{Work done}}{\text{Heat supplied}}$$

$$= \frac{C_v(T_3 - T_2) - C_v(T_4 - T_1)}{C_v(T_3 - T_2)} \quad 3.4$$

$$= 1 - \frac{(T_4 - T_1)}{(T_3 - T_2)}$$

$$= 1 - \frac{T_4(1 - T_1/T_4)}{T_3(1 - T_2/T_3)} \quad 3.5$$

Let, $V_3 = V_2 = 1$

$$\text{Compression ratio (r)} = \frac{V_1}{V_2} = \frac{V_4}{V_3}$$

$$\text{During the adiabatic expansion process (3-4)} = \frac{T_4}{T_3} = \left(\frac{p_4}{p_3}\right)^{\frac{\gamma-1}{\gamma}} = \left(\frac{V_3}{V_4}\right)^{\gamma-1} = \left(\frac{1}{r}\right)^{\gamma-1} \quad 3.6$$

$$\text{During the adiabatic expansion process (1-2)} = \frac{T_1}{T_2} = \left(\frac{p_1}{p_2}\right)^{\frac{\gamma-1}{\gamma}} = \left(\frac{V_2}{V_1}\right)^{\gamma-1} = \left(\frac{1}{r}\right)^{\gamma-1} \quad 3.7$$

$$\text{or} \quad \frac{T_4}{T_3} = \frac{T_1}{T_2}$$

$$\text{or} \quad \frac{T_2}{T_3} = \frac{T_1}{T_4} \quad 3.8$$

On substituting the value of $\left(\frac{T_2}{T_3}\right)$ from equation 3.8 in equation 3.5, we get,

$$\text{Efficiency of the cycle } (\eta) = 1 - \frac{1}{(r)^{\gamma-1}} \quad 3.9$$

The efficiency of the cycle depends on the compression ratio (r) and the ratio of specific heat (γ). As the compression ratio increases, the efficiency of the cycle increases, considering γ as constant. The compression ratio is limited between 6-10 for Otto cycle to prevent the phenomenon of knocking due to the auto-ignition of fuel.

Example 3.1: In an air standard Otto cycle, the maximum and minimum temperatures are 1673 K and 288 K. The heat supplied per kg of air is 800 kJ. Calculate;

(a) Compression ratio and efficiency.

(b) Maximum to minimum pressure in the cycles.

Solution:

Given: $T_1 = 288$ K, $T_3 = T_{max} =$ 600 kJ/min.

(a) Heat supplied $= C_v\ (T_3 - T_2)$

$$800 = 0.718\ (1673 - T_2)$$

$$T_2 = 558.77 \text{ K}$$

$$\frac{T_1}{T_2} = \left(\frac{1}{r}\right)^{\gamma-1}$$

$$r = \left(\frac{T_2}{T_1}\right)^{1/\gamma-1} = \left(\frac{558.77}{288}\right)^{1/0.4} = (1.94)^{2.5} = 5.243 \quad \text{Ans}$$

$$\text{Efficiency of the cycle } (\eta) = 1 - \frac{1}{(r)^{\gamma-1}} = 1 - \frac{1}{(5.243)^{0.4}}$$

$$= 1 - \frac{1}{1.94} = 1 - 0.515 = 0.4845$$

$$\eta = 48.45\ \% \quad \text{Ans}$$

(b) $$\frac{p_2}{p_1} = (r)^{\gamma-1} = (5.243)^{1.4} = 10.172$$

$$\frac{p_2}{p_3} = \frac{T_2}{T_3}$$

$$\frac{p_2}{p_3} \text{ can be written as ;} \quad \frac{p_2}{p_1} \times \frac{p_1}{p_3} = \frac{T_2}{T_3} = \frac{558.77}{1673}$$

$$10.172 \times \frac{p_1}{p_3} = \frac{558.77}{1673}$$

$$\text{Therefore, } \frac{p_3}{p_1} = 2.994 \times 10.172 = 30.53 \quad \text{Ans}$$

Example 3.2: In a SI engine working on the ideal Otto cycle, the compression ratio is 5.5. The pressure and temperature at the beginning of compression are 1 bar and 27 °C, respectively. The peak pressure is 30 bar. Determine;

(a) Pressure and temperature at the salient points.

(b) The air-standard efficiency.

Solution:

Given: $r = 5.5$, $p_1 = 1\text{bar}$, $T_1 = 27\ °\text{C}$, $p_3 = 30$ bar

As we know, $V_2 = V_3 = V_c$

$$V_1 = rV_2 = rV_c$$

$$\frac{p_2}{p_1} = r^\gamma = 5.5^{1.4} = 10.88$$

$$p_2 = 10.88 \times 1 \times 10^5 = 10.88 \times 10^5 \text{ N/m}^2 \quad \text{Ans}$$

$$\frac{T_2}{T_1} = r^{\gamma-1} = 5.5^{0.4} = 1.978$$

$$T_2 = 1.978 \times 300 = 593.4 \text{ K} = 320.4 \text{ °C} \quad \text{Ans}$$

$$p_3 = 30 \times 10^5 \text{ N/m}^2$$

$$\frac{T_3}{T_2} = \frac{p_3}{p_2} = \frac{30}{10.88} = 2.757$$

$$T_3 = 2.757 \times 593.4 = 1636 \text{ K}$$

$$= 1363 \text{ °C} \quad \text{Ans}$$

$$\frac{p_3}{p_4} = \left(\frac{V_4}{V_3}\right)^\gamma = \left(\frac{V_4}{V_3}\right)^\gamma$$

$$= r^\gamma = 5.5^{1.4} = 10.88$$

$$p_4 = \frac{p_3}{10.88} = 2.76 \times 10^5 \text{ N/m}^2$$

$$\frac{T_3}{T_4} = r^{\gamma-1} = 5.5^{0.4} = 1.978$$

$$T_4 = \frac{T_3}{1.978} = \frac{1636}{1.978} = 827.1 \text{ K}$$

$$= 554.1\text{°C}$$

$$\text{Efficiency of the cycle } (\eta) = 1 - \frac{1}{(r)^{\gamma-1}} = 1 - \frac{1}{(5.5)^{0.4}} = 0.4943$$

$$= 49.43 \% \quad \text{Ans}$$

3.5.2 Diesel Cycle

The Diesel cycle is an ideal thermodynamic cycle used for diesel engines. The cycle has a sequence of processes that describes the functioning of compression ignition (diesel) engines; it was first given by Rudolph Diesel in 1893. In this cycle, the air is compressed to a temperature higher than the self-ignition temperature of the fuel. Now, the fuel is injected which ignites on its own, therefore, the engines are called compression ignition (CI) engines. The heat addition takes place at constant pressure; therefore, it is also called as *constant pressure* cycle.

The air and fuel are compressed separately, and therefore no auto-ignition of the fuel, thus avoiding the limitation on compression ratio as in the case of the Otto cycle. The cylinder walls are made of perfectly non-conducting, whereas the cylinder head is made of conducting material. The p-V curve of the Diesel cycle is shown in Fig. 3.16.

- The process (1-2) is an isentropic compression of air. As entropy is constant, no heat flow takes place out of the cylinder,

$$Q_{1-2} = 0$$

- The process (2-3) is constant pressure heat addition through a hot body and combustion takes place at constant pressure.

$$Q_{2-3} = C_p(T_3 - T_2) \quad 3.10$$

- The process (3-4) is an isentropic expansion and work is obtained from the system. No heat interaction takes place with the surrounding.

$$Q_{3-4} = 0 \quad 3.11$$

- The process (4-1) is a constant volume heat rejection to the cold body, and the cycle is completed.

$$Q_{4-1} = C_v(T_4 - T_1) \quad 3.12$$

Work done during the cycle = Heat supplied – Heat rejected

$$= C_p(T_3\text{-}T_2) - C_v(T_4\text{-}T_1) \quad 3.13$$

Compression ratio $= V_1/V_2 = r$

Expansion ratio $= V_4/V_3 = r/\rho$

Cut off ratio $= \dfrac{V_3}{V_2} = \dfrac{\rho}{1}$

Efficiency of the cycle (η) $= \dfrac{\text{Output}}{\text{Input}}$

$$= \frac{\text{Work done}}{\text{Heat supplied}} = \frac{C_p(T_3\text{-}T_2) - C_v(T_4\text{-}T_1)}{C_p(T_3\text{-}T_2)} \quad 3.14$$

$$= 1\text{-}\left(\frac{C_v}{C_p}\right) \times \frac{(T_4\text{-}T_1)}{(T_3\text{-}T_2)}$$

$$= 1 - \left(\frac{1}{\gamma}\right) \times \frac{(T_4\text{-}T_1)}{(T_3\text{-}T_2)} \quad 3.15$$

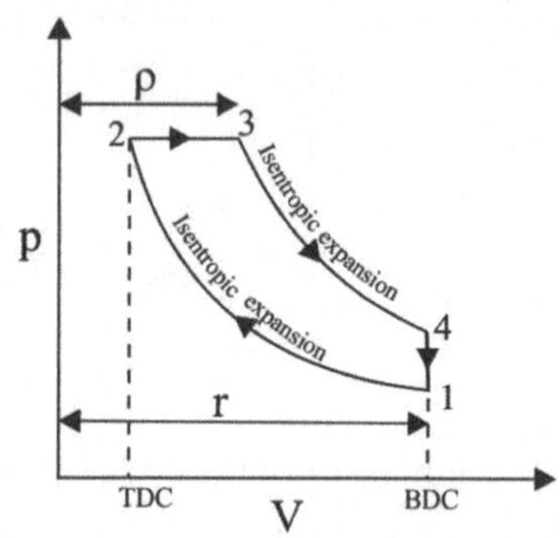

Fig. 3.16: p-V curve of Diesel cycle

From constant pressure line (2-3)

$$\frac{p_2 V_2}{T_2} = \frac{p_3 V_3}{T_3}$$

$$\frac{T_3}{T_2} = \rho \quad \text{and} \quad T_2 = \frac{T_3}{\rho}$$

From adiabatic process (3-4)

$$p_3 V_3^{\gamma} = p_4 V_4^{\gamma}$$

$$\frac{p_3}{p_4} = \left(\frac{V_4}{V_3}\right)^{\gamma} = \left(\frac{r}{\rho}\right)^{\gamma} \qquad 3.16$$

$$\frac{p_3}{p_4} = \left(\frac{T_4}{T_3}\right)^{\frac{\gamma}{\gamma-1}}$$

$$\frac{T_4}{T_3} = \left(\frac{r}{\rho}\right)^{\gamma-1}$$

$$T_3 = \frac{T_4}{(r/\rho)^{\gamma-1}} \qquad 3.17$$

From adiabatic process (1-4)

$$p_1V_1^{\gamma} = p_2V_2^{\gamma}$$

$$\frac{p_1}{p_2} = \left(\frac{V_2}{V_1}\right)^{\gamma} = \left(\frac{1}{r}\right)^{\gamma} \qquad 3.18$$

Also, $$\frac{p_1}{p_2} = \left(\frac{T_4}{T_3}\right)^{\frac{\gamma}{\gamma-1}}$$

$$\frac{T_1}{T_2} = \left(\frac{1}{r}\right)^{\frac{\gamma}{\gamma-1}}$$

$$T_1 = T_2 \times \left(\frac{1}{r}\right)^{\gamma-1} \qquad 3.19$$

$$= \frac{T_3}{\rho} \times \left(\frac{1}{r}\right)^{\gamma-1} \qquad \text{As we know that } \left(T_4 = \frac{T_1}{\rho}\right) \qquad 3.20$$

By substituting the values of temperatures in equation 3.15, we get

$$\text{Efficiency of the cycle } (\eta) = 1 - \frac{1}{\gamma}\left[\frac{\frac{T_1}{(r/\rho)^{\gamma-1}} - \frac{T_1}{\rho}\left(\frac{1}{r}\right)^{\gamma-1}}{T_3 - \frac{T_1}{\rho}}\right] \qquad 3.21$$

$$= 1 - \frac{1}{\gamma} \times \frac{1}{(r)^{\gamma-1}}\left[\frac{\rho^{\gamma-1} - \frac{1}{\rho}}{1 - \frac{1}{\rho}}\right]$$

$$= 1 - \frac{1}{\gamma} \times \frac{1}{(r)^{\gamma-1}}\left[\frac{\rho^{\gamma} - 1}{\rho - 1}\right] \qquad 3.22$$

The efficiency of a diesel engine depends on compression ratio (r), cut-off ratio (ρ) and the ratio of specific heat (γ). The efficiency of the cycle increases with a decrease in the cut-off ratio, and by using the gas having larger values of the ratio of specific heat. For the same compression ratio, the Otto cycle is more efficient than the diesel cycle. For the diesel cycle, a compression ratio is between 16-20. Due to higher values of compression ratio, the efficiency of the diesel cycle is more. In actual practice, heat addition only at constant pressure in an engine is not possible, except when an air injection system is used.

Example 3.3: In a Diesel cycle, air at 17 °C, 1 bar and 1 m³ is compressed adiabatically until the pressure is 50 bar and the volume before adiabatic expansion is 0.1 m³. Calculate;

(a) Temperature at the major points of the cycle

(b) Air standard efficiency of the cycle.

Solution:

Given: $T_1 = 17 + 273 = 290 \text{ K}, p_1 = 1 \text{ bar}, V_1 = 1 \text{ m}^3$

$$p_2 = 50 \text{ bar}, V_3 = 0.1 \text{ m}^3$$

$$\frac{T_2}{T_1} = \left(\frac{p_2}{p_1}\right)^{\frac{\gamma-1}{\gamma}} = \left(\frac{50}{1}\right)^{\frac{0.4}{1.4}} = 3.0577$$

$$T_2 = 290 \times 3.0577 \text{ K} \quad \text{Ans}$$

$$T_2 = 889.7924 \text{ K} \quad \text{Ans}$$

$$\frac{p_1 V_1}{T_1} = \frac{p_2 V_2}{T_2}$$

$$V_2 = \frac{p_1 T_2 V_1}{p_2 T_1} = \frac{1}{50} \times \frac{889.7924}{291} \times 1 = 0.06109 \text{ m}^3$$

$$T_3 = T_2\left(\frac{V_3}{V_2}\right) = 889.7924 \times \frac{0.1}{0.06109}$$

$$= 1456.527 \text{ K}$$

$$\frac{T_4}{T_1} = (\rho)^{\gamma} = \left(\frac{T_3}{T_2}\right)^{\gamma} = \left(\frac{1456.527}{889.7024}\right)^{1.4} = 1.99$$

$$T_4 = 300 \times 1.99 = 599 \text{ K}$$

$$Q_s = m\, C_p (T_3 - T_2)$$

$$= 1.005\,(1456.527 - 889.7924)$$

$$= 569.568 \text{ kJ}$$

$$Q_r = m\, C_v (T_4 - T_1)$$

$$= 1 \times 0.718\,(599 - 291)$$

$$= 221.144 \text{ kJ}$$

Efficiency of the cycle (η) = $1 - \frac{Q_r}{Q_s} = 1 - \frac{221.144}{569.568}$

$= 1 - 0.38826 = 0.6117 = 61.17\ \%$ Ans

Example 3.4: In an engine working on a Diesel cycle, inlet pressure and temperature are 1 bar and 17 °C respectively. Pressure at the end of adiabatic compression in 35 bar. The ratio of expansion i.e. after constant pressure heat addition, is 5. Calculate;

(a) Heat addition

(b) Heat rejection

(c) Efficiency of cycle

Assume: $\gamma = 1.4$, $C_p = 1.004$ kJ/kg K, and $C_v = 0.717$ kJ/kg K.

Solution:

Given: $p_1 = 1$ bar, $p_2 = 35$ bar, $T_1 = 17 + 273 = 290$ K

$$\frac{V_1}{V_2} = r = \left(\frac{p_2}{p_1}\right)^{1/\gamma}$$

$$= \left(\frac{35}{1}\right)^{1/1.4} = 12.674$$

$$\text{Cut - off ratio} = \frac{V_3}{V_2} = \frac{V_3}{V_1} \times \frac{V_1}{V_2}$$

$$= \frac{\text{Compression ratio}}{\text{Expansion ratio}} = \frac{12.674}{5} = 2.535$$

$$\frac{T_2}{T_1} = \left(\frac{p_2}{p_1}\right)^{\gamma-1/\gamma} = \left(\frac{35}{1}\right)^{0.286} = 2.76$$

$$T_2 = 2.76 \times 290 = 801.7 \text{ K}$$

Now,

$$T_3 = T_2\left(\frac{V_3}{V_2}\right) = 801.7 \times \left(\frac{V_3}{V_2}\right)$$

$$= 801.7 \times 2.535 = 2032.3 \text{ K}$$

$$T_4 = T_3\left(\frac{V_3}{V_4}\right)^{\gamma-1} = 2032.3 \times \left(\frac{1}{5}\right)^{0.4}$$

$$= 1067.6 \text{ K}$$

Heat addition = $C_p\ (T_3 - T_2) = 1.004\ (2032.3 - 801.7)$

$= 1235.5$ kJ/kg Ans

Heat rejection = $C_v(T_4 - T_1) = 0.717\ (1067.6 - 290)$

$$= 557.5 \text{ kJ/kg} \quad \text{Ans}$$

$$\text{Efficiency of the cycle } (\eta) = 1 - \frac{Q_r}{Q_s} = 1 - \frac{557.5}{1235.5} = 0.549$$

$$= 54.9\ \% \quad \text{Ans}$$

3.5.3 Dual Cycle

In actual Otto cycle (SI) engines, the combustion process does not take place at constant volume because chemical reaction needs time during combustion. Similarly, in the actual Diesel cycle (CI) engine, combustion does not take place at constant pressure because of fast, uncontrolled combustion in an engine. To overcome these, another cycle was proposed called as Dual cycle, which approximates the working of spark and compression ignition engines. The p-V diagram of the Dual cycle is shown in Fig. 3.17. In an ideal Dual cycle, the heat addition takes place at constant volume and at constant pressure, whereas compression and expansion take place isentropically.

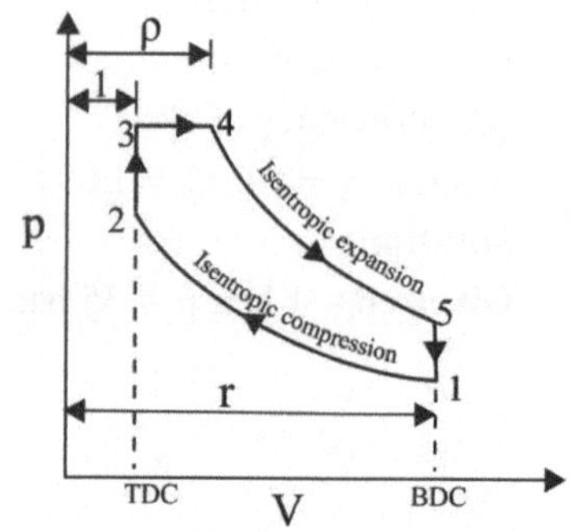

Fig. 3.17: p-V curve of Dual cycle.

Total heat supplied during the process (2-3 and 3-4) = Heat supplied at constant volume + Heat supplied at constant pressure.

$$= C_v(T_3 - T_2) + C_p(T_4 - T_3) \qquad 3.23$$

$$\text{Heat rejected during the process (5-1)} = C_v(T_5 - T_1) \qquad 3.24$$

$$\text{Air standard efficiency of the cycle} = \frac{\text{Workdone}}{\text{Heatsupplied}}$$

$$= \frac{(\text{Heat supplied - Heat rejected})}{\text{Heat supplied}} = 1 - \frac{\text{Heat rejected}}{\text{Heat supplied}}$$

$$= 1 - \frac{C_v(T_5 - T_1)}{C_v(T_3 - T_2) + C_p(T_4 - T_3)} \qquad 3.25$$

$$= 1 - \frac{(T_5 - T_1)}{(T_3 - T_2) + \gamma(T_4 - T_3)} \qquad 3.26$$

Let us consider clearance volume = 1

Then, $V_2 = V_3 = 1$

Let us consider volume at the point of cut-off = ρ

Then, $V_4 = \rho$

$$= \frac{V_1}{1} = \frac{V_5}{1}$$

and consider the explosion ratio $= \dfrac{p_3}{p_2} = \alpha$

The process (2-3) takes place at constant volume

$$\frac{T_3}{T_2} = \frac{p_3}{p_2} = \alpha \qquad 3.27$$

Also, $\frac{V_1}{V_2} = r = \frac{V_5}{V_2}$

and $\frac{V_5}{V_4} = \frac{r}{\rho}$

Now, finding out all the temperatures in terms of temperature 'T_3' and substituting in equation 3.26

From process (4-5) $\frac{T_4}{T_5} = \left(\frac{r}{\rho}\right)^{\gamma-1}$ 3.28

$$T_5 = \frac{T_4}{\left(\frac{r}{\rho}\right)^{\gamma-1}} = T_4\left(\frac{\rho}{r}\right)^{\gamma-1}$$

From process (3-4), $\frac{p_3V_3}{T_3} = \frac{p_4V_4}{T_4}$ 3.29

$$= T_4 = T_3\left(\frac{V_4}{V_3}\right) = T_3\rho \qquad 3.30$$

From process (1-2), $\frac{T_2}{T_1} = \left(\frac{V_1}{V_2}\right)^{\gamma-1} = (r)^{\gamma-1}$

$$T_1 = \frac{T_2}{(r)^{\gamma-1}} \qquad 3.31$$

From process (2-3), $\frac{p_2V_2}{T_2} = \frac{p_3V_3}{T_3}$

or $\frac{p_3}{p_2} = \frac{T_3}{T_2} = \alpha$

$$T_2 = \frac{T_3}{\alpha} \qquad 3.32$$

$$T_1 = \frac{T_3}{\alpha(r)^{\gamma-1}} \qquad 3.33$$

Again, substituting the values of T_4, T_5, T_3 and T_2 in equation (3.26), we get

$$\eta = 1 - \left[\frac{\rho.T_3\left(\frac{\rho}{r}\right)^{\gamma-1} - \frac{T_3}{\alpha(r)^{\gamma-1}}}{\left(T_3 - \frac{T_3}{\alpha}\right) + \gamma(\rho.T_3 - T_3)}\right] \qquad 3.34$$

The term T_3 will be canceled from the above equation, then

$$\eta = 1 - \frac{1}{(r)^{\gamma-1}}\left[\frac{\left(\rho^{\gamma} - \frac{1}{\alpha}\right)}{\left(1-\frac{1}{\alpha}\right) + \gamma(\rho-1)}\right] \quad 3.35$$

$$= 1 - \frac{1}{(r)^{\gamma-1}}\left[\frac{\left(\alpha\rho^{\gamma}-1\right)}{(\alpha-1) + \gamma\alpha(\rho-1)}\right] \quad 3.36$$

It can be observed from the above equation that, for the value of $\alpha > 1$, and for a given value of the 'γ' and 'ρ', the cycle efficiency increases. The efficiency of the dual combustion cycle lies between Otto and Diesel cycles, i.e., greater than the Diesel cycle and less than the Otto cycle, for the same compression ratio.

For the ideal Otto cycle, $\rho = 1$; and for the ideal diesel cycle, $\alpha = 1$.

Example 3.5: The dual cycle operates with air intake at 1 bar 300 K. The compression ratio is 16, and the maximum pressure and temperature reached in the cycle are 80 bar and 2500 K. Determine;

(a) The parameters P, V and T for each point of the cycle
(b) The thermodynamic efficiency of the cycle
(c) Net work done

Solution:

Given: $p_1 = 1$ bar, $T_1 = 300$ K, $r = 16$, $p_3 = 80$ bar, $T_3 = 2500$ K

$$p_2 = p_1\,(r)^{\gamma} = 1 \times (16)^{1.4} = 48.5 \text{ bar}$$

$$T_2 = T_1\,(r)^{\gamma-1} = 300(16)^{0.4} = 909.43 \text{ K}$$

$$T_3 = \frac{p_3 T_2}{p_2} = \frac{80 \times 909.43}{48.5} = 1500 \text{ K}$$

$$p_3 = p_4 = 80 \text{ bar}$$

$$T_4 = 2500\text{K}$$

$$\frac{T_3}{T_2} = \frac{P_3}{P_2} = \alpha = \frac{1500}{909.43} = 1.649$$

$$\frac{V_4}{V_3} = \rho = \frac{T_4}{T_3} = \frac{2500}{1500} = 1.666$$

$$\frac{T_5}{T_1} = \alpha\,\rho^{\gamma} = 1.649 \times (1.666)^{1.4} = 3.37$$

$$T_5 = 300 \times 3.37 = 1011.4167 \text{ K}$$

$$\frac{p_5}{p_4} = \left(\frac{T_5}{T_4}\right)^{\gamma/\gamma-1} = \left(\frac{1011.4167}{2500}\right)^{3.5} = (0.4045)^{3.5} = 0.0421$$

$$p_5 = 80 \times 0.0421 = 3.369 \text{ bar}$$

$$p_1 V_1 = RT_1$$

$$1 \times 100 \times V_1 = 0.287 \times 300$$

$$V_1 = 0.861 m^3/kg = V_5$$

$$V_2 = \frac{V_1}{16} = \frac{0.861}{16} = 0.0538\ m^3/kg$$

$$V_4 = \frac{RT_4}{p_4} = \frac{0.287 \times 2500}{80 \times 100} = 0.0897\ m^3/kg$$

$$V_1 - V_2 = 0.0861 - 0.0538 = 0.0872\ m^3/kg$$

$$\text{Heat Supplied} = C_v\ (T_3 - T_2) + C_p(T_4 - T_3)$$

$$= 0.718\ (1500 - 909.43) + 1.005\ (2500 - 1500)$$

$$= 424.1 + 1005 = 1429.1\ kJ/kg$$

$$\text{Heat rejected} = C_v\ (T_5 - T_1) = C_v T_1 \left(\frac{T_5}{T_1} - 1 \right)$$

$$= 0.718 \times 300(3.37 - 1)$$

$$= 510.498\ kJ/kg$$

$$W_{net} = \text{Heat supplied} - \text{Heat rejected}$$

$$= 1429.1 - 510.498$$

$$= 918.602\ kJ/kg \quad \text{Ans}$$

$$\text{Air standard efficieny of the cycle} = \frac{\text{Work done}}{\text{Heat supplied}}$$

$$= \frac{918.602}{1429.1} = 0.6427$$

$$= 64.27\ \% \quad \text{Ans}$$

3.6 INTERNAL COMBUSTION ENGINES

In a thermodynamic system, heat engines are devices used to convert the chemical energy of the fuel burnt during combustion into thermal energy and thereafter utilize this thermal energy into mechanical work done. The output power generated decides the efficiency of the engine. Heat engines are classified as internal combustion and external combustion engines.

Internal combustion (IC) engines are heat engines where the combustion of the fuel takes place in a combustion chamber (cylinder) within the engine. The high temperature and pressure gases generated expand and force the piston. The force moves the piston and other components by converting chemical energy into thermal energy which is then converted into useful work. Petrol, diesel and gas engines are common examples of IC engines. In *external combustion engines,* the combustion of fuel takes place outside the engine and the heat generated is used to generate high-pressure steam. This steam works as fluid in the cylinder of reciprocating engines or turbines.

The heat engine works on a model following an air standard cycle such as the Otto cycle, Diesel cycle and Dual engine cycle. The first Otto cycle was developed in 1876 and used to operate petrol engine-based automobiles, and thereafter many revolutions took place. Due to rapid industrialization and many vehicles on the road, the ecology and balance of the environment have changed with harmful exhaust gases. To meet the increasing demand of petrol and diesel, alternative fuels such as gases and biofuels are being used which are less polluting the environment. These days, electric vehicles are in demand to overcome environmental issues. Internal combustion engines are classified as;

Rotary engines: Open cycle gas turbine and Wankel engine

Reciprocating engines: Petrol (SI) engine and Diesel (CI) engine

3.6.1 Components of IC Engines

The major working components of IC engines are shown in Fig. 3.18 and are listed below:

1. **Cylinder:** It is the major component of an IC engine, where the compression and combustion of fuel take place. The high pressure and temperature generated guide the piston. Therefore, the material used for the cylinder is cast iron which has high compressive strength.
2. **Cylinder head:** It is the top removable part of a cylinder. It is made up of cast iron and has an opening for two ports closed by an inlet and exhaust valve, spark plug, and injector.
3. **Piston:** The piston transfers the pressure generated in the cylinder after combustion of gases to the crankshaft through connecting rod. The top portion of the piston facing high temperature and pressure is called *head or crown*. It is made of low-carbon steel or aluminum alloys and moves freely inside the cylinder.
4. **Piston rings:** For proper sealing, pistons are provided with piston rings split at one end made up of cast iron or spring steel. These are fitted in the grooves at the periphery of the piston. For a small two-stroke engine, two rings are provided, whereas for a four-stroke engine, an extra ring called an oil ring is provided.
5. **Inlet and exhaust valves:** Values are used to control the inlet and exhaust of IC engines and their number depends on a number of cylinders in an engine. Each cylinder has two valves, one for the intake of air or air-fuel mixture and the other for the exhaust of burnt gases and both valves open inwards.
6. **Inlet and exhaust manifold:** The inlet manifold is a pipe through which air or air-fuel mixture enters the inlet valve. The exhaust manifold is also a pipe through which burnt gases escape from the exhaust valve.
7. **Cylinder liner:** It is a separate sleeve fitted in the cylinder that provides wear resistance to the cylinder bore for longer life. Cast iron liners are commonly used in engines.
8. **Gudgeon pin:** It connects the piston with the smaller end of connecting rod.
9. **Connecting rod:** It is a linkage between the piston and crankshaft and transmits the pressure of gases generated inside the cylinder to produce work at the crankshaft.
10. **Crankshaft:** It is connected to the big end of connecting rod through a crank pin. It transfers the reciprocating motion of the piston into rotary motion of the output shaft. Crankshaft consists of a crank web, crank pin, and balance weights for static and dynamic balancing.
11. **Camshaft:** It operates the inlet and exhaust valve according to the timing gears through power received from crankshaft. It has a push rod, rocker arms, valve springs and a tappet.
12. **Flywheel:** During one complete cycle, the torque transmitted to the crankshaft fluctuates which changes the angular velocity of the shaft. To obtain a uniform torque throughout the cycle, an inertia mass fixed to the output shaft is called a flywheel.

3.6.2 Terminologies used in IC Engines

The different terminologies used during discussion and analysis in IC engines need to be understood. Following are the terminologies that may be observed in Fig.3.18and are discussed below.

1. **Bore:** The nominal inner diameter of the cylinder is called a bore, denoted by (d).
2. **Top dead center (TDC):** It is the topmost position attained by the piston during compression and exhaust stroke in a vertical cylinder engine.
3. **Bottom dead center (BDC):** It is the bottommost position attained by the piston during suction and expansion stroke in a vertical cylinder engine.
4. **Stroke:** It is the distance travelled by the piston moving from TDC to BDC, denoted by (L)
5. **Clearance volume:** It is the volume above the piston in combustion chamber when the piston has reached its top dead center. It is denoted by V_c and measured in a cubic centimeter (cc).
6. **Swept volume:** It is the volume swept by the piston when it moves from TDC to BDC or BDC to TDC. It is denoted by V_s and measured in a cubic centimeter. For an engine having (n) number of cylinders, then swept volume will be;

$$V_S = \frac{\pi}{4} \times d^2 \times L \times n \tag{3.37}$$

7. **Compression ratio:** It is the compression developed due to piston movement during the compression stroke, denoted by (r). It is defined as the ratio of the total volume of the cylinder ($V_{T)}$ when the piston is at BDC to the clearance volume (V_c).

$$r = \frac{V_T}{V_C} = \frac{V_C + V_S}{V_C} = 1 + \frac{V_S}{V_C} \tag{3.38}$$

8. **Engine capacity or cubic capacity:** It is the swept volume of one cylinder multiplied by a number of cylinders of an engine.

Cubic Volume = $V_S \times n$ 3.39

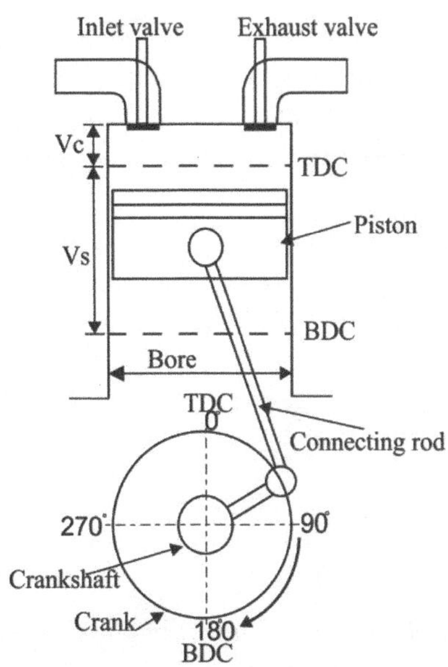

Fig. 3.18: Basic geometry and nomenclature

3.7 SPARK IGNITION (SI) AND COMPRESSION IGNITION (CI) ENGINES

Spark ignition (Petrol) engines and compression ignition (Diesel) engines are internal combustion engines. The SI and CI engines are further divided into four stroke and two stroke types.

3.7.1 Four-Stroke SI Engine

Spark ignition (Petrol) engines are internal combustion engine that operates on the principle of the Otto cycle. The piston reciprocates inside the cylinder and a mixture of air and petrol as working fluid is compressed. The spark from the spark plug ignites the mixture after compression. The one cycle is completed after four strokes of the piston or two complete revolutions of the crankshaft. The four different strokes are suction, compression, expansion and exhaust stroke. During each stroke, the crankshaft rotates by 180° and the camshaft by 90°. The working principle of four stroke petrol engine and its corresponding p-V diagram is shown in Fig. 3.19 (a-d). The different strokes are explained below.

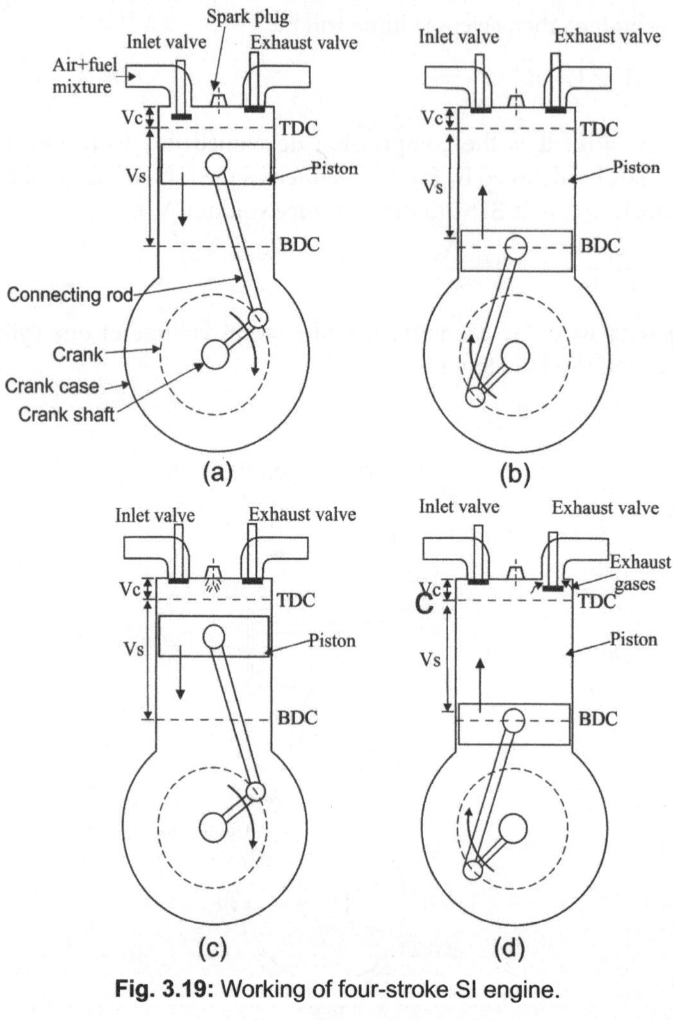

Fig. 3.19: Working of four-stroke SI engine.

(i) Suction or intake stroke

Fig. 3.19 (a) and process (0-1) in p-V diagram of Fig.3.20 show the suction or intake of a combustible mixture of air and petrol vapors into the cylinder through an inlet valve. During this stroke, the inlet valve is open and the exhaust valve remains closed. The piston initially at TDC moves towards BDC and the pressure inside the cylinder is below atmospheric. Due to the pressure difference, the mixture rushes inside the cylinder. At the end of the stroke when the piston reaches BDC, both the inlet and exhaust valves are closed, as seen in Fig. 3.19 (b). The air and petrol mixture in appropriate proportion for complete combustion is supplied by a carburettor is termed as stoichiometric ratio.

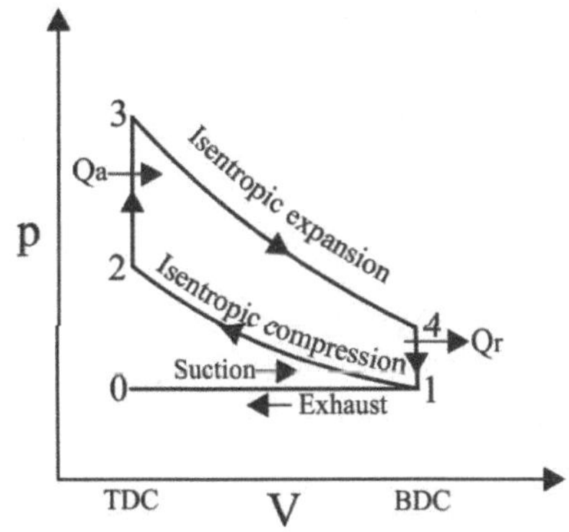

Fig. 3.20: p-V curve four stroke SI engine

(ii) Compression stroke

After the completion of the suction stroke, the air-fuel mixture in the cylinder is ready to be compressed as shown in Fig. 3.19 (b). The isentropic compression is shown in p-V diagram as the process (1-2) in Fig. 3.20. Both inlet and exhaust valves are closed and the piston moves from BDC to TDC, during this stroke pressure and temperature increase. At the end of the compression stroke at TDC the mixture is ignited with the help of a spark plug. It is assumed that combustion of fuel takes place when the piston is at TDC, and ignition of fuel is considered as a constant volume heat addition process (2-3). At the end of the stroke, pressure, and temperature raises due to heat released by the burnt mixture.

(iii) Expansion stroke

The burnt mixture after ignition, expands in the cylinder, which pushes the piston from TDC towards BDC as shown in Fig. 3.19 (c). Since power is produced in this stroke, therefore, it is also known as a power stroke. The p-V diagram in Fig. 3.20, process (3-4) shows an isentropic expansion of burnt gases. During the expansion stroke both the inlet and exhaust valves remain in the closed position. The pressure and temperature of the gases decrease and heat is released at constant volume (4-1).

(iv) Exhaust stroke

After obtaining the useful work in the expansion stroke, the burnt gases are ready to be exhausted out into the open atmosphere. At the end of the expansion stroke, the exhaust valve opens while the inlet valve remains in the closed position as shown in Fig. 3.20 (d). The process (1-0) in the p-V diagram represents the exhaust stroke. The piston moves from BDC to TDC and sweeps the burnt gases out of the cylinder. When the piston reaches TDC, the exhaust valve is also closed, with a small quantity of residual gases trapped in the clearance volume. At the end of the exhaust stroke, the cycle is completed and the next cycle begins. During all four strokes, the crankshaft makes two complete revolutions and the camshaft makes one revolution.

3.7.2 Four-Stroke CI Engine

The compression ignition (CI) engine uses diesel as a fuel and operates on the principle of the Diesel cycle. Its working is similar to the Otto cycle, but only air is compressed inside the cylinder and the compression ratio is high. After the compression stroke, fuel is injected into the cylinder, and the heat of compressed air ignites the fuel. The working principle of four stroke petrol engine is shown in Fig. 3.21 (a-d), and different strokes are explained below;

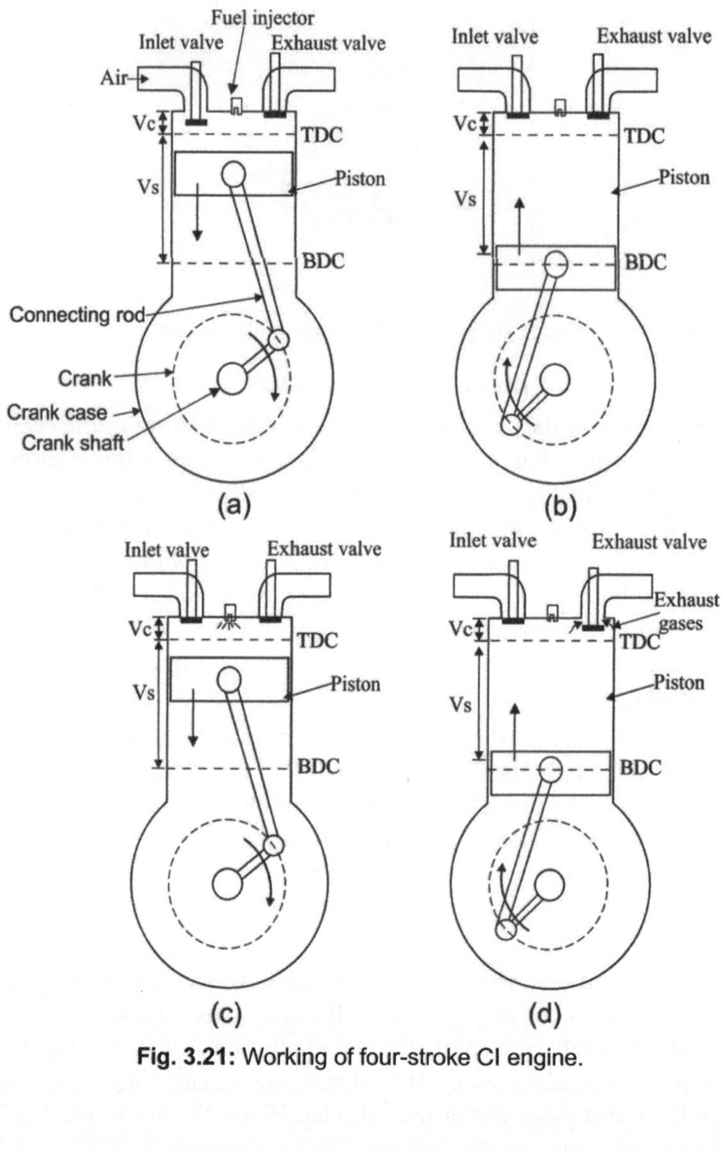

Fig. 3.21: Working of four-stroke CI engine.

(i) **Suction or intake stroke:**

During this stroke, the inlet valve is open and the exhaust valve remains closed. The piston initially at TDC moves towards BDC, as seen in Fig. 3.21 (a). The air alone is inducted inside the cylinder.

(ii) **Compression stroke:**

During this stroke both inlet and exhaust valves are closed and piston moves from BDC to TDC and as depicted in Fig 3.21 (b). The air induced into the cylinder is compressed up to the clearance volume. The temperature of the air increases and is higher than the ignition point of the fuel. The fuel is injected into the combustion chamber with the help of a high-pressure fuel pump and an injector. The injector has many fine holes for spray.

(iii) **Expansion stroke:**

At the end of the compression stroke the fuel first vaporizes and then auto ignites. Heat addition is assumed to be made at constant pressure. When the fuel injection is completed, i.e., the cut-off point, the high pressure and temperature gases expand. Both valves are closed and the piston moves from TDC towards BDC, as seen in Fig 3.21 (c).

(iv) **Exhaust stroke:**

At the end of the expansion stroke, the exhaust valve opens and the inlet valve remains closed. The piston moves from BDC to TDC and sweeps the burnt gases out of the cylinder. When the piston reaches TDC, the exhaust valve is also closed, with a small quantity of residual gases trapped in the clearance volume, as seen in Fig 3.21 (d).

During all four strokes, the crankshaft makes two complete revolutions and the camshaft makes one revolution. The compression ignition engines are made heavier than spark ignition engines for the same power output because of high pressure generated in the cycle. It has better thermal efficiency than spark ignition engines due to a higher compression ratio.

3.7.3 Two-Stroke Engine

The basic difference between two stroke and four stroke engine is the method of induction of fresh charge and expulsion of burnt product from the cylinder. In the case of four stroke engines, the suction and expulsion process take place during suction and exhaust stroke, respectively.

In the case of two stroke engine, the fresh charge is compressed in the crank case, which is thereby passed through the transfer port to fill the cylinder. This induced compressed charge in cylinder assists in expelling the burnt products through an exhaust port. The piston strokes required for the suction and exhaust process are eliminated. Hence, only two strokes are sufficient to complete the cycle, i.e., one for the compression of fresh charge and the other for the power stroke while the crankshaft makes one revolution only. Therefore, one power stroke for each revolution of the crankshaft can be attained. In two stroke engines, ports are used instead of valves. The opening and closing of ports are made through the reciprocating piston.

When the piston moves upward from BDC to TDC during the compression stroke, the pressure in crank case becomes lower than atmospheric pressure as shown in Fig. 3.22 (a). In a diesel engine, the air enters the crankcase through the one-way spring-loaded valve. At the end of the compression stroke, the piston reaches TDC, and the diesel fuel is injected. Whereas, in the case of a petrol engine, the air-fuel mixture enters the crankcase, and a spark is introduced through a spark plug at the end of the compression stroke.

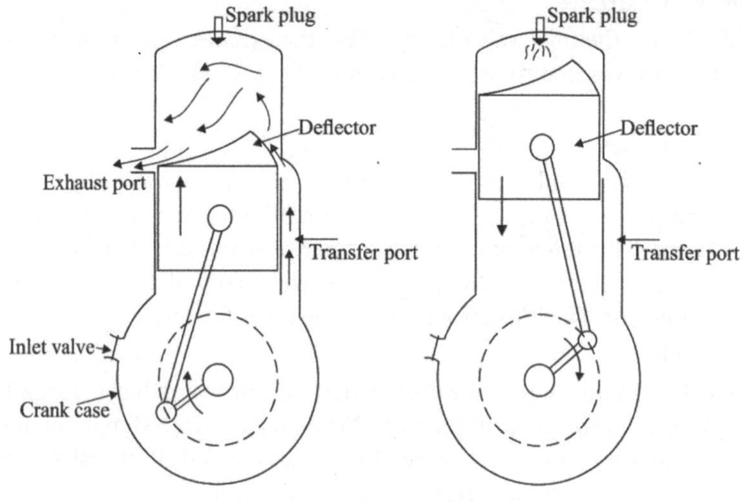

Fig. 3.22: Working of two-stroke engine.

During the expansion stroke, the high temperature and gases push the piston downwards from TDC to BDC. The pressure in crank case increases and the valve closes. During the down ward movement of the piston, the exhaust port opens first and the exhaust gases start escaping out. Simultaneously, the transfer port opens with the further downward movement of the piston allowing fresh compressed charge to enter into the cylinder and push out the burnt gases. As the piston reaches BDC both the transfer and exhaust ports are open. The fresh charge pushing the gases out of the cylinder is termed as scavenging, as shown in Fig. 3.22 (a). The piston moves upwards, the exhaust port is closed and the charge in the cylinder is compressed and the cycle repeats.

During scavenging, some of the fresh charges may go out through the exhaust port is termed as short-circuiting. To avoid this problem, the top of the piston is provided with a deflector. In the case of the SI engine, the petrol also goes out with air. Therefore, fuel is wasted and the thermal efficiency of the engine decreases. This also creates an environmental issue, as unburnt hydrocarbon comes out with the exhaust.

3.7.4 Comparison of SI and CI Engines

The basic differences between spark ignition and compression ignition engines are discussed in Table 3.2.

Table 3.2: Difference between spark ignition and compression ignition engine

Description	SI Engine	CI Engine
Cycle of operation	Works on Otto cycle	Works on Diesel cycle
Fuel used	The fuel used is Petrol has a high self-ignition temperature	The fuel used is Diesel has a low self-ignition temperature
Intake of fuel	Air and fuel mixture prepared by carburettor is induced during suction and compressed	Only air is used and compressed

Description	SI Engine	CI Engine
Ignition of fuel	The compressed mixture is ignited with a spark plug in the combustion chamber. Initially, voltage is supplied by the battery of a magneto system.	Diesel in the vaporised form is injected at the end of compression. A fuel pump and injector are required.
Compression ratio	The compression ratio is between 6 to 10.	The compression ratio is high, between 16 to 20.
Thermal efficiency	Thermal efficiency is low due to a lower compression ratio.	Thermal efficiency is high.
Weight	Light in weight due to low pressure developed	Heavy in weight due to high pressure developed
Speed	Due to light weight, speed is high.	Due to heavy weight, speed is low.
Fuel consumption	High	Low
Maintenance cost	Low	High
Applications	Light vehicles, like; automobiles, motorcycles, small engines, lawn mowers etc.	Used for high power vehicles, like; locomotives, trucks, buses, etc.

3.7.5 Comparison of Two-Stroke and Four-Stroke Engines

The basic differences between two-stroke and four-stroke engines are discussed in Table 3.3.

Table 3.3: Difference between two-stroke and four-stroke engine

Two Stroke	Four Stroke
The cycle is completed in two strokes of the piston. Crankshaft makes one complete revolution. Therefore, one power cycle is obtained in one revolution of crank shaft.	The cycle is completed in four strokes of the piston. Crankshaft makes two complete revolutions. Therefore, one power cycle is obtained in two revolutions of the crank shaft.
Two-stroke engines has ports	Four-stroke engine has valves
A lighter flywheel is required as the turning moment is more uniform.	A heavier flywheel is required for a uniform turning moment.
As it has one power stroke in one revolution, so more wear and tear occur. Therefore, more lubrication and cooling are required.	As it has one power stroke in two revolutions, so less wear and tear occur. Therefore, lesser lubrication and cooling are required.
More maintenance is required	Less maintenance is required
Scavenging takes place	No scavenging takes place.
The power produced for the same size of the engine is almost twice	The power produced for the same size of the engine is less.
For the same power output engine is lighter in weight and more compact	For the same power, the output engine is heavy in weight and bulky

Two Stroke	Four Stroke
Due to light weight and simple port mechanism, engine cost is less	Due to higher weight and valve mechanism, engine cost is more
Thermal efficiency is low	Thermal efficiency is high
Volumetric efficiency is low	Volumetric efficiency is high
Used in scooters and motorcycles, blowers, trimmers and hedge trimmers, mopeds etc.	Used in cars, buses, trucks, airplanes etc.

3.8 CALCULATION OF ENGINE POWER AND EFFICIENCY

The power of an engine is defined as work done per unit of time or rate of doing work. The rating of any engine is evaluated on the basis of output power and torque produced.

Indicated horse power (IHP): It is the power produced by the gases transmitted to the piston and measured at the crankshaft. The IHP is calculated as;

$$\textbf{IHP} = \frac{k \times p_m \times L \times A}{4500} \times \frac{N}{n} \tag{3.40}$$

$$\textbf{In SI units; } \text{Power output (kW)} = \frac{k \times p_m \times L \times A}{60} \times \frac{N}{n} \tag{3.41}$$

Where;

A = Area of piston, cm^2 (SI unit= m^2)

L = Length of stroke in metre, m

k = No. of cylinders

N = Engine speed, rpm

p_m = Indicated mean effective pressure, kgf/cm^2 (SI unit= kPa)

n = 1, for two stroke engine

= 2, for four stroke engine

For a gas engine n = f = Explosions per second or firing strokes per second.

Brake horse power (BHP): It is the actual power delivered to the driving shaft. The BHP is always less than IHP because of the power loss in friction between the moving parts and components and the power required to operate engine accessories.

$$\textbf{BHP} = \frac{2 \times \pi \times N \times T}{4500} \tag{3.42}$$

$$\textbf{In SI units; } \text{Power output (kW)} = \frac{2 \times \pi \times N \times T}{60} \tag{3.43}$$

Where;

N= Engine speed, rpm

T= Torque developed by the engine, kgf-m (kN-m)

$$\textbf{Also; } \text{Break power (bp)} = \frac{k \times p_{mb} \times L \times A}{60} \times \frac{N}{n} \tag{3.44}$$

Mechanical efficiency: It is the ratio of brake power (B.P.) to the indicated power (I.P.)

$$\eta_{mech} = \frac{B.P.}{I.P.} \qquad 3.45$$

Indicated thermal efficiency: It is the ratio of the heat equivalent to one kilo Watt hour to the heat in the fuel per I.P. (hour).

$$\eta_{ith} = \frac{\text{Heat equivalent to one kilo watt hour}}{\text{Heat in fuel per I.P., hour}} = \frac{\text{I.P.} \times 3600}{m_f \times C} = \frac{3600}{\text{isfc} \times C} \qquad 3.46$$

The ratio of $m_f / I.P.$ is termed as specific fuel consumption per I.P. hour.

Where;

I.P. = Indicated power in kW,

m_f = Mass of fuel consumed in kg per hour, and

C = Calorific valve of fuel in kJ / kg of fuel.

$$\eta_{ith} = \frac{\text{Heat equivalent to one horse power, hour}}{\text{Heat in fuel per I.H.P., hour}} = \frac{\text{I.H.P.} \times 632.3100}{m_f/\text{hour} \times CV} \qquad 3.47$$

$$\text{Heat equivalent to one horse power, hour} = \frac{75 \times 60 \times 60}{427} = 632.31 \text{ kcal/hour}$$

CV = Calorific valve of fuel in kcal / kg of fuel.

Brake thermal efficiency: It is the ratio of the heat equivalent to one kW hour to the heat in the fuel per B.P. hour.

$$\eta_{bth} = \frac{\text{Heat equivalent to one kilo watt hour}}{\text{Heat in fuel per B.P., hour}} = \frac{\text{B.P.} \times 3600}{m_f \times C} = \frac{3600}{\text{bsfc} \times C} \qquad 3.48$$

The ratio of $m_f / B.P.$ is termed as break specific fuel consumption per B.P. hour.

Where;

B.P. = Brake power in kW,

m_f = Mass of fuel consumed in kg per hour, and

C = Calorific valve of fuel in kJ / kg of fuel.

$$\eta_{bth} = \frac{\text{Heat equivalent to one horse power, hour}}{\text{Heat in fuel per B.H.P., hour}} = \frac{\text{B.H.P.} \times 632.3100}{m_f/\text{hour} \times CV} \qquad 3.49$$

$$\text{Heat equivalent to one horse power, hour} = \frac{75 \times 60 \times 60}{427} = 632.31 \text{ kcal/hour}$$

CV = Calorific valve of fuel in kcal / kg of fuel.

Relative efficiency: It is the ratio of indicated thermal efficiency to air standard efficiency.

$$\text{Relative efficiency} = \frac{\text{Indicated thermal effeciency}}{\text{Air standard effeciency}}$$

Mean piston speed: It is an important parameter in engines, and it lies between 8-15 m/s.

$$S = \frac{2LN}{60} \qquad 3.50$$

Where;
L = Length of stroke (m),
N = Engine speed (rpm)

Example 3.6: The following data refers to a single cylinder 4 stroke diesel engine;
bp = 100 kW
bmep = 850 kPa
Speed = 400 rpm
bsfc = 0.335 kg/kWh
Calorific value = 43500 kJ/kg
Stroke/bore = 1.25
Mechanical efficiency = 80 %
Determine;
(a) Bore and stroke of the engine
(b) The break thermal efficiency
(c) Indicated thermal efficiency

Solution:

$$\text{Break power }(\text{bp}) = \frac{k \times p_{mb} \times L \times A}{60} \times \frac{N}{n}$$

$$100 = \frac{1 \times 850 \times (1.25 \times d) \times \frac{\pi}{4} d^2}{60} \times \frac{400}{2}$$

$$d^3 = 0.03595 \text{ m}^3$$

$$d = 0.33 \text{ m} = 330 \text{ mm} \quad \text{Ans}$$

$$L = 1.25 \times 330 = 412.5 \text{ mm} \quad \text{Ans}$$

$$\eta_{bth} = \frac{3600}{\text{bsfc} \times C} = \frac{3600}{0.335 \times 43500} = 0.247 = 24.7\ \%$$

$$\eta_{mech} = \frac{\text{B.P.}}{\text{I.P.}}$$

$$0.8 = \frac{100}{\text{I.P.}}$$

$$\text{I.P.} = 125 \text{ kW}$$

$$\text{I.P.} = \frac{k \times p_m \times L \times A}{60} \times \frac{N}{n}$$

$$125 = \frac{1 \times p_m \times 0.4125 \times \frac{\pi}{4}(0.33)^2}{60} \times \frac{400}{2}$$

$$p_m = 1062.893 \text{ kPa}$$

$$\text{imep} = 10.62 \text{ bar}$$

$$\text{bsfc} = \frac{m_f}{BP}$$

$$m_f = 0.335 \times 100 = 33.5 \text{ kg/h}$$

$$\eta_{ith} = \frac{3600}{\frac{m_f}{IP} \times C} = \frac{3600}{\frac{33.5}{125} \times 43500} = 0.3088$$

$$= 30.88\ \% \quad \text{Ans}$$

Example 3.7: Diesel engine has a geometry compression ratio of 14:1, cut-off ratio 2, and an efficiency ratio of 0.7 when referred to as standard efficiency. The fuel consumption is 5.65 kg/h. The calorific value of the fuel is 44.5 MJ/kg. Determine; the indicated power developed.

Solution:

Given: Compression ratio (r) = $\frac{V_1}{V_2} = \frac{14}{1}$, Cut-off ratio = $\frac{V_3}{V_2} = \frac{\rho}{1}$

C = 44.5 MJ/kg, Effeciency ratio = 0.7, fuel consumption =5.65 kg/h

$$\eta = 1 - \frac{1}{\gamma} \times \frac{1}{(r)^{\gamma-1}} \left[\frac{\rho^{\gamma}-1}{\rho-1}\right]$$

$$= 1 - \frac{1}{1.4} \times \frac{1}{(14)^{0.4}} \left[\frac{2^{1.4}-1}{2-1}\right]$$

$$= 0.5926$$

$$= 59.26\ \%$$

$$\text{Effeciency ratio} = \frac{\text{Thermal effeciency}}{\text{Air standard effecency}}$$

$$0.7 = \frac{\eta_{th}}{0.5926}$$

$$\eta_{th} = 0.4148 \quad = 41.48\ \%$$

$$\eta_{ith} = 0.4148 = \frac{3600 \times IP}{5.65 \times 44500}$$

$$IP = 28.97 \text{ kW} \quad \text{Ans}$$

Example 3.8: A 2 stroke petrol engine with a cylinder diameter of 100 mm and stroke length of 120 mm has a compression ratio of 5.0. What is the

(a) Clearance volume and ideal efficiency of the engine?

If the mean effective pressure in the cylinder is 6 kgf/cm^2, find the

(b) Indicated horse power at 1000 rpm

(c) The petrol consumption per hour

(d) Relative efficiency of the engine

The indicated thermal efficiency of an engine is 24% and there are 4 cylinder. The calorific value of petrol is 10500 kcal/kg.

Solution:

Given: Compression ratio (r) = $\frac{V_1}{V_2} = \frac{5}{1}$,

$k = 4$, $p_m = 6$ kgf/cm², $d = 100$ mm, $L = 120$ mm, $N = 1000$ rpm, $n = 1$, $C = 10500$ Kcal/kg, $\eta_{ith} = 24\%$

$$\text{Swept volume } (V_s) = \frac{\pi}{4} \times 100^2 \times 120 = 942477 \text{ mm}^3 = 942.5 \text{ cm}^3$$

$$r = \frac{V_C + V_S}{V_C}$$

$$5 = \frac{942.5 + V_c}{V_C}$$

$$V_c = \frac{942.5}{4} = 235.6 \text{ cm}^3 \quad \text{Ans}$$

$$\eta_a \text{(Ideal effecciency)} = 1 - \frac{1}{(r)^{\gamma-1}} = 1 - \frac{1}{(5)^{0.4}} = 0.4746 = 47.46\ \% \quad \text{Ans}$$

$$\text{I.H.P} = \frac{k \times p_m \times L \times A}{4500} \times \frac{N}{n}$$

$$= \frac{4 \times 6 \times \frac{120}{1000} \times \frac{\pi}{4} \times 10^2}{4500} \times \frac{1000}{1} = 50.28 \quad \text{Ans}$$

$$\eta_{ith} = \frac{\text{Heat equivalent to one horse power, hour}}{\text{Heat in fuel per I.H.P., hour}} = \frac{\text{I.H.P.} \times 632.3185}{m_f \times C}$$

$$\text{fuel consumption /hr} = \frac{\text{I.H.P.} \times 632.3185}{\eta_{ith} \times C} = \frac{50 \times 632.3185}{0.24 \times 10500} = 12.616 \text{ kg/hour} \quad \text{Ans}$$

$$\text{Relative effeciency} = \frac{\text{Indicated thermal effeciency}}{\text{Air standard effeciency}} = \frac{0.24}{0.4746} = 0.5056 \quad \text{Ans}$$

In SI units

$$\text{Indicated power} = \frac{k \times p_m \times L \times A}{60} \times \frac{N}{n} = \frac{4 \times 588.6 \times \frac{120}{1000} \times \frac{\pi}{4} \times \left(\frac{10}{100}\right)^2}{60} \times \frac{1000}{1}$$

$$= 9.245 \text{ kW}$$

$$\eta_{ith} = \frac{\text{I.P.} \times 3600}{m_f \times C}$$

$$m_f = \frac{36.98 \times 3600}{0.24 \times 43961.4} = 12.61 \text{ kg/hr}$$

Example 3.9: A 4 cylinder 2 stroke petrol engine has a 75 mm bore and 90 mm stroke, operates on the constant volume cycle and has a compression ratio of 6:1. The efficiency ratio being 55%, calculate;

(a) The indicated thermal efficiency.

When running at 40 rps, the engine develops a break mean effective pressure of 350 kPa and uses 9.2 kg of fuel per hour of calorific value 44000 kJ/kg. Calculate;

(b) The break thermal efficiency

(c) Mechanical efficiency

(d) Indicated specific fuel consumption in a kg/kWh.

Solution:

Given: Compression ratio (r) $= \frac{V_1}{V_2} = \frac{6}{1}$, k = 4, p_m = 350 kPa, d =75 mm, L= 90 mm, N=1000 rpm, n = 1, m_f = 9.2 kg, C = 44000 kJ/kg, η = 55%

$$\eta_a \text{(Air standard effeciency)} = 1 - \frac{1}{(r)^{\gamma-1}} = 1 - \frac{1}{(6)^{0.4}} = 0.512$$

$$= 51.2\ \%$$

Indicated thermal efficiency= efficiency ratio x Air standard efficiency

$$= 0.55 \times 0.512 \quad = 0.2816$$

$$= 28.16\ \% \quad \text{Ans}$$

$$\eta_{ith} = \frac{\text{I.P.} \times 3600}{m_f \times CV} = \frac{3600}{\text{isfc} \times CV}$$

$$0.2816 = \frac{3600}{\text{isfc} \times 44000} = 0.2905 \text{ kg/kWh}$$

$$\text{isfc} = \frac{m_f}{IP}$$

$$0.2905 = \frac{9.2}{IP}$$

$$IP = 31.66 \text{ kW}$$

$$IP = \frac{k \times p_m \times L \times A}{60} \times \frac{N}{n} = \frac{4 \times p_m \times \frac{90}{1000} \times \frac{\pi}{4}\left(\frac{75}{1000}\right)^2}{60} \times \frac{(40 \times 60)}{1} \quad \text{[As speed is in rps]}$$

$$31.66 = p_m \times 0.0636$$

$$p_m = 497.732 \text{ kPa}$$

$$\eta_{mech} = \frac{\text{bmep}}{\text{imep}} = \frac{360}{497.732} = 0.723$$

$$= 72.3\ \% \quad \text{Ans}$$

$$\eta_{mech} = \frac{BP}{IP}$$

$$0.723 = \frac{BP}{31.66}$$

$$BP = 22.899 \text{ kW}$$

$$bsfc = \frac{m_f}{BP} = \frac{9.2}{22.899} = 0.40176 \text{ kg/kWh}$$

$$\eta_{bth} = \frac{BP \times 3600}{m_f \times CV} = \frac{3600}{bsfc \times CV} = \frac{3600}{0.40176 \times 44000} = 0.2036$$

$$= 20.36\ \% \quad \text{Ans}$$

Example 3.10: A four-stroke engine has the indicated thermal efficiency of 35% and the mechanical efficiency of 80%. The fuel consumption rate is 22 kg/h running at a fixed speed. The brake mean pressure developed is 8 bar, and the mean piston speed is 10 m/s. Considering engines as single cylinder square engines and CV= 42000 kJ/kg, calculate;

(a) The crank radius

(b) Speed of the engine

Solution:

Given: $\eta_{mech} = 80\%$, $\eta_{ith} = 35\%$, m_f = 22 kg/hr, brake mean effective pressure= 8 bar, S=10 m/s, C = 42000 kJ/kg

$$\eta_{mech} = \frac{\eta_{bth}}{\eta_{ith}}$$

$$\eta_{bth} = \eta_{mech} \times \eta_{ith} = 0.80 \times 0.35 = 0.28$$

$$= 28\%$$

$$\eta_{bth} = \frac{BP \times 3600}{m_f \times CV}$$

$$0.28 = \frac{BP \times 3600}{22 \times 42000}$$

$$BP = 71.86 \text{ kW}$$

$$BP = \frac{k \times p_m \times L \times A}{60} \times \frac{N}{n}$$

$$71.86 = \frac{1 \times p_m \times L \times \frac{\pi}{4} L^2}{60} \times \frac{N}{2}$$

$$p_m (\text{kPa}) = 6 \times 100 = \frac{71.86 \times 60 \times 2 \times 4}{3.14 \times L^3 \times N} \qquad \text{[Pressure is given in bar]}$$

$L^3N = 18.29$

$$\text{Piston speed} = S = 10 = \frac{2LN}{60}$$

$LN = 300$

$$\frac{L^3N}{LN} = \frac{18.29}{300} = 0.0609$$

$L = 0.246 \text{ m} \quad = 246 \text{ mm}$

$$\text{Crank radius} = \frac{246}{2} = 123 \text{ mm}$$

$LN = 300$

$$N = \frac{300}{0.246} = 1219.5 \text{ rpm} \qquad \text{Ans}$$

3.9 REFRIGERATION

Refrigeration may be defined in a number of ways, the A.S.H.R.A.E. (American Society of Heating, Refrigerating and Air-Conditioning Engineers) defines refrigeration as *"Cooling of a space, substance or system to lower and/or maintain its temperature below the ambient"*.

Refrigeration is the process of eliminating heat, and its practical use is to generate or maintain temperatures that are lower than the ambient. In other words, refrigeration is a process of removal of heat from an object or a place where it is not desired and transferring that heat to some other place where it makes a negligible impact. A reverse heat engine or a heat pump constitutes a refrigerator in which heat is transferred from a cold body to a hot body. The working fluid used to extract heat from a cold body and transmit it to a hot body is called as a *refrigerant*. Freons, ammonia and air are some examples of refrigerants used in various refrigeration cycles.

3.9.1 Unit of Refrigeration

The standard unit of refrigeration is expressed in *'tons of refrigeration'* commonly denoted by *TR*. One ton of refrigeration is defined as the capacity of a refrigeration system that can freeze 1 ton of water at 0 °C to the ice at 0 °C in one day (24 hours). In British Thermal Units (BTU) one ton is equivalent to 900 kg and the latent heat of ice is 335 kJ/kg.

Therefore,

$$1 \text{ TR} = 900 \times 335 \text{ kJ/day}$$

$$= \frac{900 \times 335}{24 \times 60} = 210 \ kJ/\text{min} \qquad 3.51$$

$$= \frac{210}{60} = 3.5 \ kW \qquad 3.52$$

The worldwide unit accepted for one ton of refrigeration is 210 kJ/min or 3.5 kW.

3.9.2 Coefficient of Performance (COP)

The efficiency of a refrigerator or a heat pump is expressed in terms of the coefficient of performance and it is defined as the ratio of the desired effect to the input required. The desired effect may either be the amount of refrigeration produced in case of a refrigerator or heating of space in case of a heat pump while the required input is the work done on the refrigerant in a compressor. The value of COP of the refrigerator can be greater than unity.

$$\text{COP} = \frac{\text{Desired effect}}{\text{Input required}} = \frac{Q}{W} \quad 3.53$$

where Q is the amount of heat extracted to produce the refrigeration effect and W is the amount of work done in a compressor.

The COP of a refrigerator can be expressed in terms of the amount of heat extracted from a cold body (Q_2) to the amount of work input required (W) as shown in Fig 3.23 (a).

$$\text{COP}_\text{R} = \frac{Q_2}{W} = \frac{Q_2}{Q_1 - Q_2} \quad 3.54$$

In the case of a refrigerator, the desired effect lies between the cold reservoir and the ambient temperature.

While in a heat pump, the desired effect lies in between the hot reservoir and the ambient temperature, as illustrated in Fig. 3.23 (b).

Therefore,

$$\text{COP}_\text{HP} = \frac{Q_1}{W} = \frac{Q_1}{Q_1 - Q_2} \quad 3.55$$

$$\text{COP}_{HP} = \frac{Q_2}{Q_1 - Q_2} + 1 \quad 3.56$$

$$\text{COP}_\text{HP} = \text{COP}_\text{R} + 1 \quad 3.57$$

From Eq. 3.57, it is clear that for fixed values of Q_1 and Q_2 the COP_{HP} is always greater than unity because the COP_R cannot be a negative quantity.

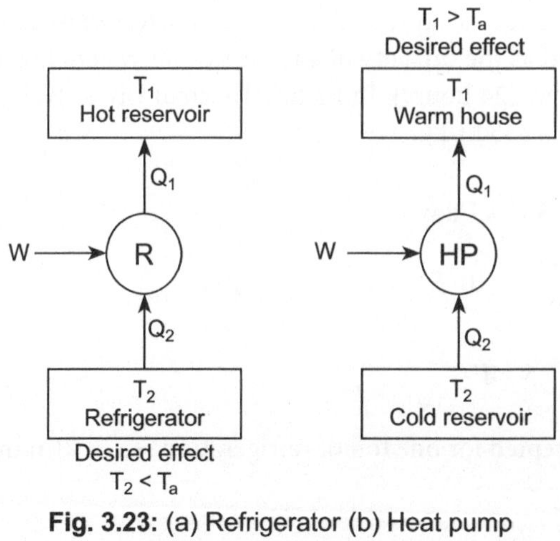

Fig. 3.23: (a) Refrigerator (b) Heat pump

3.10 TYPES OF REFRIGERATION SYSTEMS

In the present perspective, there are many types of refrigeration systems in use out of which the important ones are air refrigeration, vapour compression and vapour absorption systems. Air is the primary working fluid in an air refrigeration system, while in vapour compression system there many types of refrigerants that are used as a primary working fluid. The vapour absorption system uses ammonia as a primary coolant.

3.10.1 Air Refrigeration System

It is one of the primitive refrigeration techniques which employs air as a refrigerating medium. The air does not change its phase and remains gaseous throughout the cycle which means that the heat carrying capacity of air per unit mass is very small as compared to vapour absorption system. Air refrigeration system requires high power and hence its COP is very less therefore its application is limited to the aircraft refrigeration system.

3.10.1.1 Air Refrigeration System Working on Reversed Carnot Cycle

As it is evident that a heat engine working on the Carnot cycle has the highest possible efficiency, in a similar analogy a refrigerator working on a reversed Carnot cycle has the highest possible COP. A reversed Carnot cycle is constituted of four reversible processes in which two are isentropic compression and expansion and two are isothermal heat extraction and heat rejection. The p-V and T-s diagrams are shown in Fig. 3.24. The four processes of the cycle are explained below.

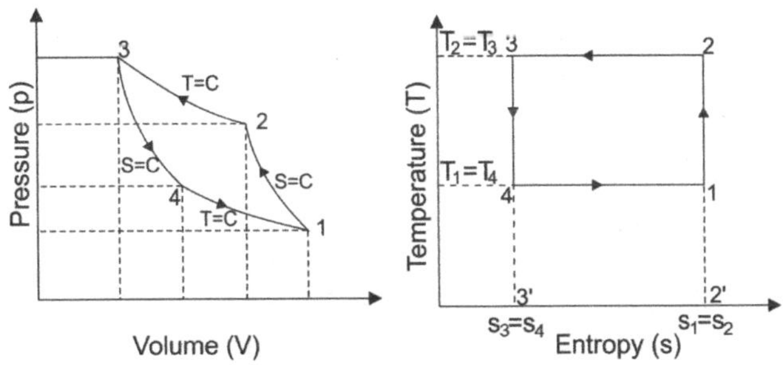

Fig. 3.24: p-V and T-s curve of Reversed Carnot cycle

(a) **Process 1-2 (Isentropic compression):** The air is compressed isentropically from pressure P_1 to P_2 simultaneously, the temperature increases from T_1 to T_2 and volume decreases from V_1 to V_2.

(b) **Process 2-3 (Isothermal compression):** During this process the air rejects its heat and is compressed isothermally ($T_2 = T_3$). The pressure increases from P_2 to P_3 and the volume further decreases from V_2 to V_3. The area under T-s plot for process 2-3 will give the heat rejected by the air during the isothermal compression.

$$Q_{2-3} = \text{Area}(2\text{-}3\text{-}3'\text{-}2'\text{-}2)$$

$$= T_3(s_2 - s_3) = T_2(s_2 - s_3) \quad 3.58$$

(c) **Process 3-4 (Isentropic expansion):** The air is now expanded isentropically and the pressure decreases from P_3 to P_4 and the volume increases from V_3 to V_4 with a corresponding decrease in temperature from T_3 to T_4. No heat is absorbed during this process.

(d) **Process 3-4 (Isothermal expansion):** Further expansion of air is achieved isothermally ($T_4 = T_1$) where the pressure decreases from P_4 to P_1 and the volume increases from V_4 to V_1. The area under T-s plot for process 4-1 will give the heat absorbed by the air during the isothermal compression.

$$Q_{4-1} = \text{Area(4-3'-2'-1-4)}$$

$$= T_4(s_1 - s_4) = T_4(s_2 - s_3) = T_1(s_2 - s_3) \quad 3.59$$

Work done during the cycle $= T_2(s_2 - s_3) - T_1(s_2 - s_3) = (T_2 - T_1)(s_2 - s_3)$ 3.60

$$\text{Therefore, } COP_R = \frac{\text{Heat absorbed}}{\text{Work required}} = \frac{Q_{4-1}}{Q_{2-3} - Q_{4-1}}$$

$$= \frac{T_1(s_2 - s_3)}{(T_2 - T_1)(s_2 - s_3)} \quad 3.61$$

$$COP_R = \frac{T_1}{(T_2 - T_1)} \quad 3.62$$

Example 3.11: The refrigerator operating on a reversed Carnot cycle has a COP of 5.5. The evaporator is at a constant temperature of -6 °C. The power required to run the refrigerator is 4 kW. Calculate;

(a) Condenser temperature of the refrigerator

(b) The refrigeration effect.

Solution:

Given: COP = 5.5, T_2 = -6 °C = -6+273 = 267 K

$$COP_R = \frac{Q_2}{W} = \frac{Q_2}{Q_1 - Q_2}$$

COP for refrigerator is given as;

$$COP_R = \frac{Q_2}{W} = \frac{Q_2}{Q_1 - Q_2}$$

$$COP_R = \frac{T_2}{T_1 - T_2}$$

$$5.5 = \frac{267}{T_1 - 267}$$

$T_1 = 315.55$ K Ans

$$COP_R = \frac{Q_2}{W}$$

$Q_2 = COP \times W$

$Q_2 = 5.5 \times 4$ kW

$Q_2 = 22$ kJ/s

$Q_2 = 1320$ kJ/min Ans

Example 3.12: Reversed Carnot cycle-based refrigeration system has a higher temperature of the refrigerant at 35 °C and a lower temperature at -15 °C. The capacity is 15 Tons. Calculate;

(a) COP of the refrigerant
(b) COP of heat pump
(c) Heat rejected from the system per hour
(d) Power required.

Solution:

Given: $T_1 = 35+273 = 308$ K, $T_2 = -15+273 = 258$ K, $Q_2 = 15$ Tons

$$COP_R = \frac{Q_2}{W} = \frac{Q_2}{Q_1 - Q_2}$$

(a) COP for refrigerator is given as;

$$COP_R = \frac{T_2}{T_1 - T_2} = \frac{258}{308-258} = 5.16 \quad \text{Ans}$$

(b)

$$COP_{HP} = \frac{Q_1}{W} = \frac{Q_1}{Q_1 - Q_2}$$

Alternate, $COP_{HP} = COP_R + 1 = 5.16 + 1 = 6.16$ Ans

(c) $$COP_R = \frac{\text{Refrigerating effect}}{\text{Work input}}$$

$$\text{Work input} = \frac{15 \times (60 \times 210)}{5.16}$$

Work input = 36627.907 kJ/hour

$$W = Q_1 - Q_2$$

$$Q_1 = W + Q_2$$

$$= 36627.907 + 15 \times 210 \times 60$$

$$= 225627.907 \text{ kJ/hour} \quad \text{Ans}$$

(d) Power required = work input in kW

$$= \frac{36627.907}{3600} = 10.174 \text{ kW} \quad \text{Ans}$$

Example 3.13: A refrigerator working on the Carnot cycle extracts 600 kJ of heat from a house, maintained at -10 °C and is discharged to the atmosphere at 35 °C. Determine the power required to run the refrigerator in kW.

Solution:

Given: $T_1 = 35+273 = 308$ K, $T_2 = -10+273 = 263$ K, $Q_2 = 600$ kJ/min.

(a) As we know that for the reversible cycle,

$$Q_1 = \frac{Q_2}{T_2} \times T_1$$

$$Q_1 = \frac{600}{263} \times 308 \text{ kJ/min}$$

$$Q_1 = 702.66 \text{ kJ/min}$$

Also, Work done = $W = Q_1 - Q_2$

$$W = 702.66 - 600 = 102.66 \text{ kJ/min}$$

$$= \frac{102.66}{60} = 1.711 \text{ kJ/s}$$

$$= 1.71 \text{ kW} \quad \text{Ans}$$

3.10.1.2 Air Refrigeration System Working on Reversed Brayton cycle (Bell-Coleman Cycle)

In this cycle, all the components are the same as in reversed Carnot cycle, with a slight difference in that the isothermal processes of Carnot cycle are replaced by constant pressure processes. Therefore, this cycle is executed with two isentropic processes and two isobaric processes. The components of the Bell-Coleman cycle are illustrated in Fig. 3.25. The processes associated with this cycle are represented on P-V and T-s plots in Fig. 3.26 (a) and (b) respectively.

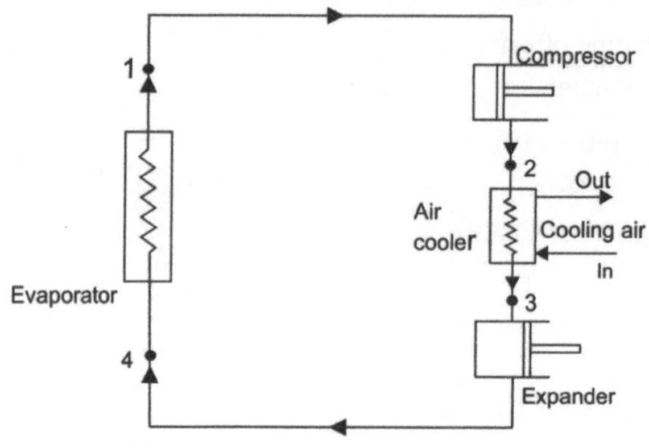

Fig. 3.25: Reversed Brayton cycle

First, the air is compressed isentropically in the compressor (process 1-2) where the pressure is raised from P_1 to P_2 and the temperature from T_1 to T_2 simultaneously the volume decreases from V_1 to V_2. Air after coming out of the compressor is then cooled at constant pressure using an air cooler (process 2-3). Here the temperature comes down from T_2 to T_3 and volume decreases from V_2 to V_3. The air is now expanded isentropically in an expander (process 3-4), due to which the pressure drops from P_3 to P_4 and the temperature of air further decreases from T_3 to T_4. The cooled air now enters the evaporator where heat is absorbed by the air at constant pressure (process 4-1) as a result, the temperature and volume of air increase from T_4 to T_1 and V_4 to V_1 respectively. The air coming out of the evaporator is again passed through the compressor and the cycle repeats.

COP of Bell-Coleman air refrigeration cycle

$$\text{COP} = \frac{\text{Heat absorbed (refrigeration effect)}}{\text{Work required}}$$

$$\text{COP} = \frac{T_4}{(T_3 - T_4)} = \frac{T_1}{(T_2 - T_1)} \quad 3.63$$

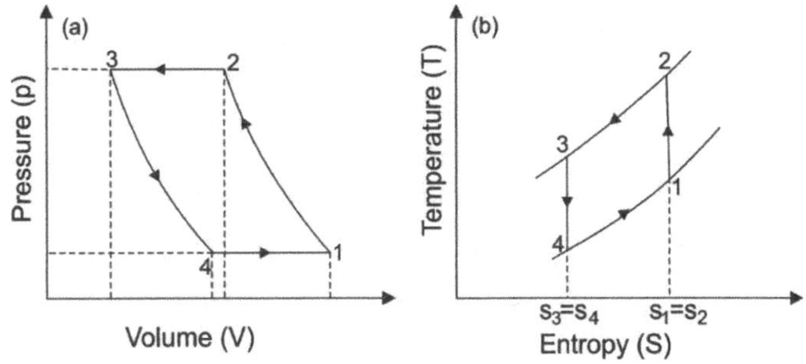

Fig. 3.26: (a) p-V plot, (b) T-s plot of Bell-Coleman cycle

3.10.2 Vapour Compression Refrigeration System

The vapour compression refrigeration cycle is a *phase change cycle* and therefore, the COP of this cycle is near to the ideal refrigeration cycle. As per the Carnot air refrigeration cycle the heat rejection and extraction at constant temperature is not possible because the air does not change its phase during the cycle. In contrast, if the working fluid is capable of changing its phase in a cycle, then constant temperature heat rejection and extraction are possible.

3.10.2.1 Working Principle

The working fluid (refrigerant) undergoes an alternative compression, condensation, expansion and evaporation in vapour compression refrigeration system. The schematic diagram of the system is shown in Figs 3.27 and its P-h and T-s plots are illustrated in 3.28 (a) and (b) respectively. The essential components of the system are discussed as follows.

1. **Compressor:** A compressor isentropically compresses the low-temperature and low-pressure refrigerant vapours coming out of the evaporator to higher temperature and pressure (process 1-2). The compression of saturated vapour (dry compression) is always desired over the compression of wet vapour (wet compression) as represented by points 1 and 1' respectively (refer to Fig. 3.28 b). The liquid droplets in wet compression may wash away the lubricating oil from cylinder walls and may cause wear and tear of the cylinder and piston. Also, the liquid droplets trapped in the cylinder head may damage valves during the compression process.
2. **Condenser:** It is a heat exchanger that provides heat transfer surface to the high-temperature and high-pressure refrigerant vapours coming out of the compressor outlet ready to dissipate its latent heat in the condenser. While passing through the condenser, the refrigerant gives up its heat which is absorbed in the evaporator plus the heat equivalent of the work done upon it by the compressor. This heat is transferred to air or water used as a cooling medium in the condenser. The process from 2-2' is the de-superheating and 2'-3 is the condensation of constant refrigerant pressure where 3 is the point of saturated liquid as shown in Fig. 3.28 (b).
3. **Expansion valve**: The primary function of the expansion valve is to allow the high-pressure liquid refrigerant coming out of the condenser to pass at a controlled rate in the evaporator after a considerable reduction in its pressure. The liquid or liquid-vapour mixture thus attains a very

low temperature capable of absorbing the latent heat of the materials kept in the evaporator, which is being cooled. Process 3-4 is the expansion process which is also called as throttling process and it takes place at constant enthalpy (refer to Fig. 3.28a)

4. **Evaporator:** The evaporator is a closed compartment in which items to be cooled are stored. Its primary function is to serve as a heat transfer surface through which the heat of the cooling space may be transferred to the refrigerant. The low-temperature refrigerant from the expansion valve passes through the evaporator tubes at constant pressure (process 4-1), as depicted in Fig. 3.28 (a). The saturation temperature of the refrigerant is lower than the surrounding temperature. Due to the temperature differential, heat is transferred from the surroundings, and the refrigerant cools the materials in the evaporator. The refrigerant evaporates while maintaining a steady temperature until all of the liquid refrigerants has been converted to saturated vapour as depicted in Fig. 3.28 (b).

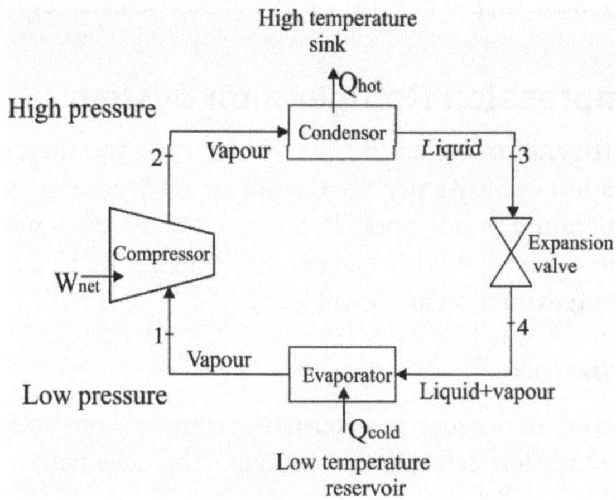

Fig. 3.27: Vapour compression refrigeration system

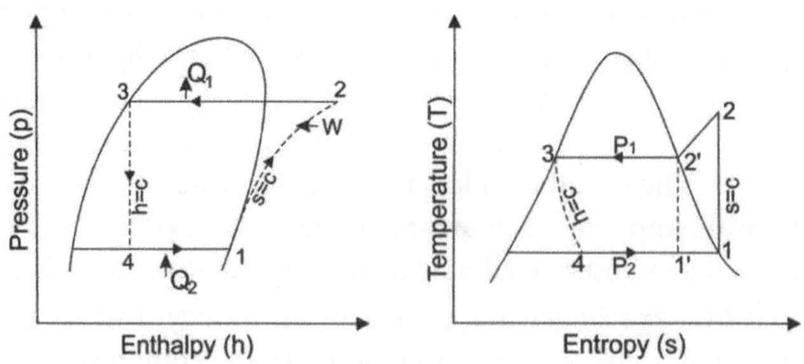

Fig. 3.28: (a) P-h plot (b) T-s plot of vapour compression refrigeration system

COP of vapour compression refrigeration system = heat rejected in the condenser/compressor work.

$= (h_2 - h_3)/\ (h_2 - h_1)$

$= (h_2 - h_4)/\ (h_2 - h_1)$ since, $h_3 = h_4$ 3.64

3.10.3 Vapour Absorption Refrigeration System

In vapour absorption refrigeration systems cooling is achieved by directly utilizing the thermal energy. The heat energy source for heating purposes in the generator can be anything like waste heat, solar energy, automobiles etc. These features make this system a very economical option in the context of energy conservation. Since there is no compressor involved in this system, the function is achieved by incorporating an absorber, pump, and generator. The working mechanism of the *aqua-ammonia* vapour absorption refrigeration system is described below, and its schematic diagram is shown in Fig. 3.29.

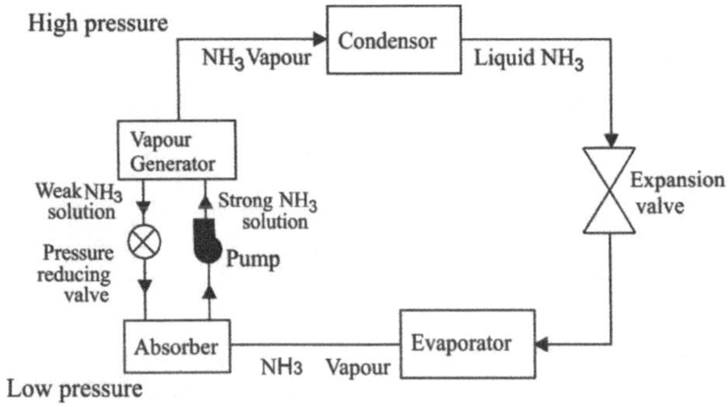

Fig. 3.29: Aqua-ammonia vapour absorption refrigeration system

3.10.3.1 *Working Mechanism of Aqua-Ammonia Refrigeration System*

Starting from the evaporator side, the low temperature and low-pressure ammonia vapour, after absorbing heat from the items present in the evaporator goes to the absorber. The absorber containing a weak solution of water-ammonia absorbs the ammonia vapour coming from the evaporator. Due to absorption, the pressure in the absorber decreases which further helps in absorption of more ammonia from the evaporator and a strong solution is formed. The strong water-ammonia (aqua-ammonia) solution from the absorber at low temperature is pumped into the generator. The ammonia vapour needs to be separated from the strong solution; therefore, the solution is heated by some external means like gas, steam, or solar energy inside the generator.

The heat addition in the generator leads to the separation of ammonia vapour from the strong aqua-ammonia solution leaving behind the weak solution. The weak solution passes through the pressure-reducing valve which is then returned back to the absorber through a heat exchanger. The provision of a heat exchanger increases the performance of the system by utilizing the heat of the weak solution coming from the generator to preheat the strong solution from the absorber which is pumped to the generator. The high-temperature and pressure-rich ammonia vapours are then passed through a condenser, where it gets condensed into liquid ammonia. The high-pressure liquid ammonia is then throttled in the expansion valve, where the pressure and temperature of ammonia decrease, which is then allowed to enter the evaporator. The liquid ammonia while passing through the evaporator absorbs the latent heat of the products available in the evaporator and comes out in the form of saturated vapour. This vapour is again passed to the absorber and the cycle repeats.

COP of vapour absorption refrigeration system =

$$= \frac{\text{heat extracted from evaporator}}{\text{(heat supplied to the genefrator + pumping work)}}$$

$$COP = \frac{Q_e}{Q_g + W_P} \quad 3.65$$

3.10.4 Comparison of Vapour Absorption and Vapour Compression Refrigeration Systems

The difference between vapour absorption and vapour compression refrigeration system are mentioned in Table 3.4.

Table 3.4: Comparison of vapour absorption and vapour compression refrigeration systems

Vapour absorption refrigeration system	Vapour compression refrigeration system
No moving parts	Contains moving parts
Less wear and tear and low maintenance cost	More wear and tear and high maintenance cost
For the same capacity, less pumping work is required	For the same capacity, more pumping work is required
It can be used in remote locations where there is no electricity supply	It can be used only at locations where electricity supply is readily available.
Considerably high refrigeration capacity from a single unit	Less refrigeration capacity from a single unit
No significant damage to the parts if liquid droplets are present in the vapour	Detrimental effect on the parts if liquid droplets are present in the vapour
COP is less	COP is high
Load variation does not affect the COP	A decrease in load decreases the COP

UNIT SUMMARY

- Steam turbines are used to convert the energy of high pressure and temperature steam into useful mechanical work. This steam is passed through the nozzles where the pressure drops resulting in an increase in the velocity.
- Types of turbines are;
  - **Impulse turbine:** The turbine works on the principle of impulse where the steam expands in the nozzle only and there is no change in pressure in the moving blade.
  - **Reaction turbine or impulse-reaction turbine:** In these turbines, pressure drop takes place in the fixed blades as well as moving blades.
- Compounding is employed to reduce the rotational speed of the impulse turbine, and is performed in three ways; Pressure compounding, Velocity compounding, and Pressure-velocity compounding.
- Condenser condenses the exhaust steam of the turbine with the help of cooling water. In direct contact type of condensers, the exhaust steam from the turbine and cooling water is allowed to mix directly and they come out as a single stream. The direct contact type condensers are; (a) Spray condenser, (b) Barometric condenser and (c) Jet condenser
- In Indirect contact type or surface condensers, there is no mixing of cooling water and steam.
- The cooling tower cools the hot water coming out of the condenser and recirculates it back to the condenser; therefore, cooling water requirements are greatly reduced. The cooling towers are mainly of two types, namely wet and dry type towers.

- Internal combustion engines are reciprocating engines, where the working fluid is fuel and air and combustion takes place inside the cylinder.
- The efficiency of Otto cycle is given as;

$$\eta = 1 - \frac{1}{(r)^{\gamma-1}}$$

- The efficiency of the Diesel cycle is given as;

$$\eta = 1 - \frac{1}{\gamma} \times \frac{1}{(r)^{\gamma-1}}\left[\frac{\rho^{\gamma}-1}{\rho - 1}\right]$$

The efficiency of Dual cycle is given as;

$$\eta = 1 - \frac{1}{(r)^{\gamma-1}}\left[\frac{(\alpha\rho^{\gamma}-1)}{(\alpha-1) + \gamma\alpha(\rho-1)}\right]$$

- For constant maximum pressure and heat input, the air standard efficiency of the gas power cycle is in the order.

  Diesel cycle > Dual cycle > Otto cycle
- For the same compression ratio and heat rejection, heat supplied is more in the Otto cycle and hence the air standard efficiency of the gas power cycle is in the order.

  Otto cycle > Dual cycle > Diesel cycle
- The spark ignition (Petrol) engine operates on the principle of the Otto cycle. The piston reciprocates inside the cylinder and a mixture of air and petrol as working fluid is compressed. The spark from the spark plug ignites the mixture after compression.
- The compression ignition (Diesel) operates on the Diesel cycle, but only air is compressed inside the cylinder and the compression ratio is high. After the compression stroke, fuel is injected into the cylinder, and the heat of compressed air ignites the fuel.
- In four strokes engine, one cycle is completed after four strokes of the piston or two complete revolutions of the crankshaft and one revolution of the camshaft. Therefore, one power cycle is obtained in two revolutions of the crank shaft.
- In a two-stroke engine, the cycle is completed after two strokes of the piston or one revolution of the crankshaft. Therefore, one power cycle is obtained in one revolution of crankshaft.
- Refrigeration is defined as the cooling of a space, substance or system to lower and/or maintain its temperature below the ambient.
- The one ton of refrigeration is equal to 210 kJ/min or 3.5 kW.
- The COP of a refrigerator can be expressed in terms of the amount of heat extracted from a cold body (Q_2) to the amount of work input required (W). Therefore,

$$COP_R = \frac{Q_2}{W} = \frac{Q_2}{Q_1 - Q_2}$$

- In a heat pump, the desired effect lies between the hot reservoir and the ambient temperature.

  Therefore, $COP_{HP} = \frac{Q_1}{W} = \frac{Q_1}{Q_1 - Q_2}$

$$COP_{HP} = COP_R + 1$$

- The air refrigeration system works on reversed Carnot cycle and the practical cycle is based on reversed Brayton or Bell-Coleman cycle.
- The vapour compression refrigeration cycle is a phase change cycle and the COP of this cycle is near to the ideal refrigeration cycle.
- Vapour absorption cycle does not have any moving part, while vapour compression system involves a compressor; therefore for the same capacity, less pumping work is required.

EXERCISE

Multiple Choice Questions

1. De-Laval turbines are mostly used.......
   (a) Where low speeds are required
   (b) For small power purposes and low speeds
   (c) For small power purposes and high speeds
   (d) For large power purposes
2. In a reaction turbine........
   (a) The steam is allowed to expand in the nozzle, where it gives a high velocity before it enters the moving blades
   (b) The expansion of steam takes place partly in the fixed blades and partly in the moving blades
   (c) The steam is expanded from high pressure to a condenser pressure in one or more nozzles
   (d) The pressure and temperature of steam remain constant
3. In an impulse turbine.......
   (a) The steam is expanded in nozzles only and there is a pressure drop and heat drop
   (b) The steam is expanded both in fixed and moving blades continuously
   (c) The steam is expanded in moving blades only
   (d) The pressure and temperature of steam remain constant
4. A condenser where circulating water flows through tubes which are surrounded by steam is known as.........
   (a) Surface condenser (b) Jet condenser
   (c) Barometric condenser (d) Evaporative condenser
5. Compounding in steam turbines is done to
   (a) Reduce the rotor speed (b) Increase the rotor speed
   (c) Keep constant rotor speed (d) None
6. In a velocity compounded steam impulse turbine, the magnitude of velocity in fixed nozzles is
   (a) Constant (b) Variable
   (c) Both (a) & (b) (d) None
7. Velocity compounding best suited for
   (a) Big turbines (b) Small turbines
   (c) Both for small & big turbines (d) None

8. Which of the following is a pressure compounded turbine
   (a) Parson's turbine (b) Curtis turbine
   (c) Rateau turbine (d) All of the above
9. Why is induced draught considered better than the forced draught
   (a) Because the power requirement is high for forced draught
   (b) Maintenance of induced draught fan is costlier
   (c) Forced draught is less efficient
   (d) Forced draught produces less amount of speed of air
10. In ________cooling arrangement, the air is blown through and the fans are located at the bottom
   (a) Forced draught (b) Induced draught
   (c) Both induced and forced draught (d) None of these
11. Which is the jet-type condenser?
   (a) Parallel flow (b) Central flow
   (c) Down flow (d) Inverted flow
12. In the steam condensing plant the pump used to send condensate to the boiler as feed water is
   (a) Air extraction pump (b) Condensate extraction pump
   (c) Circulating cooling pump (d) Boiler feed pump
13. What is the function of a baffle plate in parallel flow type low-level jet condenser?
   (a) It keeps cooling water and exhaust steam from mixing
   (b) It ensures proper mixing of cooling water and exhaust steam
   (c) It collects condensate over it for extraction
   (d) It helps the condenser wall withstand the inside pressure
14. Condensate can be used as feed water in a
   (a) Jet condenser (b) Surface condenser
   (c) Both in Jet and Surface condenser (d) None
15. What is the function of a cooling tower in a power plant?
   (a) It cools the hot water coming back from the condenser
   (b) It cools the hot water being supplied to the condenser
   (c) It heats the cold water coming back from the condenser
   (d) It heats the cold water being supplied to the condenser
16. Which of the following phenomenon is used to cool water in a cooling tower?
   (a) Evaporation (b) Radiation
   (c) Condensation (d) Conduction
17. Diesel engines, as compared to petrol engines require
   (a) Bigger flywheel (b) Same size flywheel
   (c) Smaller flywheel (d) None of the above
18. The thermal efficiency of diesel engines is about ___________
   (a) 65% (b) 50%
   (c) 80% (d) 70%

19. If the compression ratio in the I.C. engine increases, then its thermal efficiency will ________
   (a) Increase
   (b) Decrease
   (c) Remain same
   (d) None of the mentioned
20. For the same compression ratio and for the same heat added ________
   (a) Otto cycle is more efficient than Diesel Cycle
   (b) Diesel cycle is more efficient than Otto Cycle
   (c) Efficiency depends on other factors
   (d) None of the mentioned
21. For constant maximum pressure and heat input, the air standard efficiency of the gas power cycle is in the order.
   (a) Diesel cycle, Dual cycle, Otto cycle
   (b) Otto cycle, Diesel cycle, Dual cycle
   (c) Dual cycle, Otto cycle, Diesel cycle
   (d) Diesel cycle, Otto cycle, Dual cycle
22. A Carburettor is used to supply ________
   (a) Petrol, air and lubricating oil
   (b) Air and diesel
   (c) Petrol and lubricating oil
   (d) Petrol and air
23. A compression ratio of the I.C. engine is ________
   (a) The ratio of volumes of air in the cylinder before compression stroke and after the compression stroke
   (b) Volume displaced by piston per stroke and clearance volume in the cylinder
   (c) Ratio of pressure after compression and before compression
   (d) Swept volume/cylinder volume
24. If the intake air temperature of the I.C. engine increases, its efficiency will ________
   (a) Increase
   (b) Decrease
   (c) Remain Same
   (d) None Of The Mentioned
25. The thermal efficiency of a diesel cycle having a fixed compression ratio, with an increase in the cut-off ratio, will ________
   (a) Increase
   (b) Decrease
   (c) Be independent
   (d) None of the mentioned
26. Indicated power of a 4-stroke engine is equal to ________
   (a) $\frac{k \times p_m \times L \times A}{60} \times \frac{N}{2}$
   (b) $\frac{k \times p_m \times L \times A}{60} \times \frac{N}{4}$
   (c) $\frac{k \times p_m \times L \times A}{60} \times N$
   (d) $\frac{k \times p_m \times L \times A}{60} \times 2N$
27. Scavenging in a diesel engine means;
   (a) Air used for combustion sent under pressure
   (b) Forced air for cooling cylinder
   (c) Burnt air containing products of combustion
   (d) Air used for forcing burnt gases out of the cylinder during the exhaust

28. The ratio of indicated thermal efficiency to the corresponding air standard cycle efficiency is called........

(a) Net efficiency
(b) Efficiency ratio
(c) Relative efficiency
(d) Overall efficiency

29. A two stroke cycle engine gives......the number of power strokes as compared to the four stroke cycle engine, at the same engine speed

(a) Half
(b) Same
(c) Double
(d) Four times

30. In a four-stroke engine, the working cycle is completed in.......

(a) One revolution of the crankshaft
(b) Two revolutions of the crankshaft
(c) Three revolutions of the crankshaft
(d) Four revolutions of the crankshaft

31. The break power of an engine is always......the indicated power

(a) Equal to
(b) Less than
(c) Greater than
(d) None of the above

32. In a diesel engine, the fuel is ignited by

(a) Spark
(b) Injected fuel
(c) Heat resulting from compressing air that is supplied for combustion
(d) Ignition

33. A 75 cc engine has the following parameter as 75 cc

(a) Fuel tank capacity
(b) Lubrication oil capacity
(c) Swept volume
(d) Cylinder volume.

34. Piston rings are usually made of......

(a) Cast iron
(b) Aluminium
(c) Carbon steel
(d) Babbitt

35. One tonne refrigerating machine means.......

(a) One tonne is the total mass of the machine
(b) One tonne of refrigerant is used
(c) One tonne of water can be converted into ice
(d) One tonne of ice when melts from at 0^0C in 24 hours; the refrigeration effect produced is equivalent to 210 kJ/min

36. During a refrigeration cycle, heat is rejected by the refrigerant in a.......

(a) Condenser
(b) Compressor
(c) Evaporator
(d) Expansion valve

37. Air refrigerator works on......

(a) Carnot cycle
(b) Reversed Carnot cycle
(c) Bell-Coleman cycle
(d) Both b and c

38. The refrigerant, commonly used in vapour absorption refrigeration systems, is.......
(a) Sulphur dioxide
(b) Ammonia
(c) Freon
(d) Aqua-ammonia

39. The vapour compression refrigerator employs the following cycle
(a) Rankine
(b) Carnot
(c) Reverse Rankine
(d) Reverse Carnot

40. In a vapour compression system, the lowest temperature during the cycle occurs after.......
(a) Compression
(b) Condensation
(c) Expansion
(d) Evaporation

41. In vapour compression refrigeration cycle, the condition of the refrigerant is very wet vapour.......
(a) Before entering the expansion valve
(b) Before entering the compressor
(c) After passing through the condenser
(d) After passing through the expansion or throttle valve

42. The C.O.P. of domestic refrigerator.......
(a) Is less than 1
(b) Is more than 1
(c) Is equal to 1
(d) None of the above

43. The evaporative type of condenser has
(a) Water in pipes surrounded by steam outside
(b) Steam and cooling water mixed to give the condensate
(c) Steam in pipes surrounded by water
(d) None of the above

44. The natural draft or hyperbolic towers have been used for
(a) Large capacity of water
(b) Small capacity of water
(c) High efficiency
(d) Low capital cost

45. In a Jet condenser, condensate and cooling water are
(a) Mixed fully
(b) Mixed partially
(c) Not mixed
(d) None

Answers for Multiple Choice Questions							
1. (c)	2. (b)	3. (a)	4. (a)	5. (a)	6(a)	7. (b)	8. (c)
9. (a)	10. (a)	11. (a)	12. (d)	13. (b)	14. (b)	15. (a)	16. (a)
17. (a)	18. (d)	19. (a)	20. (a)	21. (a)	22. (a)	23. (d)	24. (b)
25. (b)	26. (a)	27. (d)	28. (c)	29. (c)	30. (b)	31. (b)	32. (c)
33. (c)	34. (a)	35. (d)	36. (a)	37. (d)	38. (d)	39. (d)	40. (d)
41. (d)	42. (b)	43. (c)	44. (a)	45. (a)			

Short and Long Answer Type Questions

1. What is a turbine? How are steam turbines classified?

2. What is the difference between an impulse and a reaction turbine.
3. What is the purpose of compounding steam turbines?
4. Explain different types of compounding with neat sketches.
5. What do you understand by steam condensers? Explain their functions.
6 Explain the key components of a condenser.
7. Differentiate between surface and jet condensers.
8. What is the purpose of a cooling tower? What are its different types?
9. What is the difference between natural draught and mechanical draught cooling towers? Which is preferred?
10. Differentiate between direct and indirect contact type condensers.
11. Explain the working of wet and dry-type cooling towers.
12. What is the basic difference between external and internal combustion engines.
13. Compare SI and CI engines based on different parameters.
14. What are the fundamental differences between 2 stroke and 4 stroke engines.
15. Explain with a neat sketch the working of 4 stroke petrol engine.
16. Prove the efficiency of the ideal Otto cycle using standard air;

$$\eta = 1 - \frac{1}{(r)^{\gamma-1}}$$

Where, r and γ have their usual meanings.

17. Explain with a neat sketch the working of 2 stroke diesel engine.
18. Define the following terms:
    (i) Brake power
    (ii) Indicated power
    (iii) Brake thermal efficiency
    (iv) Indicated thermal efficiency
    (v) Compression ratio
    (vi) Swept volume
19. What is the difference between a refrigerator and a heat pump?
20. Explain with a neat sketch the principle and working of the vapour compression refrigeration system.
21. Explain with a neat sketch the principle and working of vapour absorption refrigeration system.
22. Prove that $COP_{HP} = 1 + COP_{Ref}$
23. Prove the efficiency of the ideal Diesel cycle using standard air;

$$\eta = 1 - \frac{1}{\gamma} \times \frac{1}{(r)^{\gamma-1}}\left[\frac{\rho^{\gamma}-1}{\rho - 1}\right]$$

Where, r, ρ and γ have their usual meanings.

Numerical Problems

1. One Ton refrigerator machine works on reverse Carnot cycle. It requires 1.75 kW to maintain a low evaporator temperature of 250 K. Find the COP of the machine, if working as
   (a) Refrigerator
   (b) Heat pump
   (c) The temperature at which heat is rejected.

   [2,3, 375 K]

2. A refrigerating machine operating under Bell- Coleman cycle uses air as working fluid under the following conditions.

   Refrigerator temperature= 150 K, expander temperature= 300 K. The air temperature at the entry of the refrigerator is 50 K less than the refrigerator temperature. Pressure in the refrigerator= 2 bar. Calculate;
   (a) Refrigeration effect
   (b) Network
   (c) COP of the machine
   (d) Cooler pressure

   [52.5 kJ/kg, 450 K, 0.5, 93.5 bar]

3. A reversed Carnot cycle is used for making ice at -5 ^oC from water at 25 ^oC. The temperature of the brine is -10 ^oC. Calculate the quantity of the ice formed per kWh of work input. Assume the specific heat of the ice as 2 kJ/kgK, the latent heat of the ice is 335kJ/kg, and the specific heat of water is 4.18 kJ/kgK.

   [60.17kg]

4. A reversible heat pump is used for heating an apartment in winter. The heat is absorbed from the atmosphere at 0 ^oC to maintain the temperature of the apartment at 18 ^oC. If 5 kW-hr of electrical energy is needed to operate the reversible heat pump, calculate the heat supplied to the apartment in kJ.

   [307058.8 kJ]

5. Following particulars are related to a petrol engine;

   Diameter of cylinder = 8.0 cm

   Stroke (s) = 11.0 cm

   Clearance volume = 80 cm3.

   Calorific value of a fuel = 10500 kcal/kg (43961.4 kJ/kg)

   Consider Cp for air = 0.237 and Cv= 0.163

   Calculate;
   (a) Actual thermal efficiency
   (b) Air standard efficiency
   (c) Relative efficiency

   [30.66 %, 60.57 %, 50.61 %]

6. In a compression ignition engine, working on a dual combustion cycle has a compression ratio of 16 end pressures and temperature at the start of compression is 1.033 kgf/cm^2 (101.33 kPa) and

300 K, respectively. The heat supplied is 100 kcal (418.68 kJ) per kg of air during constant volume heating and 120 kcal (502.416 kJ) during constant pressure heating. Estimate;

(a) Pressure and temperature at each point of the cycle

(b) Cycle efficiency

[909.42 K, 1494.2 K, 8074.6 kPa, 65.92%]

7. One kilogram of air at a pressure of 1.033 kgf/ cm^2 and temperature 300 K is compressed adiabatically until the pressure becomes 7.75 kgf/ cm^2 in an auto cycle engine. At the end of the compression stroke 100 kcal (418.68 kJ/kg) heat is added at constant volume. Determine;

   (a) Compression ratio of an engine

   (b) Temperature after compression

   (c) Temperature after heat addition

[4.167, 540.12 K, 1128.35 K]

8. A petrol engine consumes 0.26kg of fuel per BHP per hour with a calorific value 10600 kcal/kg. The compression ratio is 5.8. If the mechanical efficiency is 80%, and assuming Cp = 0.24 and Cv= 0.172, determine;

   (a) Break thermal efficiency

   (b) Indicated thermal efficiency

   (c) Air standard efficiency

   (d) Relative efficiency on the basis of BHP

[22.97%, 28.7%, 50.48%, 0.455]

9. The mean effective pressure of an ideal diesel cycle is 8 bar. If the initial pressure is 1.03 bar under a compression ratio is 12, determine the cut-off ratio and the air standard efficiency. Assume the ratio of specific heat for air = 1.4.

[2.38, 54.66%]

10. In an engine working on a diesel cycle, the ratio of the weight of air and fuel supply is 50: 1. The temperature of the air at the beginning of compression is 60 °C, and the compression ratio is 14:1. Consider the calorific value of fuel= 42000 kJ/ kg, Cp for air = 0.237 and C_v= 0.163. Determine the ideal efficiency of the engine.

[60 %]

11. The 4-stroke petrol engine at full load drivers 50 kW. It requires 8.5 kW to rotate it without load at the same speed. Find the mechanical efficiency; at full load, half load, and the quarter road. Also, find out the volume of fuel consumed per second at full load if the thermal brake efficiency is 25%, given that the calorific value of fuel = 42 MJ/kg and the specific gravity of petrol is 0.75. Estimate the indicated thermal efficiency.

[85.47%, 74.62%, 59.52%, 6.34 x 10^{-6} m^3/s, 29.25%]

12. Find out the speed at which a 4-cylinder engine using natural gas can develop a break power of 50 kW working under the following conditions;

    Air-gas ratio 9:1,

    Calorific value of a fuel = 34 MJ/m^3,

    Compression ratio = 9:1

    Volumetric efficiency = 70%

Indicate thermal efficiency = 35%
Mechanical efficiency = 80% and
Total volume of the engine = 2 litre

[4500 rpm]

13. The output of a single cylinder 4 stroke internal combustion engine is measured by a rope brake dynamometer. The diameter of the break pulley is 750 mm and the rope diameter is 50 mm. The dead load on the tight side of the rope is 400 N and the spring balance reading is 50 N. The bore is 150 mm and stroke is 190 mm. The engine consumes 4 kg/h of fuel at a rated speed of 1000 rpm. The calorific value of the fuel is 44 MJ/kg. Calculate;
    (a) The brake specific fuel consumption
    (b) Brake mean effective pressure
    (c) The brake thermal efficiency

    If the mechanical efficiency is 80%, calculate;
    (d) Indicated power
    (e) Indicated mean effective pressure
    (f) Indicated specific fuel consumption
    (g) Indicated thermal efficiency

[14.66 kW, 523.949 kPa, 29.98%, 18.328 kW, 654.93 kPa, 37.48 %]

PRACTICAL

1. Study of impulse and reaction turbine through working models.
2. Study of various types of condensers and cooling towers.
3. Performance test of four stroke SI and CI engines
4. Performance test of two stroke SI and CI engines.
5. Determination of COP of vapour compression refrigeration system.

KNOW MORE

- Steam turbine utilizes the kinetic energy of super-heated steam to move the blade, therefore various components of velocity (absolute, flow and whirl velocity) play a key role in the computation of axial and tangential forces and power. Therefore, it becomes essential to draw a velocity diagram for the analysis.
- The turbine speed is a function of electrical load, which may vary, and will affect the frequency, which is not desirable. Therefore, governing of the turbine is employed through throttle, nozzle and by-pass governing to maintain the constant speed.
- The condenser is the major component of the steam power plant, where a vacuum is maintained to make the steam flow easier, and more work output may be extracted.
- A sudden rise in the pressure during combustion in an SI engine accompanied by a hammering-like sound in an engine is called detonation or knocking. This is affected by; supercharging, high compression ratio, increase in load, increased inlet temperature, and spark plug location.

- In case of CI engine, delay period short to avoid knocking. The delay period is increased by factors such as; low compression ratio, the low cetane number of fuel, low combustion temperature, and poorly atomized fuel.
- Supercharging is the process of increasing inlet air density in the CI engine, which increases the power output.
- The Wankel engine is a type of IC engine where four processes are achieved by using an eccentric rotary design to convert pressure into rotating motion.
- Subcooling in a refrigeration system increases the COP, provided that no further energy is to be spent to achieve extra cold coolant requirements.
- The refrigerants are low boiling point substances. Freon (F-12) is a very suitable refrigerant for domestic refrigerators and air conditioning systems. Other refrigerants are F-22, Ammonia, R-22, R-32 etc.
- Vapour compression system is used in household refrigerators, whereas vapour absorption is mainly used in ice plants.

REFERENCES AND SUGGESTED READINGS

1. R. Yadav, Steam and Gas Turbines and Power Plant Engineering, Central Publishing House, Allahabad, India, 2009.
2. M.P. Poonia and S.C. Sharma. Basic Mechanical Engineering – Khanna Publishing House, New Delhi, India, 2005.
3. P. N. Ananthanarayanan, Basic Refrigeration and Air Conditioning, McGraw Hill Education Pvt. Ltd. Delhi, India, 2013.
4. V. Ganesan, Internal Combustion Engine, McGraw Hill Education, Delhi, India, 2017.
5. R. K Rajput, A Textbook of Automobile Engineering, Laxmi Publications Pvt. Ltd. New Delhi, India, 2007.

4 Materials and Manufacturing Processes

UNIT SPECIFICS

Through this unit we have discussed the following aspects:

- Understand the fundamentals of materials related to mechanical engineering;
- Classification of materials and their properties;
- Fundaments of sand casting, sand properties, sand moulding process;
- Understand patterns, process of making patterns, allowances and their types;
- Demonstration of basic metal forming techniques in detail and their working principles;
- Demonstration of basic metal joining techniques and their applications;

The practical applications of the topics are discussed for generating further curiosity and creativity as well as improving problem solving capacity.

Besides giving a large number of multiple choice questions as well as questions of short and long answer types marked in two categories following lower and higher order of Bloom's taxonomy, assignments through a number of numerical problems, a list of references and suggested readings are given in the unit so that one can go through them for practice. It is important to note that for getting more information on various topics of interest some QR codes have been provided in different sections which can be scanned for relevant supportive knowledge.

After the related practical, based on the content, there is a "Know More" section. This section has been carefully designed so that the supplementary information provided in this part becomes beneficial for the users of the book. This section mainly highlights the initial activity, examples of some interesting facts, analogy, history of the development of the subject focusing the salient observations and finding. Timelines from the development of the concerned topics to the present, applications of the subject matter in our day-to-day real life or industrial applications on a variety of aspects, and finally inquisitiveness and curiosity of the topics of unit are also highlighted.

RATIONALE

This fundamental unit materials and manufacturing processes helps students to get a primary idea about different engineering materials and their classification which is important for an engineer to select a most suitable material for a desired application. It explains different material and their properties of mechanical engineering. Material science is an interdisciplinary field related to discovery, design and development of new materials particularly solids and understanding their properties and application. All these are discussed at length to develop the subject.

Manufacturing is always important to mankind and considered as backbone of social and economic development of any country. Manufacturing involves transforming of raw materials into finished goods to be used for some purpose with the systematic use of labour, machinery, tools and biological or chemical processing. It also denotes the fabrication or assembly of components into finished products on a larger scale.

Fundamentals of basic material processing techniques such as sand casting, gas and arc welding and some metal forming and shaping processes are discussed in detail to develop an understanding of subject. This subject combines engineering physics with materials science, to design, analyse, manufacture, and maintain mechanical systems. The practical applications of subject are related to the selection, design, developments of materials, and through different processing to convert for application in almost every segment of our daily life.

PRE-REQUISITES

- Basic Science

UNIT OUTCOMES

List of outcomes of this unit is as follows:

U4-O1: To describe importance of materials and their properties in engineering.

U4-O2: To classify materials according to their types.

U4-O3: To understand and apply the principles of metal casting to produce components.

U4-O4: To understand the mechanism of deformation for different metal forming processes.

U4-O5: To identify different welding processes and their application.

Unit-4 Outcomes	EXPECTED MAPPING WITH COURSE OUTCOMES (1- Weak Correlation; 2- Medium correlation; 3- Strong Correlation)					
	CO-1	CO-2	CO-3	CO-4	CO-5	CO-6
U4-O1	3	-	-	3	3	2
U4-O2	3	-	-	3	3	1
U4-O3	3	-	-	3	3	2
U4-O4	3	1	1	3	3	1
U4-O5	3	-	1	3	3	1

4.1 INTRODUCTION

Materials are available around us and are used to make different objects. These objects are made up of different materials such as wood, metal, glass, plastic, clay for different applications. Therefore, materials are having significant impact in our life from ancient time for the purpose of household goods, food, clothing, transportation, and in almost every segment of our daily life. These materials possess different properties and characteristics based on structure.

Material science is an interdisciplinary field related to discovery, design and development of new materials particularly solids and understanding their properties and application. It also includes fabrication process used to mould metals into different shapes. The field is also termed as *materials science and engineering* having emphasis on engineering aspects of building useful products. The four interdependent essential elements of material science and engineering are processing, structure, properties, and performance or application as shown in Fig. 4.1.

Processing → Structure → Properties → Performance

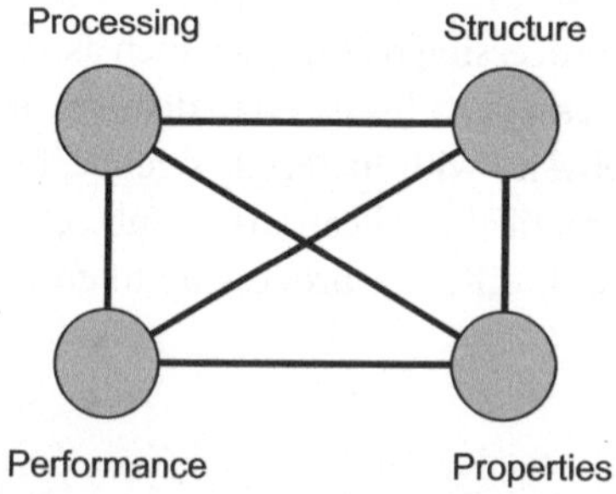

Fig.4.1: Elements of material science

Processing includes mining of ore, purifying and shaping the material, it may also include secondary processing like *forging, rolling, milling, cutting, extruding,* and *polishing* to develop final material. Structure means the arrangement of different constituents at different length scales based on atomic bonding and chemical composition. Material structure can be classified as: macrostructure, microstructure, substructure, crystal structure, electronic structure and nuclear structure. Structure decides the properties/ characteristics, and performance/ application for specific purpose.

Materials acts as base for manufacturing and service technologies; therefore, engineers of other branches depend mainly on materials scientists and engineers to provide advanced materials for manufacturing of products for their particular design and application. The engineers should have better understanding of material systems for selecting right material for applications such as of cars with better mileage, aeroplanes, and advanced computers having hard drive capacities, smaller electronics devices, sensors and equipment used to harvest renewable energy.

With the advancements in technologies materials with higher strength to density ratio were developed that resulted in a wide variety of new products, from body implants to light weight tennis racquets.

4.1.1 Historical Perspective

The progress of human civilization is witnessed by improvement and simplification in their standard of living. It can be noticed that materials have an impact on human civilization and significantly controlled human activities over thousands of years. The ancient civilizations are named after advances in materials as they have transformed and shaped the civilizations;

Stone → Bronze → Iron → Advanced materials

In the beginning from Stone Age, natural materials were used such as stones, clay, skin, and wood for various useful purposes to make weapons, instruments, shelter etc. In Bronze Age, people started using copper and its alloy (Bronze) which could be casted and hammered in various shapes.

The Iron Age began about 3000 years ago, where Iron was available in abundance on earth which has dramatically changed the life and is continued till today. With the advancements in science and technology, intelligent design of advanced materials, ceramics and polymers have simplified the human life.

4.2 CLASSIFICATION OF MATERIALS

Every engineer should have basic knowledge of various materials based on cost and performance to use them for different engineering application with desirable properties. Therefore, classification of materials is important for selecting a suitable material based upon the chemical properties, crystal structure, condition of operation, machinability and other parameters. Presently, engineering materials are grouped into four basic classes; metals, ceramics, polymers and composites as shown in Fig 4.2.

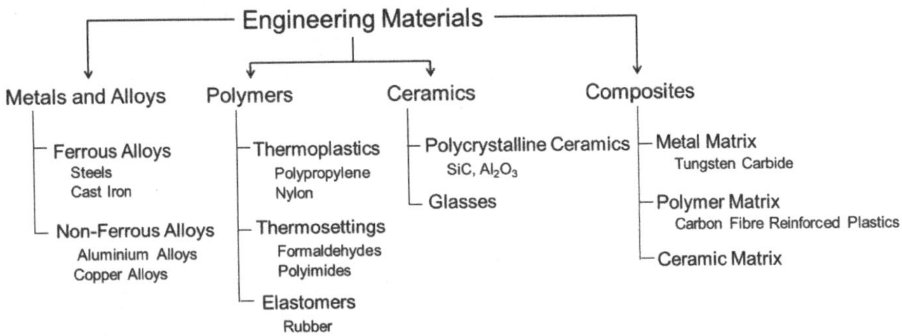

Fig. 4.2: Classification of materials

4.2.1 Metals and Alloys

Metals in pure form are soft, ductile and corrosive and are not useful. For practical applications, metals are combined with different elements to impart desired properties called as *alloys*. Metals may be composed of one or more metallic elements (iron, aluminium, copper, gold and nickel) and non-metallic elements in a small amount (carbon, nitrogen and oxygen). In metals and alloys, atoms are arranged in a particular order are called as *crystalline materials*. Metals have valence electrons due to which they have good electrical and thermal conductivity, opacity, lustre and malleability. The most common alloy steel is a mixture of iron and carbon, along with chromium and manganese. Metals may be further classified as *ferrous* and *non- ferrous* metals.

Ferrous metals: Ferrous metals contain iron and steel. Iron such as cast iron wrought iron, steel is the main constituents in ferrous metals. Ferrous metals are magnetic in nature and possess high tensile strength which makes them suitable for structural applications. The most common ferrous metals are cast iron, wrought iron, alloy steel, and carbon steel.

Non - ferrous metals: Metals other than iron are called non-ferrous, includes aluminium, copper, lead, tin, titanium and zinc, and alloys such as brass. These have light weight, high conductivity, corrosion resistance and non-magnetic properties.

4.2.2 Polymers

Polymer is a class of natural or synthetic organic substances based on carbon, hydrogen and other non-metallic elements. They consist of long chain of large molecules (macromolecules) with many smaller repeating units (monomers) which can be tailored for different uses. The term polymer is used to define *plastics* which are synthetic polymers, whereas natural polymers are *rubber* and *wood*. They are soft and light in weight as compared to metals. Polymers have a higher corrosion resistance and can be transformed into different shapes on application of heat and pressure.

Thermoplastic polymers can be melted easily allowing them for recycling as they have molecules with covalent bond within each molecule and secondary Van der Waals forces between them. Nylon, polyethylene, polyvinyl chloride, etc. are examples of thermoplastics.

Thermosetting (Thermosets) polymers have long molecular chains attached to each other with covalent bonds. Owing to typical bonding type, polymers are electrical and thermal insulators (e.g., epoxy, phenolics, etc.)

Elastomers are long molecular chain that exists in the form of amorphous linear bonding. They have weak intermolecular force, low Young's modulus **and higher** strain to failure as compared with other materials. Examples of elastomers are natural rubbers, polybutadiene, polyurethane elastomers, and nitrile rubbers.

4.2.3 Ceramics

Ceramic materials are the inorganic and non-metallic compounds, which may be oxides, nitrides, and carbides. They are crystalline and glassy. Ceramics are harder, abrasion resistant, possess high strength with very good chemical and thermal stability. Conventional ceramic materials are made of clay, silica, feldspar and dolomite, and are used to make pottery, tableware, sanitaryware, tiles, structural clay products, refractories and blocks. The modern ceramic material includes aluminium oxide (or alumina, Al_2O_3), silicon dioxide (or silica, SiO_2), silicon carbide (SiC), silicon nitride (Si_3N_4). They are high-purity compounds produced through a series of specialized manufacturing processes and are used for various industrial application due to their permeability, magnetism, insulation and conduction.

4.2.4 Composites

Composite materials are multiphase materials composed of two or more different materials with different physical and chemical properties viz., metals, ceramics and polymers. The objective is to achieve a good combination of properties which cannot be achieved by any individual components. It also incorporates better properties of each and every individual component. Wood and bone are a natural composite.

Composite materials are categorized based to matrix materials; like metal-matrix, polymer-matrix, and ceramic-matrix. Concrete is a common example of artificial composite material that contains small rocks and cement to hold it. To prepare specialized jobs like lightweight brake disc, SiC particles are embedded in Al-alloy matrix.

Glass fibres are polymer materials like epoxy or polyester combined with glass fibers. Here the glass fibers are strong and stiff, whereas polymer is ductile in nature. Similarly in carbon fiber, polymers are combined with carbon fiber. This class of materials have improved strength and stiffness, are lighter and resistant to electricity and corrosion as compared to pure material. Probably, they will replace steel in future.

4.3 PROPERTIES OF MATERIAL

Material properties determine the behaviour under specific conditions. The selection of correct material is important for an engineer to develop a product suitable for a given application to perform within a given range of properties. Different material possess different properties and materials are combined together to achieve a good combination of properties than individual ones. Properties are affected by their chemical composition and rearrangement in internal structure, like grain size or crystal structure.

Mechanical processes or metal working and heat treatment techniques might plays an important role in changing internal structure (arrangement of atoms), thus changes the properties like density, strength and electrical conductivity. Increase in strength may be achieved, but at the expense of ductility, so process optimization is important that provides the best material for a given application. In automobile sector steel is being replaced with carbon fibre reinforced plastic (CFRP) which significantly reduces the vehicles weight by 60% due to higher strength to weight ratio. Thus, engineers must have basic idea of material properties. There are many material properties tested in laboratory, but we will concentrate on very few which are important and grouped as follows:

4.3.1 Dimensional Properties

Dimensional properties are although not listed as property in handbooks, but form important criteria for selection of materials. It reflects, size, shape and tolerances on materials and microscopic surface morphologies including surface layout, roughness etc. The surface roughness and waviness are measured by using a *profilometer*.

4.3.2 Physical Properties

Material used for various industrial applications possess definite physical characteristics which can be found without changing the identity of the material. In general, physical properties of materials include lustre, density, thermal and electrical conductivity, melting point and boiling point.

4.3.3 Chemical Properties

Properties that describe the way a substance changes into a completely different substance. This property becomes evident during a chemical reaction when changing a substance's chemical identity. Examples of chemical properties include chemical composition, structure and reactivity. *Composition* means the percentage of other materials added to make alloy. *Structure* refers to the microstructure of the material seen under microscopes including phase structures and grain boundaries. The structure defines materials response during treatment. For example, when iron combines with oxygen in the presence of water forms rust.

4.3.4 Mechanical Properties

Mechanical properties are those characteristics of a material that become important when material is subjected to external force. These properties are important in selecting materials for structural applications and machine components. Mechanical properties are determined by the range of usefulness of the material and service under specified conditions. They usually relate the elastic and plastic behaviour of the material upon loading. The knowledge of stress, strain, elasticity and deformation up to failure are essential to realize and understand the concept of mechanical properties.

Some important mechanical properties are hardness, yield strength, tensile strength, elongation, reduction in area, modulus of elasticity, impact, fatigue and creep strengths etc. These properties can be measured with hardness test, tensile test, bend test, impact test, fatigue test and creep test. Tensile test forms the basis for almost all the mechanical properties. The stress-strain curve obtained from tensile test reveals many important properties which are helpful for design engineers. The stress-strain curve for ductile material (mild steel) is shown in Fig. 4.3. A test specimen of circular, square or rectangular cross-section may be used for testing using *universal testing machine*. When a gradually increasing axial load is applied on a uniform cylindrical bar specimen, it elongates with decrease in diameter. Then, the resistance offered by the material to this load is called *stress*. The load divided by original area of cross-section of specimen is defined as the *stress*.

$$stress = \frac{Load}{Area} \qquad 4.1$$

During loading, the material deforms, and the change in length is observed. The change in length divided by original length is defined as the *strain*. This deformation may be either elastic or plastic.

$$strain = \frac{Change\,in\,length}{Original\,length} \qquad 4.2$$

Initially with increase in load, stress is proportional to strain. Point 'a' is known as *proportionality limit*, beyond this point, the curve is not linear. The point 'b' is elastic limit, if the specimen is stressed beyond this point, it undergoes plastic deformation. A certain amount of strain is retained in it, and on unloading, it does not regain its original position. In case of most of the materials, the limit of proportionality and elastic limit are assumed to be the same.

With further increase in load, the strain increases upto point 'c'. Point 'c' is known as *upper yield point*, beyond this point, there is an increase in strain, with no increase in stress and the material is said to yield. Point 'd' is known as *lower yield point*, and again the stress increases with strain, and material is said to be *strain hardened*. The stress is maximum at point 'e' known as *ultimate tensile stress*, at this point cross-sectional area decreases and *neck* formation takes place. Beyond this point, specimen elongates and fractures at point 'f'. Now, the properties of materials can be explained as;

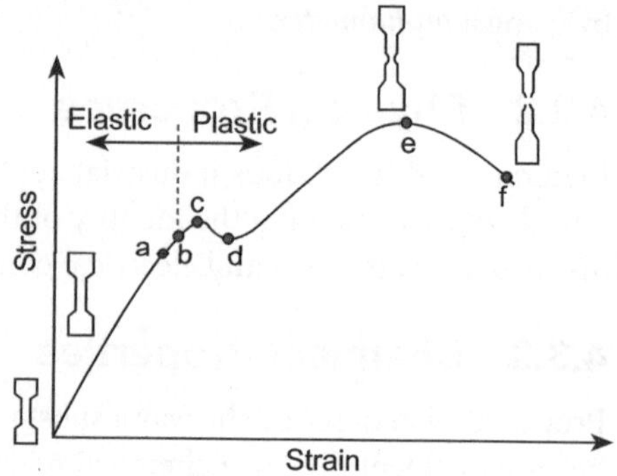

Fig. 4.3: Stress – strain curve for mild steel

Strength: Strength enables a material to resist deformation under the action of external load. A stronger material will be able to withstand greater load before breakdown or yielding. Materials have different strengths according to the nature of the applied load. For example; concrete is stronger at compressive load, but has poor tensile strength.

Elasticity: When an object is subjected to external load, it deforms. Elasticity is the ability of material or an object regains its original shape on removal of external load. Heat-treated springs and rubber are examples of elastic materials.

Plasticity: When a material is deformed beyond elastic limit the material, it undergoes some permanent deformation without fracture. Thus, plasticity may be defined as the ability of a material to permanently change the shape and it does not regain the original shape on removal of external load. Plasticity is necessary for forgings, in stamping images on coins and ornamental work in materials.

Yield strength: Yield strength is a stress where a material is in a plastic deformation region. The strength corresponding to point 'c' is the yield strength.

Tensile strength: Tensile strength or ultimate tensile strength is the maximum stress on an engineering stress-strain curve. At this point 'e', materials are plastically deformed but they may not be broken.

Hardness: Hardness is the resistance of materials being permanently deformed and also indicates resistance to abrasion, scratching and indentation. The materials which are difficult to deform or penetrate exhibits a greater hardness characteristic. Metals are heat treated through different techniques to increase hardness, specifically of tools like hammer, scriber and cutting tools etc. There are different types of hardness measurements tests like, Rockwell hardness, Brinell hardness, Vickers hardness, Micro-hardness and Rebound test.

Ductility: It is the ability of a material to change shape or deform plastically without fracture when tensile force is applied. It usually denotes stretching along its length, which enables the material to be drawn out into thin wires. Ductile material shows large scale of deformation prior to fracture and is measured as percentage elongation or percentage reduction in area during tensile test. Examples of ductile material are mild steel, copper, aluminium, nickel, zinc, tin, etc.

Brittleness: Materials that can withstand little or no plastic deformation before fracture are called as brittle materials. Temperature has a big effect on ductility of material. The rise in temperature will increase the ductility of material, while at cold temperature ductility will decrease and material become more brittle in nature. Brittle material shows high compressive strength but lower tensile strength, examples are cast iron, glass, ceramic and graphite etc.

Toughness: Toughness is defined as the ability of a material to absorb energy before it is fractured. It is related to the response of material to sudden blows or shocks. It can also be explained as the amount of energy required to create or propagate a crack. Mild steel being tough material absorbs a large amount of energy, whereas a brittle material such as glass does not absorb any energy before fracture.

Malleability: Malleability is the ability of materials to be hammered or rolled into sheets of other sizes and shapes when subjected to compressive force before fracture. Malleability is related to ductility and sometimes called workability, examples are aluminium, copper, tin, lead etc.

4.4 CASTING

Different quality products are made available to satisfy the needs, desires and demand of a consumer. The improvement in product quality at reasonable cost may be achieved by proper design, selection of material and manufacturing aspects. Manufacturing is always important to mankind and considered as backbone of social and economic development of any country. Manufacturing involves transforming of raw materials into finished goods to be used for some purpose with the systematic use of labour, machinery, tools and biological or chemical processing. It also denotes the fabrication or assembly of components into finished products on a larger scale.

The industrial revolution has changed the manufacturing from manual to mechanized system to increase production and efficiency. There are many material processing techniques available to manufacture parts, each has its advantages and disadvantages, and are categorized as;

- **Casting:** Sand casting, investment casting, permanent moulding.
- **Material removal process:** Turning, milling, boring, laser beam machining, etching.
- **Forming or shaping process:** Rolling, forging, extrusion, drawing.
- **Fabrication or consolidation process:** Gas welding, arc welding, soldering and brazing.

Casting is one of the oldest manufacturing processes, where molten material, usually metal is used. This molten metal is then poured into a mould cavity that takes the form of the finished part, and is allowed to cool and solidify into the desired shape. The solidified object is removed from the mould is called *casting*.

Castings are made in specific factories or industries called as *foundries*. Casting process is known for its use for preparing copper and bronze ornaments, weapons, tools and utensils from time as old as Indus Valley Civilization. The steps in casting process are;

- Preparation of pattern and mould
- Melting and pouring of metal
- Solidification and removal
- Cleaning, finishing and inspection

The casting obtained may be cleaned, finished and machined to required dimensions. It is inspected to get a defect free casting. There are different types of casting process depending on type of mould, pattern and material to be casted.

Advantages and limitations of casting process

In casting, tools required are simple and inexpensive. Both ferrous and non-ferrous materials can be casted easily. During casting molten metal can flow into thin and small sections of the mould, so intricate and complex shapes can be made. Size and weight of the casting is not a limitation, a casting of around 200 tons can be made. Casting objects have high compressive strength and are suited for mass production.

However, in sand casting the dimensional accuracy and surface finish achieved is poor, therefore operations such as cleaning and finishing are required. Therefore, die casting process, investment casting process, vacuum-sealed moulding process, and shell moulding process are developed. Handling of molten metal poses safety hazards to workers. Many casting defects may occur resulting in poor product quality.

Applications

Many parts used in industry can be made such as automobile cylinder block, clutch plate, brake drum, machine tool beds, housings, crank shafts, cam shafts, differential housing, valves, brackets, rollers, turbine blades etc.

4.4.1 Casting Terminologies

Before going for detailed study of sand casting process, the terminologies used should be known to us, and is represented in Fig. 4.4.

Mould Box: The mould box or flask is used to hold the sand mould in proper alignment. The sand is rammed into the box to make sand mould which has a cavity of the object to be casted, and the metal is poured into it to get desire shape. It is made up of metal and has parts called as cope, drag and cheek.

- **Cope:** The upper part of mould box flask is called as cope.
- **Drag:** The lower part of mould box is called as drag.
- **Cheek:** When the mould box is made of more than two parts, then the central part is called as cheek. This is used when the casting to be made is complex.

Pattern: Pattern is replica of part to be casted. It is made by either wood, metal or any suitable material such as wax. It is kept in the moulding box and sand is rammed around it, which after ejection forms a cavity in the mould.

Pouring basin: It is a funnel or cup shape cavity at the top of the mould. The molten metal is poured into it which through the sprue and gating system passes in the mould cavity.

Sprue: A tapered shaped passage through which the molten metal passes from the pouring basin to the runner and reaches the mould cavity. Sprue also controls the flow of metal into the mould.

Runner: Runner is a channel through which molten metal passes from the sprue to the gating system.

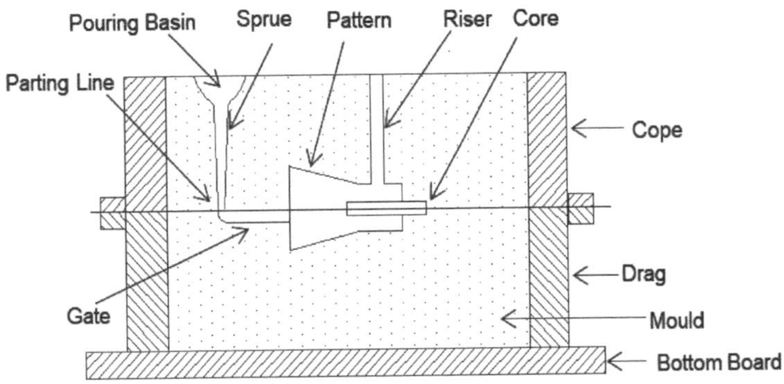

Fig. 4.4: Schematic illustration of a sand mold

Riser: Riser is a column of molten metal used as reservoir to feed the casting when pouring of molten metal is stopped. It feeds the casting during solidification and cooling to avoid shrinkage which may reduce the size of object. Riser compensates the volumetric shrinkage and classified as top riser, blind riser, side riser etc.

Ingate: It is the channel through which molten metal actually enters into the mould cavity.

Core: It is a baked sand structure used to cast hollow cavity. It is placed at proper location into mould cavity to create a hollow part in the casting.

Chaplets: These are metal inserts used to provide support and hold the core at proper location in the mould cavity. It protects the core from deflection due to the flow of metal into the casting.

Chills: Chills are solid metal inserts in the mould to create local chilling. They have high thermal conductivity and equalize the rate of directional solidification throughout the casting.

Vents: These are small passages made in mould to facilitate the escape of the gases during solidification of casting.

Parting line: It is the dividing line between the two halves of moulding box or between spilt patterns.

Facing sand: It is fine grained carbonaceous material sprinkled on inner surface of the mould cavity to get a good surface finish of the casting. It comes in direct contact with the molten metal, so it should have high refractoriness and strength.

Moulding sand: It is a well prepared fresh refractory material used for making the mould and surrounds the pattern. It is a prepared mixture of silica with 10 to 20% clay and about 8% water with some organic additives to get the sound casting.

4.4.2 Moulding Tools

The different tools used in preparing a mould are discussed below:

Sprue: It is used to make sprue hole to pour the molten metal in the mould (Fig. 4.5 a).

Mallet: It is a wooden hammer used to loosen the pattern in the mould so that it can be withdrawn without damage to the mould (Fig. 4.5 b).

Swab: It is made of flax or hemp and is used for applying water to the mould around the edge of the pattern. This prevents the sand edges from crumbling, when the pattern is removed from the mould (Fig. 4.5 c).

Rammer: It is used for packing/ramming the moulding sand in the moulding box (Fig. 4.5 d).

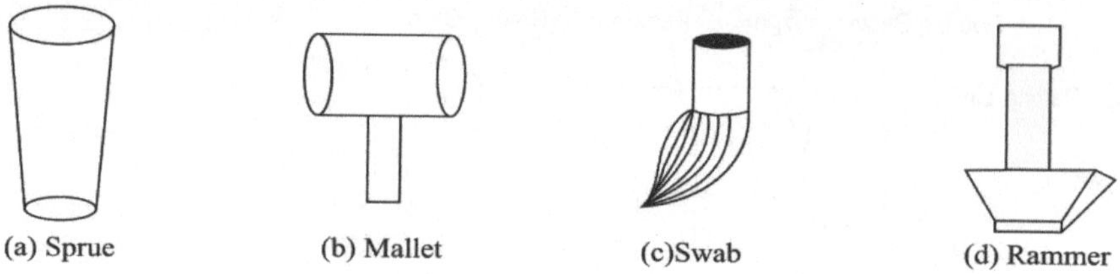

Fig. 4.5: Moulding tools: sprue; mallet; swab and rammer

Vent rod: It is used for making vent holes in the sand mould so that the gases released during casting process, can be escaped out from mould easily (Fig. 4.6 a).

Riddle: It is a metal sieve used for removing foreign materials such as stones, nails etc. from the moulding sand (Fig. 4.6 b).

Lifter: It is used to remove the loose sand from the mould. It is also used to finish the bottom and sides of the mould (Fig. 4.6 c).

Slick: It is used for repairing and finishing of surfaces and to round the corners of the mould (Fig. 4.6 d).

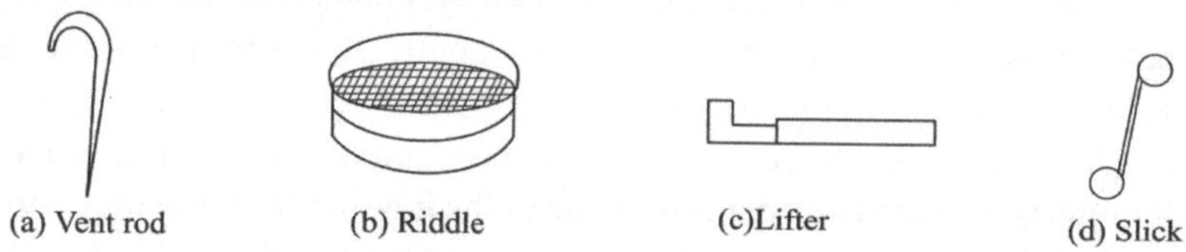

Fig. 4.6: Moulding tools; vent rod; riddle; lifter and slick

Shovel: It is one of the foundry tools which have a broad metal blade with a long wooden handle. It is used to carry moulding sand from the sand pit to the moulding box and is also used for mixing and adding sufficient water (tempering) the sand (Fig. 4.7 a).

Trowels: It is used for smoothening the surfaces of mould and also to repair the damaged portion of mould (Fig. 4.7 b).

Ladle: A ladle is a pouring equipment used to carry molten metal from furnace into moulds to produce the casting (Fig. 4.7 c).

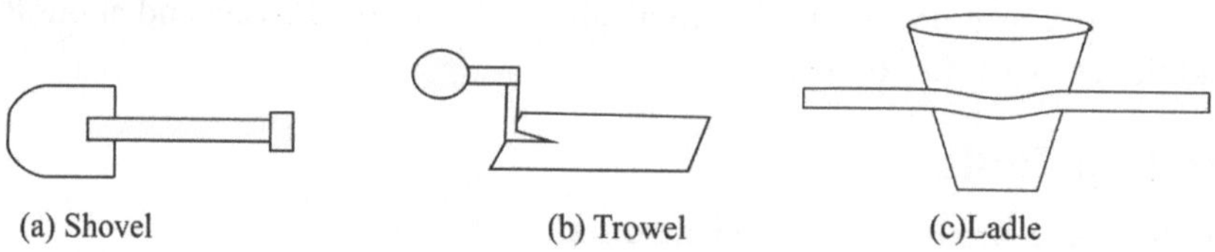

Fig. 4.7: Moulding tools; Shovel; Trowel and ladle

4.4.3 Patterns and materials

During casting process, patterns are used which is the replica of the object to be made by casting with few dimensional modifications or allowances. It forms the mould cavity when moulded in sand, which on filling with molten material and after solidification reproduces a pattern is called casting. Patterns play a very important role, so care must be taken in deciding allowances and pattern material. A pattern with proper design and features gives smooth surface and reduces casting defects. Table 4.1 shows different types of pattern materials with their advantages and disadvantages. Selection of pattern material depends on following factors;

- The dimensional accuracy, finish required and size of the casting.
- Size of pattern and complexity.
- Number of moulds to be made with a pattern.
- It should be Water resistant.
- Cheaper and light weight.
- Long lasting and smooth surface.
- Type of moulding, like, sand moulding, shell moulding, and plaster moulding.

Table 4.1: Different pattern materials and their advantages and disadvantages

Pattern material	Advantages	Disadvantages
Wood is used when size of pattern is large and less number of castings is to be produced. Commonly used woods are teak, deodar, mahogany, pine, etc.	Light in weight, cheap and easily available. A good surface quality can be achieved and can be kept for long time by applying varnish. It can be shaped into various forms and intricate designs.	Dimension changes due to absorption of moisture and may crack or split. Low resistance to abrasion. Low strength and not suitable for large scale production.
Metal **pattern**s are used for large number of castings and when high dimensional accuracy is required. Commonly used metals are steel, brass, aluminium, grey cast iron etc.	Does not absorb moisture, and used for large quantity of casting. Stronger and longer life. Good surface finish and machinability.	Patterns has tendency to rust. Expensive as compared to wood, thus not suitable for small quantity production. It requires machining.
Plastic is used for pattern making. Commonly used plastic pattern are thermosetting resins, epoxy, PVC, polyurethane foam, etc.	Light in weight and reasonably low cost. High strength, high resistance to wear and corrosion Dimensionally stable.	They are fragile and do not work when subjected to severe shock conditions i.e., machine moulding. Poor abrasion resistant as compared to metal.
Plaster of Paris is used to make plaster patterns.	Can be easily cast into complex shapes. It expands on solidification. Easily workable.	Can be used for small castings.
Wax pattern is not taken out like other patterns, the mould is inverted and then heated. The molten wax flows out from the mould. It is used in investment casting. Commonly used waxes are bee wax, shellac, paraffin wax etc.	Do not cause any distortion on mould while removing the pattern. It provides good surface finish and dimensional accuracy.	Equipment and process are costly. Requires high level of skilled operators.

4.4.4 Types of Patterns

A Pattern is a model or the replica of the casting to be made, except for the various allowances pattern exactly resembles the casting. Patterns can be made in two or three pieces, depending on shape and size of casting, whereas castings are made in a single piece. The quality of casting made is highly affected by the planning of the pattern. During pattern making, care must be taken because properly constructed patterns minimize overall cost of the casting. Selection of pattern depends on features of casting process and number of castings to be made. Various types of patterns used are discussed below;

(i) Solid or single piece pattern

Single piece pattern is the simplest pattern made from one piece without any joint. This is inexpensive and used for small scale production of large size simple castings, such as stuffing box of steam engine, regular surfaces and rectangular blocks. It does not create any problem during withdrawal from mould. The important characteristic of this pattern is that one surface is considered as flat portion and this flat surface is used for parting plane. In absence of flat surface, moulding process becomes complicated. Here, pattern is expected to lie either in cope or in drag (Fig. 4.8).

(ii) Split or two- piece pattern

Split patterns are used when withdrawal of casting from mould become difficult, mainly for intricate castings. It is also used when the depth of casting is high. The shape of casting will decide the location of split plane. Split plane and parting line are same. In split pattern one half is moulded in drag while the other half is moulded in cope. The cope part of the pattern has dowel pins, whereas drag half of the split pattern has holes. The dowel pins matches exactly with the holes to align two parts properly (Fig. 4.8).

(iii) Multi piece pattern

In case the castings have complicated designs and difficult to cast by split pattern, then multi piece type patterns may be used. Three or more piece patterns are considered as multi piece pattern. For three- piece pattern, moulding box with three parts is used, the top part is called as cope, bottom part is drag and the middle part is termed as cheek (Fig. 4.8).

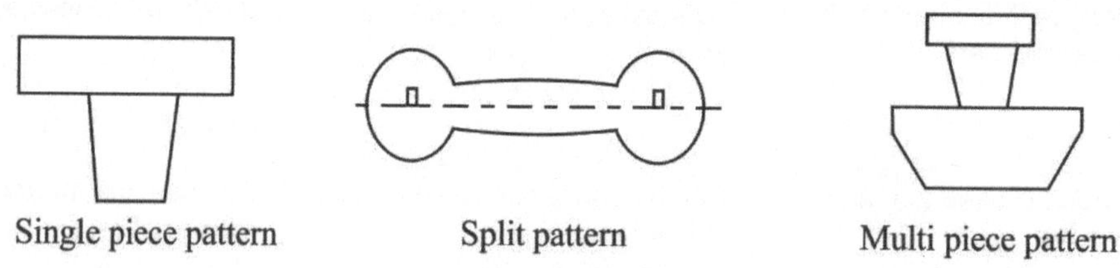

Fig. 4.8: Types of patterns: Single piece; Split piece; Multi piece patterns

(iv) Match plate pattern

Match plate pattern is a split type pattern, the two halves (cope and drag) are placed on opposite sides of metallic plate termed as match plate. The gates and runners are mounted on match plate. Locating pins are used to accurately place the match plate between cope and drag. This type of pattern is used for mass production of small precision castings, by placing several patterns on a single match plate (Fig. 4.9).

(v) **Gated pattern**

These type of patterns are used to make several small size castings with a single multi cavity mould made by joining number of pattern with a common gating and runner system. For joining different loose piece patterns gates are used (Fig. 4.9).

(vi) **Sweep pattern**

Sweep pattern is used when a mould of larger size symmetrical casting is to be produced in shorter time. A sweep is a wooden section of proper shape corresponding to the shape of mould to be made which is rotated about an axis to shape the mould cavity having shapes of rotational symmetry. It is economical as it avoids the need of full pattern for large casting (Fig. 4.9).

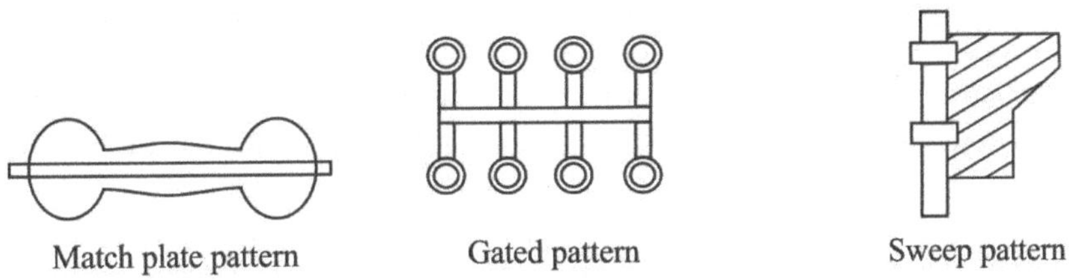

Fig. 4.9: Types of patterns: Match plate; Gated; Sweep patterns

(vii) **Loose piece pattern**

It becomes difficult to remove solid pattern from the mould which is having complex contour above or below the parting plane. Therefore, complicated portion of pattern are kept as loose pieces, and are attached to main portion of pattern with dowel pins for proper alignment. Once the main portion of pattern is removed, there is sufficient space available for the ease in removal of loose piece. It requires skilled and trained worker as the shifting may occur during ramming and it is expensive also (Fig. 4.10).

(viii) **Cope and drag pattern**

Cope and drag pattern is like a split pattern. This pattern has cope and drag portion of pattern with gating and runner system on separate wooden or metallic plate. Cope and drag portion of pattern are moulded separately and are kept in proper alignment with alignment pin. After moulding, the two separate parts are brought together to form complete mould cavity. This arrangement is used for the mass production of large and heavy castings (Fig. 4.10).

(ix) **Follow board pattern**

Follow board is used where patterns are weak and they may break or sag after the application of ramming force. The wooden board acts as a base and supports the pattern in drag box only (Fig. 4.10).

(x) **Segmental pattern**

The working principle of segmental pattern and sweep patterns are similar. For designing required shape or structure of mould they both employ a part of pattern. Segmental pattern is a part or segment used for moulding large circular castings such as gears, wheels, rims etc, thus avoiding the use of solid heavy pattern of same size. A vertical spindle is placed in drag box and pattern is mounted on it. Spindle is rotated to make the complete mould (Fig. 4.10).

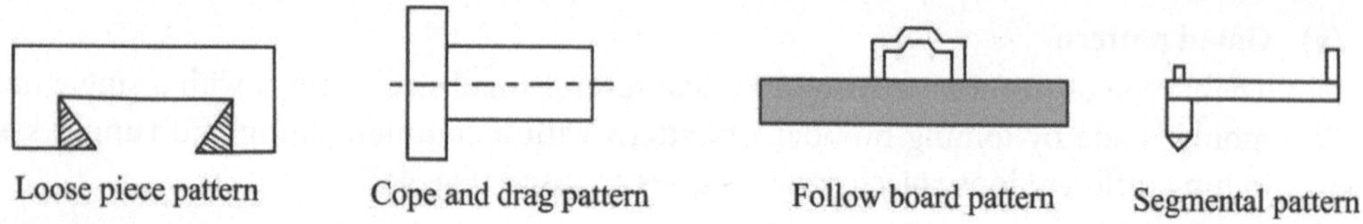

Fig. 4.10: Types of patterns: Loose piece; Cope and drag; Follow board; Segmental pattern

(xi) Skeleton pattern

In order to fabricate patterns for larger size castings with simple geometries, lot of material is wasted. Therefore, skeleton patterns are used that consists of a wooden frame and rib that outlines the shape of the part to be cast which is hollow form of pattern. The hollow part is filled with sand and clay and rammed to get required outer surface. The skeleton pattern is made in two halves; one is for the cope and other for the drag. A stickler board is used to remove the surplus sand that comes out of the spaces between the ribs. The wooden frame highlights the area to be casted. The skeleton pattern is used to cast small number of large-sized castings such as hollow cast iron pipes, boxes, valve bodies and bends (Fig. 4.11).

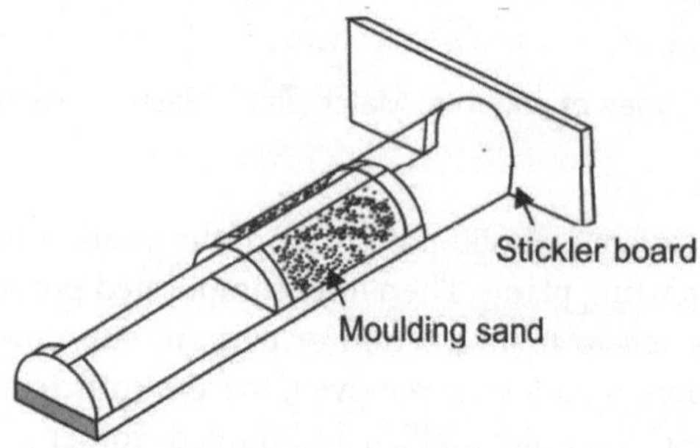

Fig. 4.11: Skeleton pattern

4.4.5 Pattern Allowances

As discussed above, pattern exactly matches or is a replica of object to be cast, but with slightly larger dimensions. This difference is because it carries certain allowances to compensate shrinkage, machining, draft and distortion due to metallurgical or mechanical changes that takes place during casting. The different types of allowances are discussed below (Refer to Fig. 4.12 and Fig. 4.13).

(i) Draft allowance

In Draft allowance a taper is provided on all the surfaces of pattern that are parallel to the direction of pattern withdrawal, so that pattern can be removed easily without damaging the sides of mould. The surfaces are tapered by about 1 to 2 degrees (Fig. 4.12). If this taper is not provided, it will damage the mould and affect the casting. Draft allowance is provided on internal and external surfaces and is more for internal surfaces. The amount of taper depends on the shape and size of pattern, mould materials, moulding methods, and depth of mould cavity.

(ii) Machining allowance

The products after and casting process has a poor surface finish, and dimensions are also inaccurate. The casting gets oxidized in the mould and during heat treatment it forms scales.

In order to remove roughness and get exact dimensions, casting is machined. Thus, extra material is provided on casting due to which a pattern has to be made oversized. The amount by which the pattern is made larger than basic size is termed as machining allowance.

The machining allowance depends on metal to be casted, shape and size of the casting, machining operation, degree of surface finish required and the moulding process. It varies from 2 mm to 15 mm depending on the size and material of the pattern. Investment casting is precision casting and do not require machining (Fig. 4.12).

(iii) Shrinkage allowance

Most of the metal castings shrink or contract on cooling. Metals shrink at different rates because shrinkage depends on linear coefficient of thermal expansion which is the property of the metal to be casted (Fig. 4.12). It also depends on pouring temperature, size of casting, design aspects and moulding materials and methods. The shrinkage takes place in three ways;

- Liquid shrinkage occurs when liquid metal starts cooling upto freezing temperature.
- Solidifying shrinkage occurs when phase changes from liquid to solid.
- Solid shrinkage occurs when solid casting cools from freezing temperature to room temperature.

The first two shrinkage types are compensated by proper design of risers which feeds the liquid metal to the casting. Whereas, the solid shrinkage is compensated by making the oversized pattern equal to shrinkage allowance. The shrinkage in casting is volumetric, but it is calculated as linear measure in mm/meter. Shrinkage allowance for different materials is given in Table 4.2.

Table:4.2: Shrinkage allowance for various metals

Metal	Grey cast iron	Steel	Brass	Copper	Zinc	Aluminium	Magnesium	Lead
Shrinkage allowance (mm/meter)	7 to 10.5	20	16	16	24	16	18	24

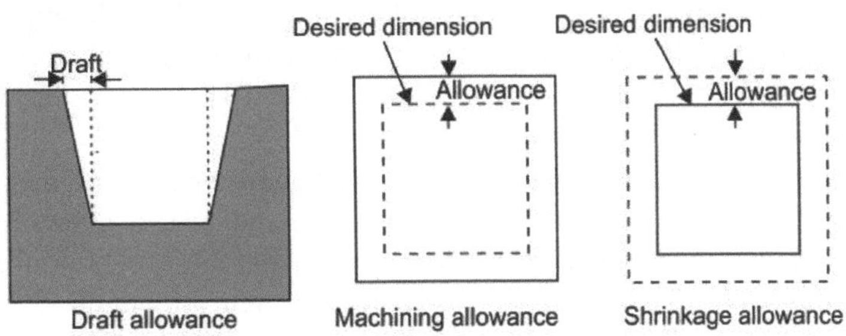

Fig. 4.12: Draft, Machining and Shrinkage allowance

(iv) Shake or rapping allowance

When the pattern is to be withdrawn from the mould sand, it is slightly shaken or rapped by striking it. This facilitates easy withdrawal of pattern but enlarges the mould cavity, resulting in increased casting size (Fig. 4.13). To avoid this, pattern size is reduced, i.e., a negative allowance is provided to the pattern in order to remove it from the sand and this will increase the dimension of the casting slightly. These small changes in the dimensions of the pattern in the casting process are called the shaking or rapping allowance. This allowance is provided to the dimensions which are parallel to the parting line.

(v) **Distortion allowance**

During cooling process of metal casting, stresses are developed. These stresses are due to irregular shape of casting as all the parts do not shrink uniformly. Sections having long flat portions as well as U' or 'V' shape sections results in uneven cooling of the casting. The stresses developed may cause to deformation or bending of the casting. To prevent this behaviour, camber is provided in the opposite direction so that, when bending takes place due to uneven thickness of metal, the casting may become straight. This change in pattern to avoid bending of casting during solidification is called as distortion or camber allowances (Fig. 4.13).

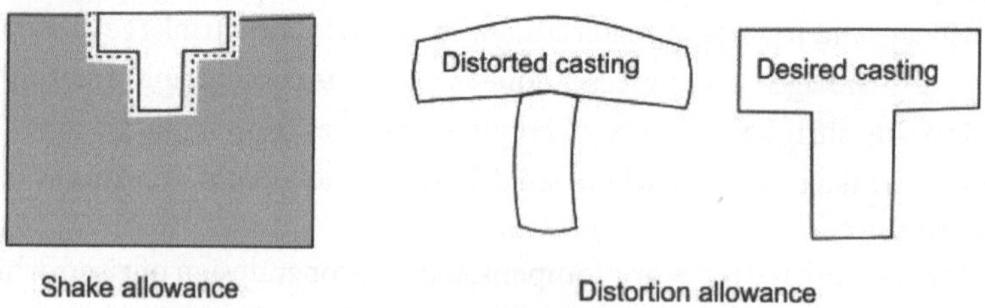

Fig. 4.13: Shake and Distortion allowance

4.4.6 Sand Moulding Process

Sand moulding is most widely used expandable moulding process which employs multiple use of patterns for casting of all types of materials due to process flexibility, heat resistance property and low cost equipment. The simplest sand moulding is a green sand casting process that provides desired shape to the object of any size by pouring molten metal in a sand mould. The step wise procedure for sand moulding is given below and complete process is presented in Fig. 4.14.

- First of all, pattern is made as per the object, which may be of wood or any other material. Generally, pattern is split type with contraction, drag and other allowances. Facing sand is sprinkled over the board to provide a non-sticky layer. Then drag part of the pattern is placed on the mould board. One half of moulding flask called drag is placed on the moulding platform or wooden mould board around pattern.
- Well prepared moulding sand is then poured in the drag over pattern till the drag is completely filled. It is then uniformly rammed to tightly pack the sand. Proper ramming must be done to allow gases to escape and obtain a defect free casting. After uniform ramming, the excess sand is removed with strike off bar. Vent holes are made with vent wire in the drag to the full depth of the moulding box to allow gases to escape during pouring and solidification of casting.
- The finished drag is now turned to make upside down to expose the pattern. Second half of the pattern called cope half is then placed over the drag part of pattern using locating pins for proper alignment. The dry parting sand is sprinkled over it.
- A sprue pin and riser pin are placed at appropriate distance from pattern for making passage. The moulding sand is sprinkled, thoroughly rammed and vent holes are made in the same manner.

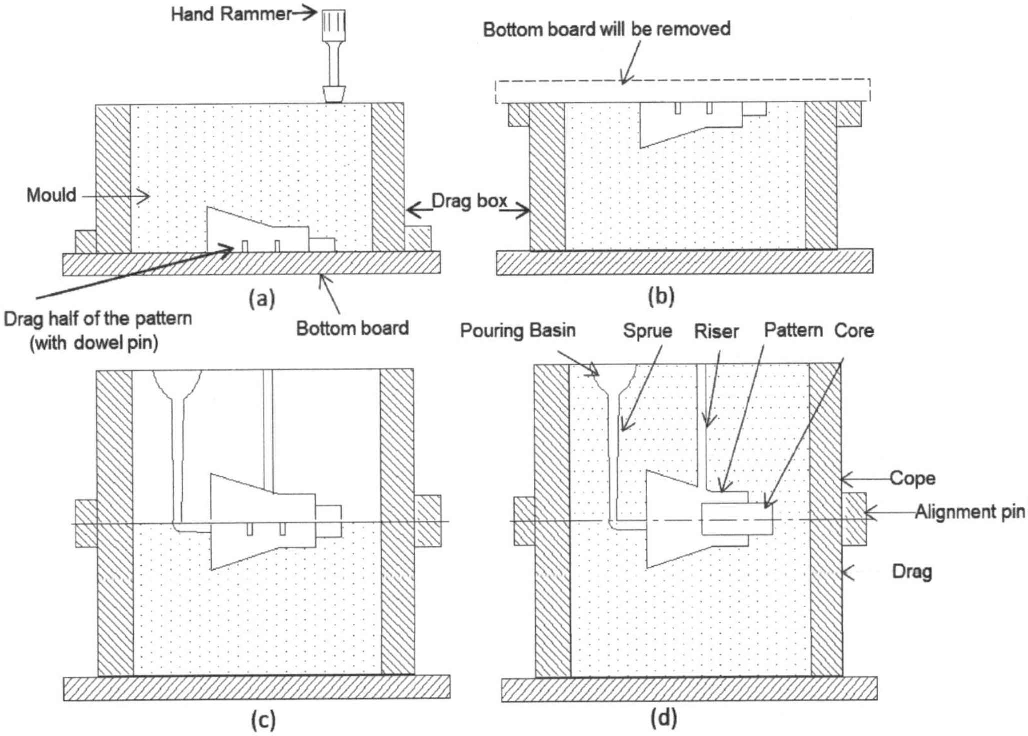

Fig.4.14: Schematic illustration of sand moulding process

- The sprue pin and riser are withdrawn, and a pouring basin is cut at the top of sprue. The cope and drag parts are separated and excess sand is blown off. Now, the pattern from the cope and drag halves are removed using draw spikes.
- Runners and gates are made in the mould by cutting the parting surface with a gate cutter without damaging the mould. Gate cutter is made of sheet metal bent to the desired radius.
- If, any hollow part is required, then core is placed with core prints for making a central hole in the mould cavity in the drag.
- Cope is placed on the drag again by properly aligning the two parts with pins. Mould is now assembled and ready for pouring.
- The metal is melted in furnace and molten metal is carried to the mould with ladle. It is then poured into mould through sprue and continued till the cavity and riser are filled sufficiently. Metal is allowed to cool and solidify.
- The solidified casting is obtained on breaking the mould, it is then cleaned and sent for further processing and inspection.

4.4.7 Types of Sand Mould

The different types of sand moulding are discussed in the given below:

(i) Green sand mould

Green sand mould is most common type of mould containing sufficient moisture in it. It has a mixture of silica, clay with water and additives. Clay and water act as a binding material to strengthen the mould. It is used for simple and rough castings of both ferrous and non-ferrous metals and used for making grills, weights, moulding boxes etc. Mould is soft and may be damaged during handling, so shape and size is not accurate. Once the casting solidifies, mould and casting can be separated.

(ii) Dry sand mould

When the moisture present in the green sand is removed, it is known as dry sand. The mould produced by dry sand has greater strength, rigidity and thermal stability. Resin binders are used for providing strength. Dry sand moulds are usually used for large and heavy steel casting like larger rolls, gear housing etc. It reduces the defects caused by moisture and provides better surface finish.

(iii) Skin-dried mould

This is the mould with dry sand facing and green sand backing. It is cheaper, but not stronger as compared to dry sand mould. This type of mould is used for the larger castings of both ferrous and nonferrous alloys.

(iv) Loam sand mould

Loam sand is a mixture of sand and clay, and also contains fire clay or ganister. The mould is made up of porous bricks cemented with load mortar. The mould is baked with forced hot air. It is suitable for large and heavy casting such as cylindrical or conical shapes cylinders, large bells etc. with improved surface finish.

(v) Metal mould

These are permanent mould used in the die casting of alloys having low melting temperatures and poured into metallic moulds. These moulds have longer life but costly, and gives high production rate with better surface finished castings.

4.4.8 Moulding Methods

Moulding methods may be classified as follows;

(i) Bench moulding

In this of moulding, all the operations are performed on a bench of convenient height and is used for moulding of small jobs. Two flasks, namely cope and drag are necessarily used.

(ii) Floor moulding

In this type of moulding, all the operations are performed on foundry floor. Floor moulding is used for medium and large size castings.

(iii) Pit moulding

Pit moulding is usually preferred for large castings, where moulds are made in a pit. The pit acts as a drag. The walls and the bottom of the pit are prepared to enable easy escape of gases. One box is generally required to complete the mould.

(iv) Machine moulding

Hand moulding is uneconomical and is not efficient for mass production of the casting. The machine moulding is advantageous because of significant savings observed in labour cost and working time with high dimensional accuracy and close tolerances. This also saves the machining cost and other post processing operations. Moulding machine helps in making the moulds at a faster rate and eliminates requirement of skilled operators. The major operations carried out with moulding machines are ramming, gating, and pattern withdrawal. Many of the mould making operations are carried out with moulding machines.

4.4.9 Properties of Moulding Sand

Sand used for moulding is a mixture of clean, uniformly grained high quality silica sand, clay and water with some additives such as, wood flour, dextrin, sea coal and molasses. The ratio is 70-85% silica, 10 to 20% clay, about 8% water with 1-5% additives. This proportion may vary for ferrous and non-ferrous castings. The grain size and distribution of silica sand affects the sand properties, it may vary the permeability of sand. Clay with water acts as a binding agent and provides strength and plasticity to the sand.

Bentonite and kelonite are popular clay types. Water is absorbed by the clay and coats the surface of clay which helps in bonding and developing the strength. Additives are added to provide desired properties in sand such as wood flour which increases collapsibility and flowability, dextrin and molasses increases strength and hardness, sea coal increases surface finish. All these ingredients are mixed together for uniform distribution. Following properties are required to be controlled during preparation of mould to achieve a sound casting of desired quality.

(a) **Porosity**

Porosity is the ability of the moulding sand to allow gases to escape out, which are generated during pouring of molten metal into the green sand mould. If, these gases are not allowed to pass, it would be entrapped inside the casting and will develop gas holes. This property is affected by shape and size of the sand particles, amount of the clay, binding material and moisture contents.

(b) **Cohesiveness**

Cohesiveness is the property of sand to hold its particles together, and defines the strength of the moulding sand. Clay and bentonite help to retain the shapes of the mould during handling.

(c) **Adhesiveness**

Adhesiveness is the property of sand due to which the sand particle sticks to the sides of the moulding box. Adhesiveness of sand enables the proper lifting of cope along with the sand.

(d) **Flowability**

Flowability is the property of moulding sand due to which it behaves like a fluid to free flow and fill the mould all around the pattern. It varies with moisture content and grain size of sand.

(e) **Collapsibility**

Collapsibility is the property of sand due to which the mould automatically gets collapsed after the solidification of casting, thus prevents damage of casting. The mould should break into small particles with minimum force after the casting is removed from it.

(f) **Refractoriness**

Refractoriness is the ability of moulding sand to withstand high temperature of molten metal without cracking and fusion. A granular shaped particle of silica sand improves the refractoriness. Poor refractoriness may burn mould during pouring of molten metal.

4.5 METAL WORKING

Metal working or forming is a group of manufacturing processes where the change in shape and size of the metal can be achieved by plastic deformation. Selection of appropriate deformation process will affect the strength and application of the product developed. Metal is first shaped into ingots, blooms, billets and slabs through casting, these shapes are then converted into structural sections by plastic deformation through different metal working techniques. The stresses induced during deformation are higher than yield sterngth of the material but lower than the fracture strength. Metal working is therefore defined as re-shaping and fabricating the metal objects without adding or removing any material, but by deformtion through a combination of heat and pressure. The input enrgy is used in improving the strength of material by *strain hardening*.

Metal working or forming is suitable for mass production due to expensive machinery and specialized tooling, without wastage of any material. The deformation of material depends on its ductility. Higher the ductlity of material, more is the deformation. Metal working can be grouped into hot working and cold working.

4.5.1 Cold Working

In cold working, the plastic deforamtion of the metals is carried out at room temperature or slightly above it, but below the recrystallization temperature. When the deformtion by cold working is increased, the material becomes harder with significant deformation of the crystal structure. The dislocations are produced in the crystal sytructure with deformation, dislocations pile up and increases the strength of the metal called as *work hardening*. In cold working residual stressess are induced, these stresses are undesirable and proper heat treatment may neutralize them without significant change in properties and structure.

Advantges

1. Good surface finish and better dimensional accuracy can be achieved.
2. Strength, fatigue and wear properties are improved.
3. Propeties and structure do not change on heat treatment.
4. Handling is easy and economical.
5. No decarburization of metal occurs, no loss of material due to low working temperature.

Diasadvantages

1. Ductility is reduced.
2. More power is required for deformation due to work hardening.
3. Residual stresses are setup, thus need heat treament.
4. Bigger parts cannot be deformed easily.
5. High cost of equipment and tooling.

4.5.2 Hot Working

In hot working, the plastic deforamtion of the metals is carried out at temperature above their recrystallization temperature. The large amount of plastic deformation can be imparted without significant strain hardening. Hot working lowers the strength of material and improves the ductility. Therefore, less stress is required for deformation and products having larger dimensions can be deformed easily. At this temperature, the deformed grains and strain hardening produced during plastic deformation is eliminated and allows the metal to recover and recrystallize. Thus new strain free grains or crystals are formed. In hot working, the temperature of metal working is important, as any extra heat left after working leads to grain growth.

Advantges

1. Metal becomes ductile, therefore load required for deformation is also reduced.
2. Elevated temperatures increase diffusion which can remove or reduce chemical inhomogeneities
3. No strain hardening, so no residual stress is induced.
4. Pores may reduce in size or closes completely during deformation
5. High production rate.
6. Higher reductions are possible.

Diasadvantages

1. At elevated temperatures scaling or oxidation of the object takes place.
2. Close tolerances can not be achieved due to thermal contraction and warping from uneven cooling.
3. Poor surface finish
4. Handing of material is difficult.
5. Lower life of equpments and tooling.

4.6 EXTRUSION

Extrusion is a metal forming process in which billet of ductile material undergoes plastic deformation by the application of compressive force. This causes the metal to pass through an orifice of a steel die of desired shape. The material acquires the shape of the cross-section of the die. Piston or plunger is used to apply compressive load, and is widely applicable for the manufacturing of pipes and rods. Extrusion is commonly used for deformation of non-ferrous ductile metals such as aluminium, copper, magnesium and lead. For the deformation of other metals, force required is high. It can be carried out in both hot and cold state.

Characteristics of metal extrusion

1. Die design, billet temperature, and extrusion speed controls the process.
2. Variety of shapes can be extruded using hot extrusion.
3. Lubrication is used to reduce wear and friction, and to improve the life of die.
4. Cold extrusion gives better surface finish and close tolerances.
5. Hot extrusion reduces the force required for deformation and improves the ductility of the material.

Extrusion can be divided into direct and indirect extrusion.

4.6.1 Direct Extrusion

Direct extrusion is also called as forward extrusion in which metal is forced to flow in the direction of motion of the punch as shown in Fig. 4.15. The die is mounted at one end of the cylindrical chamber. This process requires higher force due to higher friction between billet and container.

4.6.2 Indirect Extrusion

Indirect extrusion is also called as backward extrusion in which metal flows in the direction opposite to the direction of hydraulic ram as presented in Fig. 4.16. The die is mounted at the end of the hydraulic ram, and the ram moves towards the billet placed inside the chamber to force the material to pass through the die.

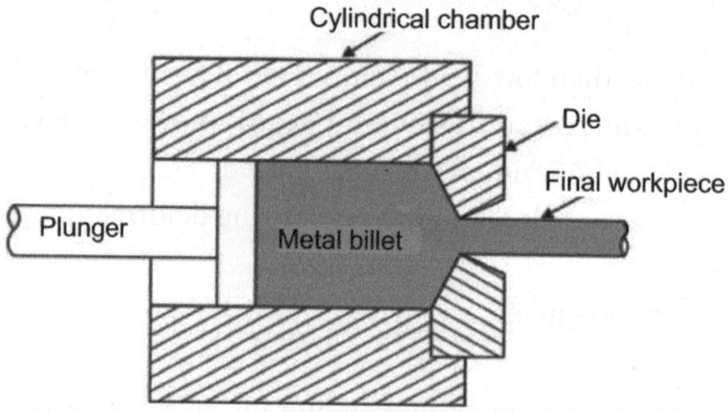

Fig. 4.15: Schematic of direct extrusion process

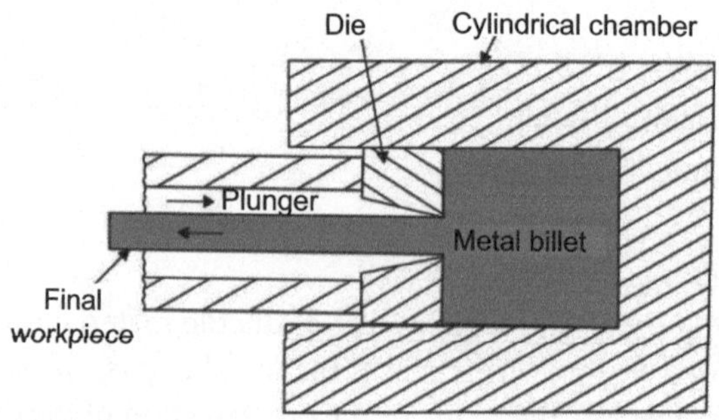

Fig. 4.16: Schematic of indirect extrusion process

4.7 FORGING

Forging is a deformation process in which the metal is shaped between two dies by applying a compressive force, either impact load or gradual load. Impact load may be applied using forging hammers, whereas gradual pressure is achieved with forging press. Forging can be carried out in both hot and cold state. The process is used to shape a variety of high-strength components including engine crankshafts, connecting rods, gears, aircraft structural components, jet engine turbine parts etc. Forging is commonly used for the deformation of stainless steel, titanium, carbon steel, and other metal alloys .

Characterstics of forging

(i) The forged parts have good strength and ductility and offers resistance towards impact and fatigue loads.

(ii) Forging improves the structure of the metal and hence its mechanical properties.

(iii) Forging usually produce little or no scrap

(iv) It is suitable for medium to large scale production.

(v) Higher equipment cost for large forging presses

(vi) Finishing procedure might be necessary.

Forging may be divided into open and close die forging as discussed below (Fig. 4.17 and Fig. 4.18)

4.7.1 Open Die Forging

Open die forging is the simplest forging process where circular workpiece heated above recrystallization temperature is placed between two flat dies compress with repeated blows of hammer as shown in

Fig. 4.17. Open die forging is also known as *upset forging* which allows the metal to flow without any restriction in the sideward direction thus increases the diameter while the workpiece height is reduced. The workpiece is frequently rotated after each pass, therefore skills of the operator is important in achieving accuracy and dimensions.

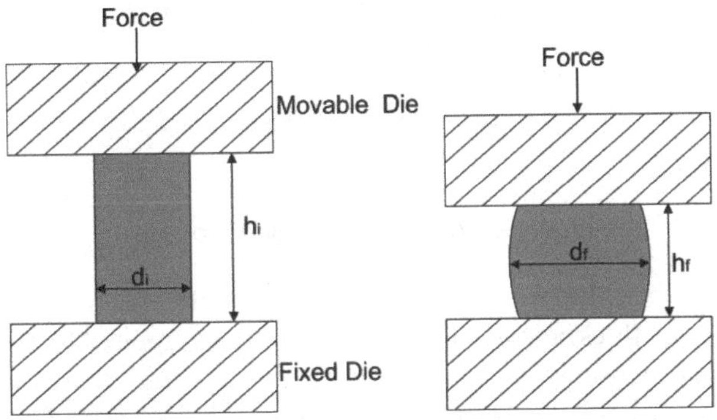

Fig. 4.17: Schematic of open die forging process

4.7.2 Close Die Forging

Close die forging is also called as *impression die forging* is performed in dies which has the impression that will be imparted to the work piece through pressure as shown in Fig. 4.18. The heated workpiece is placed in cavity of bottom die that is close to the shape and size of the final forged object.

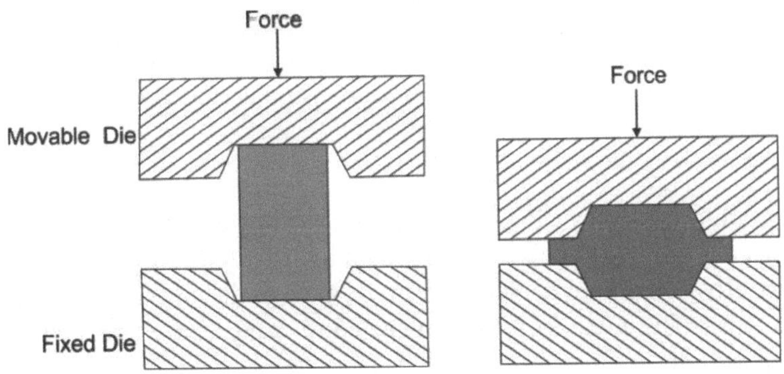

Fig. 4.18: Schematic of close die forging process

The upper die is allowed to move downwards by pneumatic and mechanical press that completely or partially envelops the workpiece. The heated workpiece deforms and fills the cavity, and excess material is squeezed out of the impression to form flash. In case of multi pass operation, several die cavities are used for shape change after each successive stage.

4.8 ROLLING

Rolling is the most widely used industrial process to reduce the thickness of material at lower cost and higher productivity. It is a metal forming process, in which the thickness of the stock of material is reduced by compressive force exerted by a set of rollers moving in opposite direction as shown in Fig. 4.19. After each pass the gap between rollers is reduced till the desired thickness is achieved.

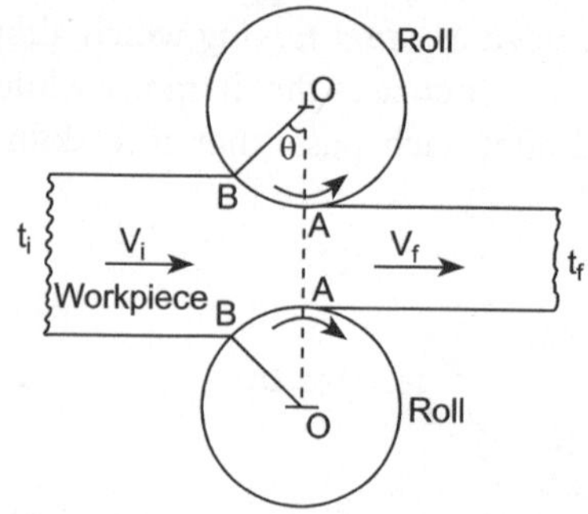

Fig. 4.19: Schematic of rolling process

This process decreases the thickness and increases the length without changing the width of the workpiece. The decrease in the thickness is called a *draft*. Rolling is applied to produce plates, sheets and strips of varying thickness which may be used for further processing such as welding, machining and other metal forming operations. Rolling can be carried out in both hot and cold state. During rolling the initial coarse grain structure is fragmented into thin elongated grains in rolling direction. When hot rolling is carried out, the elongated grains got recrystallized into fine grain structure. The grain growth takes place until recrystallization is over. Steel, magnesium, aluminium, copper, and their alloys are the commonly rolled materials.

Characteristics of rolling process

1. Uniform dimensions of the components can be obtained.
2. It is economical process and suitable for large scale production.
3. Close tolerance is possible for the components in the cold rolling.
4. In hot rolling scale formation leads to poor dimensional accuracy and surface finish.

4.9 DRAWING

Drawing is a metalworking process by which metal is stretched or elongated using tensile force through a mould or die. The metal is inserted through die or mould in order to draw or pull. The wire, rod or tube is stretched to reduce cross-section to achieve a desired shape of die and increases in length. Although the stress applied is tensile, but material is also subjected to compressive stress within the die thereby deforming material plastically.

The process is usually performed at room temperature (cold working) to ensure good surface finish, close tolerances, and improved material properties. But, it may also be performed at higher temperature (hot working) to reduce the drawing force. A wide range of metals on which drawing operation can be performed are aluminium, brass, bronze, cold rolled steel, copper and titanium etc.

Characteristics of metal drawing

1. Seamless products can be produced.
2. The process can be automated easily, thus producing high volumes of products with different cross-sections.
3. Requires less labour for handling purpose

Drawing is classified as deep drawing (sheet metal drawing) and rod drawing, wire and tube drawing, which are discussed below.

4.9.1 Rod and Wire Drawing

It is metal working process in which the work piece in the form of cylindrical bar or rod is pulled through a converging die. The main difference between rod and wire drawing is the size of the starting material. The schematic of rod and wire drawng is shown in Fig. 4.20.

Rods are drawn from rounds of larger diameter, whereas wires are drawn from rods. Wires may be drawn upto 0.05 mm diameter, using a series of drawing dies, and the subsequent die is to have die of smaller diameter than the previous one. Rods and wire have a wide range of industrial applications including shafts, structural components, blanks for bolts and rivets, cables, wires, nails, screws, rivets and many more.

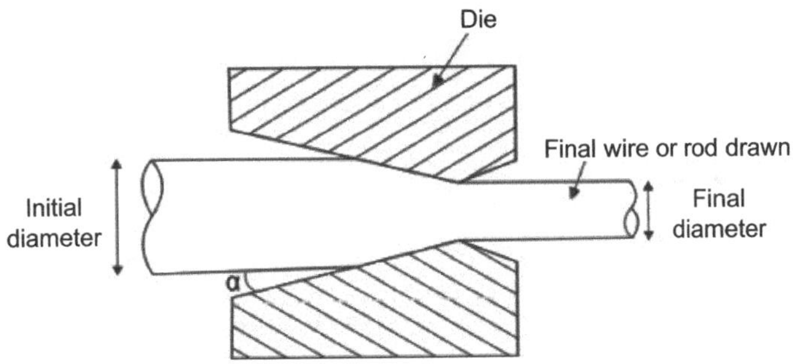

Fig. 4.20: Schematic of rod and wire process

4.9.2 Tube Drawing

It is a metal working process which reduces the diameter or wall thickness of the seamless tubes and pipes by drawing them through a die with mandrel. The mandrel controls the inside diameter and wall thickness of the tube which is shown in Fig. 4.21.

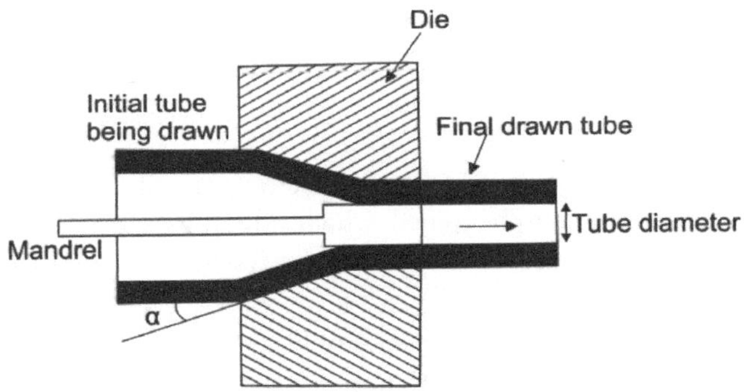

Fig. 4.21: Schematic of tube drawing process

4.9.3 Deep drawing

It is a sheet metal forming process in which the sheet metal blank is radially drawn into concave shaped (cup shaped) forming die by mechanical action of punch. The drawing tools consist of a punch, die and a blank-holder as shown in Fig. 4.22. The blank holder is used to clamp the blank on the die to facilitate controlled material flow into the die radius while the punch pushes the sheet metal.

The blank is subjected to a combined action of tensile and compressive stress. In deep drawing, the blank holder closes after the metal sheet blank has been inserted. It is a shape transformation process, and considered deep drawing when the depth of the part drawn exceeds the diameter. It is used for the fabrication of cooking utensils, containers, sinks, automobile parts, such as panels and gas tanks. The amount of drawing performed is measured with the drawing ratio and is roughly mentioned as the ratio of the diameter of the blank to the diameter of the punch.

Characteristics of deep drawing

1. Parts fabricated are uniform and seamless.
2. Material hardening during deformation may improve the strength of parts formed.
3. Parts drawn have a high degree of consistency throughout forming operation.
4. Process is faste, efficient and produces parts in larger quantity.
5. Process can be automated easily, and produces parts with closed end, thus eliminating the need of secondary processes.

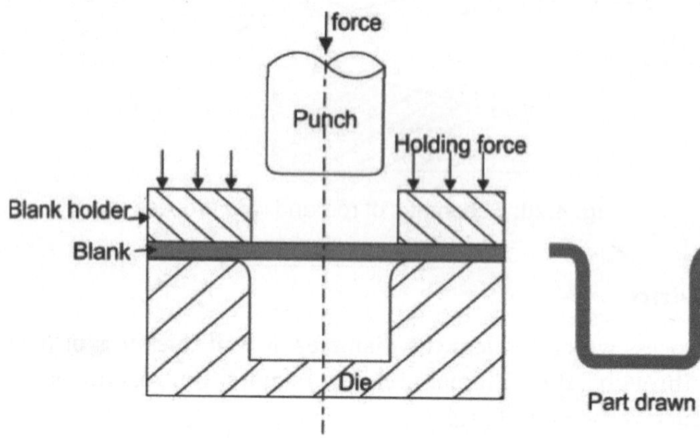

Fig. 4.22: Schematic of deep drawing process

4.10 WELDING

Welding, brazing, soldering, and adhesive bonding of materials are all examples of joining. They form a permanent bond between the pieces to be joined, which cannot be easily detached by applying force. They are mostly used to assemble numerous elements into a system.

Welding is a metal joining procedure in which two or more pieces are bonded or merged at their contacting surfaces using heat or pressure. Sometimes welding is done solely through the application of heat, with no pressure applied; in other cases, both heat and pressure are applied; and in still other cases, only pressure is applied, with no external heat applied. In some welding processes, a filler material is added to facilitate coalescence (joining)

4.10.1 Types of Welding

Welding processes can be broadly classified into (i) Fusion (non-pressure) welding and (ii) Solid state welding (pressure welding).

Fusion welding: The base metals are melted during the fusion-welding process by applying heat. To speed up the welding process and strengthen the welded joint, filler metal is frequently added to the molten pool throughout numerous fusion welding methods. A fusion welding process is referred to as *autogenous weld* when filler metal is not employed. The most common types include laser welding, electron beam welding, gas welding, resistance welding, and arc welding.

Solid state welding: This technique involves applying pressure, either alone or in conjunction with heat. The temperature in the operation is lower than the melting point of the metals being welded even when heat is applied (unlike in fusion welding). There is no filler metal used. The three main types are ultrasonic welding, friction welding, and diffusion welding.

4.10.2 Classification of Welding Processes

The field of welding is very broad. Several welding processes are available to meet the demand for joining a wide range of materials in numerous ways; each has its own special qualities, advantages, and disadvantages in comparison to others. More than 100 distinct welding techniques are commonly utilised nowadays. It's challenging to put them all into one group. Any succinct classification will have certain gaps where a few processes from a certain group may be missing. This opens the door for numerous classifications of welding techniques, which are discussed below.

Arc welding

Arc welding is a welding procedure that uses electricity to generate enough heat to melt metal, and the melted metal pool when cools, result in a binding of the metals. It is a type of welding in which an electric arc is created between a metal stick ("electrode") and the base material to melt the metals at the point of contact using a welding power supply. Arc welders can employ either direct current (DC) or alternating current (AC), as well as consumable or non-consumable electrodes. The arc welding can be of following types:

- Gas tungsten arc welding (TIG) or (GTAW)
- Gas metal arc welding (MIG) or (GMAW)
- Shielded metal arc welding (SMAW)
- Submerged arc welding
- Plasma arc welding
- Flux cored arc welding (FCAW)

Resistance welding

Resistance welding, also known as electric resistance welding (ERW), is a method of joining metals by applying pressure and carrying a high electric current through the metal combination to heat up the welding joint and melt the metals, fusing them together. Resistance welding can be done in following ways:

- Spot welding
- Seam welding
- Projection welding
- Resistance butt welding

Gas welding

One of the earliest types of heat-based welding is gas welding. By heating the ends of the materials, it fuses metals together by causing them to melt. This is made possible by the large number of gases that burn at very high temperatures. Gas welding allows for the fusing of both nonferrous (not containing iron) and ferrous metals, plus no requirement of electricity to get the welds going. The gas welding processes are as follows:

- Oxy-acetylene welding
- Oxy-hydrogen welding
- Air -acetylene welding
- Pressure Gas welding

Thermo chemical welding process

Thermo chemical welding is a type of fusion welding in which the metals are joined together using the heat generated by exothermic chemical processes. For connecting metals, there are primarily two types of thermo chemical welding processes.

- Thermite welding
- Atomic hydrogen welding

Radiant energy welding process

When a stream of electrons or an electro-magnetic radiation beam impacts the workpiece during a radiant energy welding process, heat is produced at the point of welding. Welding can be done in a vacuum or at low pressure and the major types are:

- Electron beam welding
- Laser beam welding

Although, there are many ways in which we can perform the welding processes depending upon the need and application, but there are some welding processes which are most widely used all over the world. These processes are discussed in details in the following section.

4.11 ARC WELDING PROCESS

It is a fusion welding procedure in which metals are melted and joined using the heat energy created by the arc between the work and an electrode. When the electrode makes contact with the work and then quickly separates to maintain the distance, an electric arc is formed. This arc produces a temperature of 5500°C and above. This temperature is hot enough to melt most metals. In the weld zone, the molten metal, which consists of base metal and filler, solidifies. The power source moves along the weld line to produce a seam weld. The schematic of arc welding is shown in Fig. 4.23.

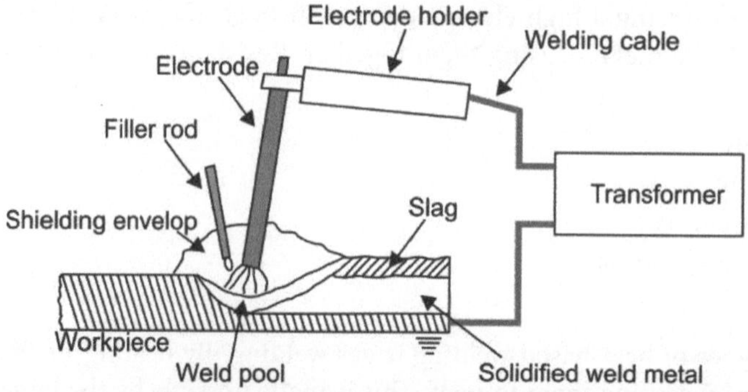

Fig 4.23: Schematic of arc welding process.

An electric current from a welding power supply, either alternating current or direct current is utilised to generate an electric arc between the electrode and the metals to be connected. As the weld is formed, the flux coating of the electrode disintegrates, emitting vapours that act as a shielding gas and forming a layer of slag that protects the weld region from ambient contamination. When current is carried between two metallic electrodes separated by a small distance, an electric arc is formed. When an electrode initially contacts the plate, a huge short circuit is formed, and current continues to flow across the air gap in the form of a spark. Air is ionised as a result of this spark and a very high temperature is generated. The addition of electrode material gives volume and strength to the welded joint.

4.12 ELECTRODES

There are two kinds of electrodes used: consumable and non-consumable.

Consumable electrodes

These electrodes are available in form of rod or wire with a length of 200 to 450 mm and a diameter of less than 10 mm in arc welding. During the welding process, the electrode is consumed by the arc and added to the weld joint as filler metal. Consumable electrodes will be changed on a regular basis as they are consumed during welding process. This is a disadvantage to the welder and lowers the production rate.

Non-consumable electrodes

During arc welding, the electrodes are not consumed. Despite this, some depletion occurs as a result of vaporisation. Here, the filler metal must be provided as a separate wire fed into the weld pool. The comparison of consumable and non-consumable electrodes is given in Table 4.3.

Table 4.3: Comparison of consumable and non-consumable electrodes

Consumable electrode	Non-consumable electrode
During welding, a consumable electrode itself melts and deposits on the weld bead.	The weld bead is not melted or deposited by a non-consumable electrode. It is not damaged during the welding process.
A consumable electrode fills the root gap by acting as filler and supplying the necessary filler material.	Electrode that is not consumable does not provide filling. Filler material must therefore be supplied separately.
A sizeable percentage of the electrode is integrated into the weld bead after welding.	The electrode is unaffected after welding (except a small erosion).
Since filler is applied naturally, this sort of electrode does not permit autogenous welding. It can be utilised for both homogeneous and heterogeneous welding processes.	It permits autogenous, homogeneous, and heterogeneous welding in all three of its forms.
To preserve chemical compatibility, the electrode material must be chosen based on the parent components.	Because non-consumable electrodes do not operate as fillers, the electrode material is unrelated to the parent materials to be welded.
Since welding consumes electrode material, it is typically preferred to replace the electrode frequently. However, the electrode size and filler deposition rate affect the frequency of replacement.	A non-consumable electrode has a longer lifespan because welding does not use it up. Additionally, frequent replacement is not desirable (it helps improving productivity).

Consumable electrode	Non-consumable electrode
Arc welding methods that use a consumable electrode include: Shielded Metal Arc Welding (SMAW) Gas Metal Arc Welding (GMAW): MIG and MAG Flux-cored arc welding (FCAW) Submerged arc welding (SAW) Electroslag welding (ESW) Electro-gas welding (EGW)	Arc welding methods that uses a non-consumable electrode include: Tungsten Inert Gas welding (TIG) or Gas Tungsten Arc Welding (GTAW) Atomic Hydrogen Welding (AHW) Carbon Arc Welding (CAW))

4.12.1 Electrode Coating

Electrode coating is covered in a layer of 1 to 3mm thickness with a relatively high-quality coating. The following are the primary roles of electrode coating:

- Improving arc stability by supplying particular chemicals that have this ability by ionising the arc path
- Create a protective gaseous atmosphere to prevent the molten metal from absorbing oxygen, hydrogen, and nitrogen.
- Provide protective slag over hot metals
- Provide flux, which aids in the removal of oxides and other impurities from molten metals
- Reduce weld metal spatter - when the covering burns off faster than the core.
- Functions as a deoxidizer
- Reduce the weld's cooling rate (due to the protective layer of slag) to prevent hardening.
- Coatings typically act as electrical insulators, making it impossible to employ electrodes in confined spaces like grooves, etc.

4.13 GAS WELDING

This procedure involves burning a mixture of different fuels and oxygen to perform welding e.g., oxyacetylene welding.

4.13.1 Oxyacetylene Gas Welding

Welding is carried out in this case by a flame created by the combustion of oxygen and acetylene. The flame is produced by a torch. A flux-coated filler rod is occasionally used to prevent oxidation and provide a better joint. The schematic of gas welding is given in Fig. 4.24.

The chemical interaction between oxygen and acetylene occurs in two steps, as illustrated below.

$$C_2H_2 + O_2 = 2CO + H_2 + \text{Heat (First stage; inner core)} \quad 4.3$$

The results of the first reaction are flammable, and the results of the second reaction are,

$$2CO + H_2 + 1.5O_2 = 2CO_2 + H_2O + \text{Heat (second stage; outer envelope)} \quad 4.4$$

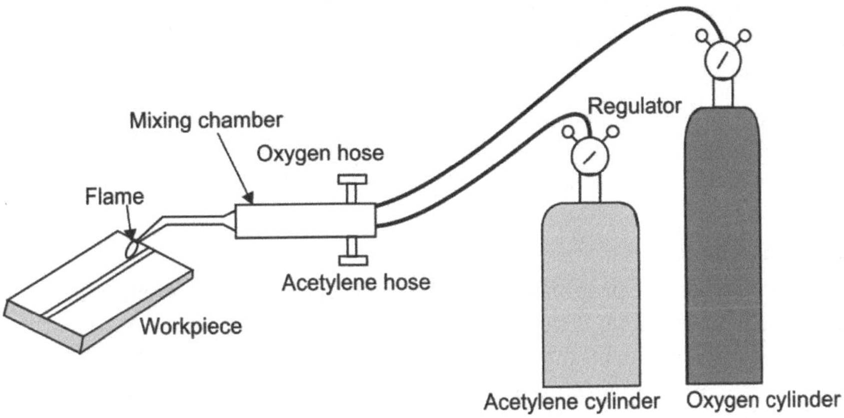

Fig 4.24: Schematic of gas welding process.

4.13.2 Gas Welding Equipments

The main equipment used in performing gas welding are discussed below (Refer Fig. 4.24)

Oxygen Cylinder: is composed of steel and has a capacity range of 2.25 to 6.3 m^3 painted in black. At 21°C, the cylinders are filled with oxygen at a density of roughly 150 kg/cm^2. It also includes a safety valve. A wheel that operates a valve opens and closes the cylinder. To protect the valve, a protector cap is installed on the top of a cylinder.

Acetylene Cylinder: Acetylene cylinders are likewise composed of steel that has been coated maroon. At a pressure of 18-20 kg/cm^2, gas is filled. The cylinder has a capacity of around 10m^3. The cylinder houses the regulator valve and the safety valve. On the bottom of the cylinder, there are also safety plugs. The acetylene filled into the cylinder is dissolved in acetone.

Hoses: Oxygen and acetylene are supplied from the oxygen and acetylene cylinders to the welding torch via hoses in oxy-acetylene gas welding. The colour coding is used to identify the pipe that is transporting the gas. The blue hose transports oxygen, whereas the red hose transports acetylene.

Regulators: The working pressure of oxygen is 1 kg/cm^2, while that of acetylene is 0.15 kg/cm^2. The working pressure varies according to the thickness of the welded work components.

Other equipment are pressure gauges, welding torch, check valve and non-return valve.

4.13.3 Types of Flames

Flame plays an important part in the formation of a weld joint, and the weld qualities are heavily dependent on it. There are three forms of oxy-acetylene flames, which are commonly referred to as:

1. Neutral
2. Carburizing (or "excess acetylene")
3. Oxidizing (or "excess oxygen")

1. Neutral flame

The type of flame created is determined by the oxygen-to-acetylene ratio in the gas mixture that exits the torch tip. It is formed when the oxygen-to-acetylene ratio in the mixture leaving the torch is nearly equal. It is called *neutral* because it has no chemical effect on the metal being welded. It will neither oxidise the weld metal, nor will it raise the carbon content of the weld metal. The addition of a little extra oxygen results in a dazzling whitish cone surrounded by a translucent blue envelope, which is known as a *neutral flame* (it has a balance of fuel gas and oxygen) (3500 °C).

When oxygen and acetylene are blended in a 1:1 ratio, a neutral flame is formed, as illustrated in the Fig. 4.25 (a). The exterior shell has a temperature of 1260°C, whereas the inner core has a temperature of approximately 3500°C. It is mainly used for welding steels, aluminium, copper and cast iron.

2. **Carburising Flame**

   It is produced when the proportion of acetylene in the mixture is greater than that required to produce the neutral flame. When used on steel, it raises the carbon content of the weld metal. When oxygen is introduced into the flame, it rapidly transforms into a long white inner area (feather) surrounded by a translucent blue envelope, which is known as a *carburizing flame* (3000 °C) is illustrated in Fig. 4.25 (b). The carburising flame is advantageous for welding high carbon steel and can endure alloys such as nickel and monel.

3. **Oxidizing Flame**

   It develops from the combustion of a mixture containing more oxygen than is required for a neutral flame. Some of the metal being welded will oxidise or burn. When more oxygen is introduced, the cone darkens and gets more pointed, while the envelope shortens and becomes more furious, is shown in Fig 4.25 (c). The hottest temperature is around 3400 °C. Oxidation welding flames are commonly used for welding metals like zinc, copper and manganese steel. It is also used in the welding and brazing of brass.

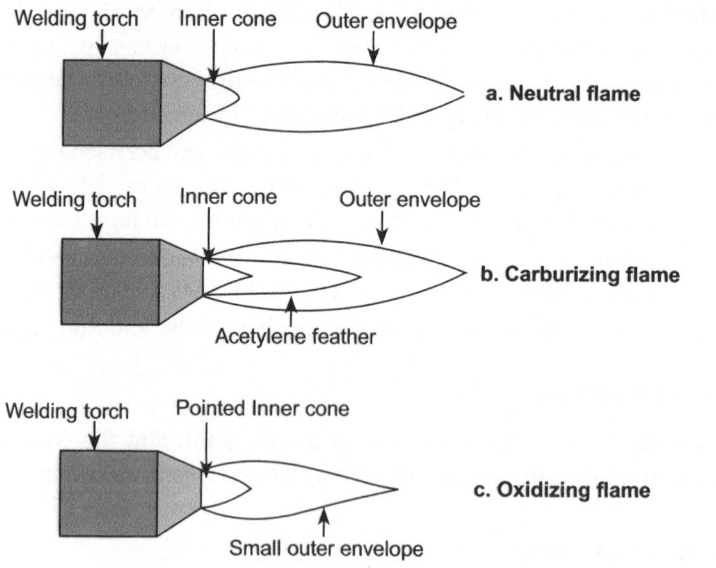

Fig 4.25: Types of flames (a) Neutral, (b) Carburizing, (c) Oxidizing flame

4.14 SOLDERING

Soldering is a method of attaching metal pieces to establish a mechanical or electrical bond. It commonly employs a low melting point metal alloy (solder), which is melted and applied to the metal pieces to be joined, where it links to the metal parts and establishes a connection when the solder hardens. It differs from welding in that the items being linked are not heated and are usually not the same substance as the solder.

Soldering is a typical method of connecting electrical components and wires. Although it can be used for plumbing, sheet metal fabrication, and automotive radiator repair, the procedures and materials are not the same as those used for electrical work. The applications of soldering include printed circuit board (PCB) manufacture, pipe joining (copper pipe) and jewellery manufacture.

Flux: Both solder wire and solder paste contain flux for electrical soldering. This helps to clean the surfaces being soldered and prevents the hot solder from oxidising. The flux composition varies depending on whether it is a paste or wire, leaded or unleaded solder. Rosin is a flux that is commonly used in solder wire.

When the solder is heated, most fluxes emit vapours that are likely dangerous to your health. A well-ventilated workspace may suffice for occasional soldering, but for prolonged or recurrent exposure, a fume extractor should be utilised. Solder flux can also cause solder to spatter, therefore wear eye protection when soldering.

4.15 BRAZING

Brazing is a metal bonding method that requires the use of heat and the inclusion of a filler metal. This filler metal, which has a lower melting point than the joining metals, is either pre-placed or supplied into the joint while the parts are heated. Capillary action allows the filler to flow into the joint in brazing pieces with narrow clearances.

The molten filler used for brazing has a temperature that exceeds 430° C. The filler metal remains below that temperature in a related procedure known as soldering. Brazed joints are typically more durable than soldered connections.

Brazing may be done on almost any metal, and the variety of accessible brazing alloys is expanding as new alloys and service requirements are released. Heating with a torch in air is sufficient if the joint is thoroughly fluxed. Comparison between welding, soldering and brazing is mentioned in Table 4.4.

Table 4.4: Comparison between Welding, soldering and brazing

Welding	Soldering	Brazing
The welding zone must be heated to 3800 °C.	Temperatures of up to 450 °C are required.	It has the potential to reach 600 °C.
These are the most powerful joints utilised to support the weight. The strength of a welded junction may exceed the strength of the base metal.	These are the weakest of the three joints. Not designed to support the weight. In general, it is used to make electrical connections.	These are more powerful than soldering but less powerful than welding. These can support the strain to some extent.
Workpieces to be bonded must be heated until they reach their melting point.	There is no need to heat the workpieces.	The workpieces are heated but not melted.
Because of heating and cooling, the mechanical characteristics of the base metal may vary at the junction.	There has been no change in the mechanica1properties after joining.	May alter the mechanical properties of the joint, however the effect is minor

Welding	Soldering	Brazing
There is a large cost involved, as well as a high level of expertise necessary.	The costs involved are cheap and less skill is required.	The cost and skill requirements are in the middle.
In most cases, heat treatment is essential to remove the unwanted effects of welding.	There is no need for heat treatment.	There is no need for heat treatment.

UNIT SUMMARY

- Material science is an interdisciplinary field related to discovery, design and development of new materials particularly solids and understanding their properties and application.
- Engineering materials are classified as; metals and alloys, polymers, ceramic and composites.
- The different properties to describe the material behaviour are; dimensional, physical, chemical and mechanical properties.
- The material behaviour is important for mechanical engineering and different mechanical properties are; Strength, elasticity, plasticity, ductility, malleability, toughness, hardness, brittleness etc.
- In casting, the molten metal is poured into a mould cavity that takes the form of the finished part, and is allowed to cool and solidify into the desired shape.
- The pattern is the replica of casting to be made. The different pattern allowances are; machining, shrinkage, draft, distortion and shake allowance.
- Various mould sand properties are; porosity, adhesiveness, cohesiveness, flowability, collapsibility and refractoriness.
- In cold working, the plastic deformation of the metals is carried out at room temperature or slightly above it, but below the recrystallization temperature
- In hot working, the plastic deformation of the metals is carried out at temperature above their recrystallization temperature.
- Extrusion is a metal forming process in which billet of ductile material undergoes plastic deformation by the application of compressive force causing the metal to pass through a die of desired shape. The different types are direct and indrect extrusion.
- Forging is a deformation process in which the metal is shaped between two dies by applying a compressive force, either impact load or gradual load.
- In rolling, the thickness of the stock of the material is reduced by compressive force exerted by a set of rollers moving in opposite direction.
- Drawing is a metalworking process by which metal is stretched or elongated using tensile force through a mould or die. Rod and wire drawing, tube drawing and deep drawing are different types of drawing process.
- Welding is a metal joining procedure in which two or more pieces are bonded or merged at their contacting surfaces using heat or pressure. Different types of welding are; arc welding and gas welding.

EXERCISE

Multiple Choice Questions

1. Metals are superior to ceramics in which of the following properties.
   (a) Hardness
   (b) Ductility
   (c) Toughness
   (d) Yield Strength
2. Ceramics are excellent refractory materials. A material is said to be refractory if it meets the following criteria:
   (a) Resists the flow of electric current
   (b) Has high refractive index
   (c) Retains its strength at high temperatures
   (d) All of the above
3. Which of the following statements about polymers is correct.
   (a) They have very high molecular mass
   (b) They do not have a linear stress-strain curve
   (c) They have high strength to mass ratio
   (d) All of the above
4. The property of a material to resist any elastic deformation is termed as:
   (a) Stiffness (b) Hardness
   (c) Malleability (d) Strength
5. Permanent material deformation with regard to time caused by constant load and fluctuating temperature is referred to as:
   (a) Elasticity (b) Isotropy
   (c) Hardness (d) Creep
6. A material's resistance to any external force is referred to as:
   (a) Stiffness (b) Malleability
   (c) Strength (d) Hardness
7. Which type of metal is used in casting process.
   (a) Liquid (b) Solid
   (c) Gas (d) Plastic
8. The casting can be defined as pouring of molten metal into a mold and taking it out after it becomes vapor.
   (a) True (b) False
9. Chaplets are metal inserts used to provide support and hold the core at proper location in the mould cavity.
   (a) True (b) False

10. In the molding sands, which of the following are the major properties.
    (a) Permeability
    (b) Sand Grains
    (c) Size of the grains
    (d) Permeability, sand grains and size of the grains
11. ____________ pattern could fit entirely in the cope or in the drag.
    (a) Single piece
    (b) Match plate
    (c) Two piece
    (d) Gated Pattern
12. ________ is used to make the large castings of axi-symmetrical shapes.
    (a) Loose piece pattern
    (b) Skeleton pattern
    (c) Sweep pattern
    (d) Match plate pattern
13. Uncontrolled permeability in the green sand mould leads to ____________
    (a) Blow Holes
    (b) Chills
    (c) Shrinkage
    (d) Tearing
14. The allowance provided to take care of the contraction of casting is known as.
    (a) Draft allowance
    (b) Shrinkage allowance
    (c) Machining allowance
    (d) Shake allowance
15. In order to reduce the chances of damage due to withdrawing of a pattern from the mould is done by giving ________
    (a) Draft allowance
    (b) Shrinkage allowance
    (c) Distortion allowance
    (d) Shake allowance
16. Which of the following is not improved by cold working of metals.
    (a) Hardness
    (b) Toughness
    (c) Surface finish
    (d) Corrosion resistance
17. In hot working of material, the continuous reformation of grains can be possible in material.
    (a) True
    (b) False
18. Tubes for shaving cream and tooth paste are made by.
    (a) Forward extrusion
    (b) Backward extrusion
    (c) Impact extrusion
    (d) All of the above
19. Which characteristic of material is used in forging process.
    (a) Characteristics of elasticity of material
    (b) Characteristics of ductility of material
    (c) Characteristics of plasticity of material
    (d) None of the above
20. Good surface finish and better dimensional accuracy can be achieved in.
    (a) Cold working process
    (b) Hot working process
    (c) Both a and b
    (d) None of the above

21. Electric arc welding uses.
    (a) DC Supply
    (b) AC supply
    (c) Both DC & AC supply
    (d) None
22. Types of electrodes used in arc welding.
    (a) Consumable electrodes
    (b) Non-consumable electrodes
    (c) Both a and b
    (d) None
23. Oxidizing flame is a flame which is obtained by supplying.
    (a) Less volume of acetylene and more volume of oxygen
    (b) More volume of acetylene and less volume of oxygen
    (c) Equal volume of acetylene and oxygen
    (d) None of the above
24. The process of joining two pieces in which a non-ferrous alloy is introduced in liquid state between the piece of metals and allowed to solidify, is known as:
    (a) Welding
    (b) Brazing
    (c) Lancing
    (d) Riveting
25. A soldering iron 'bit' is made of________
    (a) Brass
    (b) Tin
    (c) Steel
    (d) Copper

Answers for Multiple Choice Questions							
1. (b)	2. (c)	3. (d)	4. (a)	5. (d)	6. (c)	7. (a)	8. (b)
9. (a)	10. (d)	11. (a)	12. (c)	13. (a)	14. (b)	15. (a)	16. (d)
17. (c)	18. (c)	19. (a)	20. (c)	21. (c)	22. (c)	23. (a)	24. (b)
25. (d)							

Short and Long Answer Type

1. Explain various engineering materials and their applications.
2. Differentiate between elasticity and plasticity.
3. Draw stress strain curve for mild materials.
4. What are the steps involved in preparation of casting?
5. Differentiate between pattern and casting
6. Discuss briefly various pattern materials.
7. In what conditions split pattern are advantageous?
8. Write the application of core prints and chaplets.
9. List various types of allowances provided on pattern.
10. Differentiate between green and dry sand moulding.
11. Enumerate various types of commonly used patterns.
12. Differentiate between forward and backward extrusion.
13. Differentiate between hot and cold working.

14. Differentiate between open and close die forging.
15. Define rolling process?
16. Differentiate between solid state and fusion welding.
17. Differentiate between gas welding flames.
18. Explain the functions of electrode.
19. Explain various material properties.
20. What properties are desirable for moulding sand to produce sound castings?
21. What are the major limitations of sand casting and how are they overcome?
22. Explain with neat sketch any five types of patterns.
23. Distinguish between moulding sand, facing sand and baking sand.
24. Discuss briefly why draft allowances are important for patterns?
25. How are sand moulding method classified?
26. Explain briefly with sketch various pattern allowances.
27. List advantages, disadvantages and applications of extrusion process.
28. Explain briefly the following processes;
    (i) Wire drawing
    (ii) Tube drawing
    (iii) Deep drawing
29. Differentiate between welding, soldering and brazing.
30. Explain briefly following processes with neat sketch;
    (i) Rolling
    (ii) Forging
    (iii) Extrusion
31. Differentiate between consumable and non-consumable electrodes.
32. Give the classification of welding process.
33. Explain the working principal of arc welding process, and its applications.
34. Write short note on electrodes used in arc welding.
35. With the help of neat sketches, explain gas welding flame and their applications.
36. Explain soldering and brazing process and their applications.

PRACTICAL

1. Material identification of, say, 10 common items kept in a box.
2. To prepare mould of a given pattern requiring core.
3. Prepare a funnel with metal sheet.
4. Hands-on practice on arc welding.
5. Hands-on practice on gas welding.

KNOW MORE

- The mechanism of plastic deformation of materials is slip where atoms move by many interatomic distances in one crystal plane over the atoms of another crystal planes.
- The different advanced materials have been developed to obtain superior properties and performance includes; semiconductors, biomaterials, smart materials, and nano materials.
- Dies used in metal forming are; simple, compound, progressive and transfer dies.
- Press working is a chipless manufacturing process by which different components are manufactured through sheet metal.
- The advanced casting techniques are; permanent moulding, die casting, centrifugal casting, investment casting, shell moulding etc.
- Resistance welding in process of joining two metallic pieces pressed together by applying high electric current, which develops heat around the point of contact.
- The allied welding processes were developed for different applications such as; thermit welding, laser beam welding, electron beam welding, ultrasonic welding, friction welding, explosive welding etc.

REFERENCES AND SUGGESTED READINGS

1. A. Ghosh and A. K. Mallik. Manufacturing Science, Affiliated East West Press, Delhi, India, 2019
2. M.P. Poonia and S.C. Sharma. Basic Mechanical Engineering – Khanna Publishing House, New Delhi, India, 2005.
3. S. Kalpakjian and S. R. Schmid. Manufacturing Processes for Engineering Materials, Pearson Education, New Delhi, India, 2009
4. P. N. Rao. Manufacturing Technology (Vol. 2), McGraw Hill, New Delhi, India, 2015.
5. B. S. Raghuvanshi. Workshop Technology (Vol. 2), Dhanpath Rai and Sons, New Delhi, India,2003.
6. M. P. Groover. Fundamentals of Modern Manufacturing: Materials, Processes and systems, John Wiley and Sons, USA, 2010.
7. R. K. Rajput. Manufacturing Technology, Laxmi Publications, New Delhi, India, 2007.

5 Machine Tools and Machining Processes

UNIT SPECIFICS

Through this unit we have discussed the following aspects:

- Different machine tools used in workshop, like lathe, shaper, planer, milling and drilling;
- Different types of lathes and operations performed with lathe;
- Milling machine, their types and operations performed with milling machine;
- Shaper and planer machine, their mechanism and operations performed with them;
- Drilling machine types and operations performed with them;
- Grinding machine and operations performed.

The practical applications of the topics are discussed for generating further curiosity and creativity as well as improving problem solving capacity.

Besides giving a large number of multiple choice questions as well as questions of short and long answer types marked in two categories following lower and higher order of Bloom's taxonomy. Assignments through a number of numerical problems, a list of references and suggested readings are given in the unit so that one can go through them for practice. It is important to note that for getting more information on various topics of interest some QR codes have been provided in different sections which can be scanned for relevant supportive knowledge.

After the related practical, based on the content, there is a "Know More" section. This section has been carefully designed so that the supplementary information provided in this part becomes beneficial for the users of the book. This section mainly highlights the initial activity, examples of some interesting facts, analogy, history of the development of the subject focusing the salient observations and finding, timelines starting from the development of the concerned topics up to the recent time, applications of the subject matter for our day-to-day real life or/and industrial applications on variety of aspects, and finally inquisitiveness and curiosity topics of the unit.

RATIONALE

This fundamental unit of machine tools and machining processes helps students to get a primary idea about different machine tools used in workshop, their materials, parts, types, specifications nd mechanism of operation. Students will also learn different operations performed and types of surfaces generated with these machine tools and their suitability for any particular operations. This knowledge is important for an engineer to select a most suitable manufacturing process out of many alternatives to produce an object with accurate dimensions for a particular application as per the need of customer. All these are discussed at length to develop the basic and fundamental knowledge of the subject.

Manufacturing is the process of converting raw materials or parts into finished products with the use of man, machine and material and adds value to the product. All the machine tools like; lathe, shaper, milling and drilling makes use of different cutting tools to remove material in form of chips. With this, the raw material is converted into finished products or the part is manufactured with required shape, dimension, accuracy and surface finish. Whereas, grinding is an abrasive machining process uses fine abrasive particles and is basically a finishing process. Fundamentals of basic machine tools and the operations performed are discussed in detail to develop an understanding of subject. This subject combines manufacturing with materials science, to design, analyse, manufacture, and maintain mechanical systems. The practical applications of subject are related to the selection of proper machine tool to machine different materials through different operations used for different engineering applications.

UNIT OUTCOMES

List of outcomes of this unit is as follows:

U5-O1: To describe importance of machine tools, their characteristics, classification and types of lathes, different parts, operations performed and specifications of lathe.

U5-O2: To define working principles of shaper and planer machine, quick return mechanism, their parts and specifications.

U5-O3: To understand the working principle of milling machine and drilling machine, types, their parts and specification and operations performed.

U5-O4: To define the grinding machine, working, grinding wheel, selection of abrasive particles and operations performed.

U5-05: With this knowledge students will be able to select a particular machine tool to machine a defined material.

Unit-5 Outcomes	EXPECTED MAPPING WITH COURSE OUTCOMES (1- Weak Correlation; 2- Medium correlation; 3- Strong Correlation)					
	CO-1	CO-2	CO-3	CO-4	CO-5	CO-6
U5-O1	3	-	-	-	3	1
U5-O2	3	-	-	-	3	1
U5-O3	3	1	1	-	3	2
U5-O4	3	-	1	-	3	2
U5-O5	3	-	-	-	3	2

5.1 INTRODUCTION

Machine tools are power driven devices used to hold and support the cutting tool or workpiece to remove the unwanted material from the workpiece in order to produce required shape and size. In this process of material removal, motion is provided to the cutting tool or workpiece or both relative to each other through power driven devices. They form or shape the different parts by machining the material in the form of chips as shown in Fig. 5.1. The shape generated depends upon the characteristics of machine tool used. Earlier hand operated tools were used to cut and shape the material for the manufacturing of utensils, weapons, furniture, wagons etc.

With the industrial revolution, these products were manufactured by power driven machine tools. The commonly used machine tools are lathe, shaper, planer, drilling and milling generally available in any machine shop. *Machine tools are considered as mother tools because these tools will be able to produce themselves.* These machines have greater flexibility in functioning, as they have various attachments to improve it and thus improves productivity.

With industrial automation, semi- automatics and automatic machine tools were developed, which are capable of producing dimensionally accurate and interchangeable parts in large quantities. The numeric control (NC) machines were having tape or punched cards to control the motion of machine tools. Later, computers have been installed to execute the part program to control and command the machine tools. This enables it to changeover to variety of parts by changing the control program and software, to reproduce the same part for any number of times with higher efficiency and reliability with minimum human intervention.

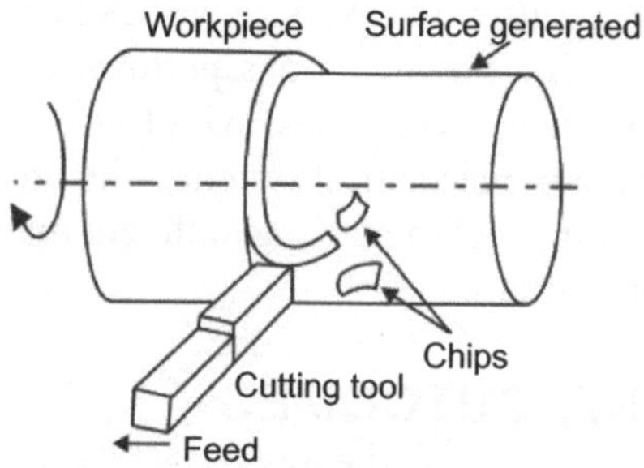

Fig. 5.1: Mechanism of Material removal

A variety of machine tools are available to perform the desired operations of machining to produce the part. The selection of any machine tool depends on;

- The workpiece material
- Type of cutting operations
- Number of parts to be machined
- Degree of accuracy and surface finish required.

5.2 CHARACTERISTICS OF MACHINE TOOLS

All type of machine tools has both the work holding and tool holding devices. The relative motion between them will decide the cutting speed and how fast the metal is removed.

The *feed rate* is the distance travelled by the tool in one complete revolution of workpiece, whereas the distance travelled by tool bit into the workpiece is called *depth of cut*. The cutting speed and feed rate will decide material removal rate, power requirements and surface finish and accuracy of part generated. A machine tool shapes the component by either generating or forming techniques. In generating, a profile is generated by manipulating the relative motion between the cutting tool and work piece. The shape developed in workpiece will not be identical to the cutting tool.

In case of centre lathe, a cutting tool is moved along the axis of rotating workpiece, thereafter a cylindrical surface is generated and this operation is called *turning*. Whereas, in forming, the contour or shape of the tool is impressed into the workpiece. The accuracy of the shape formed depends on the accuracy of form tools.

5.2.1 Classification of Machine Tool

Machine tools are classified on the basis of machining method, type and size of object, surface generated or purpose, accuracy required and methodology adopted to control. They are classified as follows;

(i) **General purpose machine tools**: These machine tools are designed to perform a large variety of machining operations with help of attachments on variety of workpieces having different shapes and sizes.

(1) Lathe machine
(2) Shaper machine
(3) Milling machine
(4) Drilling machine
(5) Planer machine
(6) Sawing machine

(ii) Special purpose machine tools: These machine tools are designed to perform some of the specified operation for mass production where a machine has capability to produce identical parts.

(1) Capstan and turret lathe
(2) Cam shaft grinder
(3) Gear cutting
(4) Copying lathe
(5) Thread rolling machines
(6) Rotary machines

5.3 LATHE MACHINE

A lathe is an oldest and versatile machine tool available in all machine shops. It holds and supports the workpiece between two rigid supports i.e., chuck *(live center)* which revolves and tail stock *(dead center)*. A cutting tool is mounted on a tool post, which is held against the revolving work.

Lathe is used to remove the extra material from the workpiece in the form of chips to achieve the required shape and size by rotating the workpiece against the cutting tool. The machining is performed with the cutting tool fed parallel or at right angles to the axis of the rotation of the workpiece.

5.3.1 Main Parts of Lathe

The main parts of lathe as illustrated in Fig 5.2 are discussed below;

- **Bed:** It is the main part of lathe and acts as the base on which all other parts of the machine are mounted. Bed is made up of cast iron and has a box section column with cross ribs. It has two longitudinal guide ways precisely made to ensure proper alignment of other parts. The tailstock slides on guideways maintaining its alignment with headstock. The bed should have sufficient strength to withstand various cutting forces during operation.
- **Headstock:** It works as a housing for gears box, pulleys and spindles and transmits the power to different parts of lathe. It is fixed on the left end side of the bed and is called as *live center.* The spindle is made hollow with Morse taper to fix live centre in spindle hole to locate the workpiece axis. The gear box provides the wide range of speeds to machine different materials. Head stock is also threaded to facilitate the mounting of chuck which hold the workpiece
- **Tailstock:** It is fixed on the right end side of the lathe bed in an alignment with live centre. It slides on the guideways to support and accommodate different lengths of workpiece by the use of centres. It may also hold the tools like, drill bit, tap and reamer etc. The dead centre is attached in front of the spindle, which has a tapered hole. A hand wheel is provided for the lateral movement and an adjusting screw to fix the position.
- **Carriage:** The carriage is mounted between headstock and tailstock with the function of supporting, guiding and feeding the tool against the workpiece during machining. It supports different parts; saddle to support cross slide, compound slide, tool post and apron. Cross- slide provides cross feed to the cutting tool perpendicular to the axis of workpiece and tool post for holding the cutting tools. The compound slide supports the tool post and its base is graduated in degrees to swivel in desired angles. This arrangement is used to make angular cuts, short tapers as well as to position the tool. Apron has gears and clutches to transmit motion from

feed rod to the carriage through lead screw. The lead screw is engaged with carriage by the use of half nut during thread cutting operation.

- **Feed mechanism:** It provides longitudinal, cross and angular feed to the cutting tools. Carriage is the part of feed mechanism. It consists of feed reverse lever, gear box, lead screw, feed rod and apron.

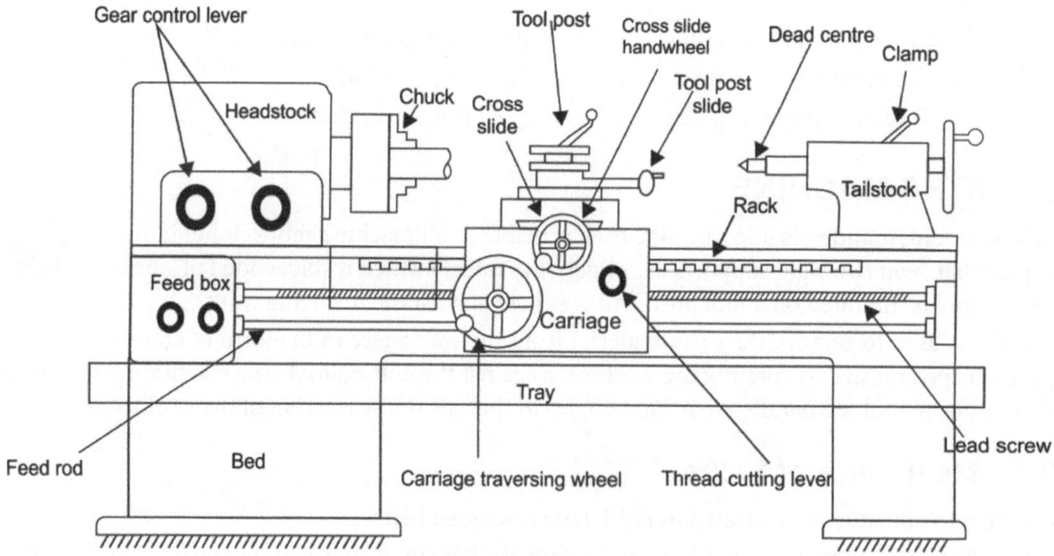

Fig. 5.2: Schematic view of centre lathe

- **Chuck:** Chuck is the device used to hold the workpiece of shorter length and larger diameter which is difficult to hold between centres. It is attached to the spindle nose of the headstock. Chuck rotates with the spindle, which rotates the workpiece. The different types of chuck used are;
  - Three jaw chuck
  - Four jaw chuck
  - Combination chuck
  - Magnetic chuck
  - Air or hydraulic chuck
  - Collet

Out of all these chucks, *three jaw chuck or self-centring chuck* is most common and quick way of centring. By rotating any of the jaw, all the three jaws will move radially inwards or outwards simultaneously by the same distance. In case of *four jaw chuck*, all the jaws can be moved separately and adjusted at desired distance. This helps in holding irregular and eccentric workpiece. Combination chuck is combination of both the principles and provided with four jaw chuck. Magnetic chuck holds the workpiece centrally by developing a strong electromagnet. For magnetic chuck, workpiece should be of iron and steel, and it is not used to make heavy cuts. In case of hydraulic or air type of chuck, air or hydraulic pressure is used to press the jaws against the workpiece. Collet is used for small jobs for very firm grips.

5.3.2 Lathe Specification

Lathes are available in different sizes, which are specified on the basis of number of parameters. The following elements must be specified to identify the capabilities of lathe as shown in Fig. 5.3.

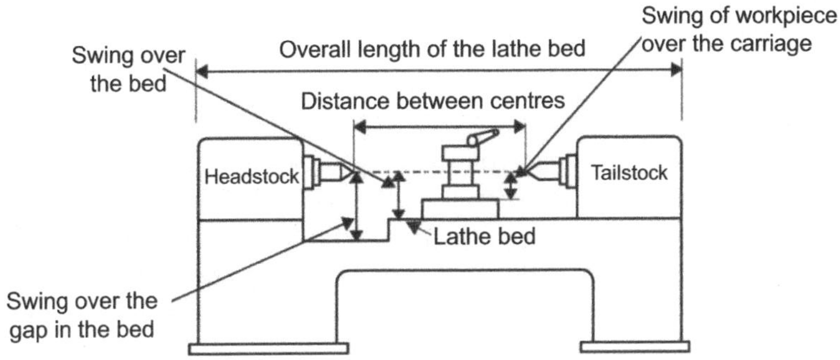

Fig. 5.3: Specifications of lathe

Distance between centres: It defines the maximum length of workpiece held between the centres for turning.

Swing over the bed: It defines the maximum diameter of the workpiece which can be turned over the lathe bed.

Overall length of the lathe bed: It is the total horizontal distance of the lathe bed from head stock to the end of tail stock. In addition to the above mentioned specifications,

Swing of workpiece over the carriage: It defines the maximum diameter of the workpiece which can be turned over the lathe carriage.

Swing over the gap in bed: In many lathes, a portion of lathe bed in front of chuck can be removed to hold the workpiece of larger diameter. Therefore, this section of bed is removed and a *gap* is left. The swing over the gap refers to accommodate the largest diameter of workpiece for machining.

Following specifications are also required to define the lathe.

- Diameter of the hole through the lathe spindle for turning the workpiece.
- Horse power of the motor
- Number of spindle speeds
- Feed range
- Diameter of lead screw
- Lead screw threads
- Pitch of lead screw

5.3.3 Lathe tools

For the general purpose work, lathe makes use of single point cutting tool which removes the material with one cutting edge of the tool, but for special operations multi point cutting tools may also be used. The cutting edges are prepared by grinding the tool. The varieties of tools are available, but their use depends on surface to be generated on the workpiece. These tools are made up of high carbon steel, high speed steel, cemented carbide, diamond tips and ceramics etc. depending on type of operation. Lathe cutting tools may be classified as;

- Turning tools (left hand or right hand type)

- Facing tools (left hand or right hand type)
- Chamfering tools
- Form tools
- Parting tools
- Threading tools
- Knurling tools
- Boring tools

Depending on the primary cutting edge, the direction of tool movement during machining is defined. The right hand cutting tool removes the material during movement towards leftward. It is derived from right hand palm, when placed over the tool, the thumb represents direction of tool feed. Opposite to that, a left hand cutting tool removes material, during movement towards rightward as shown in Fig. 5.4. Other tools used in lathe are discussed in the subsequent section of operations performed on lathe.

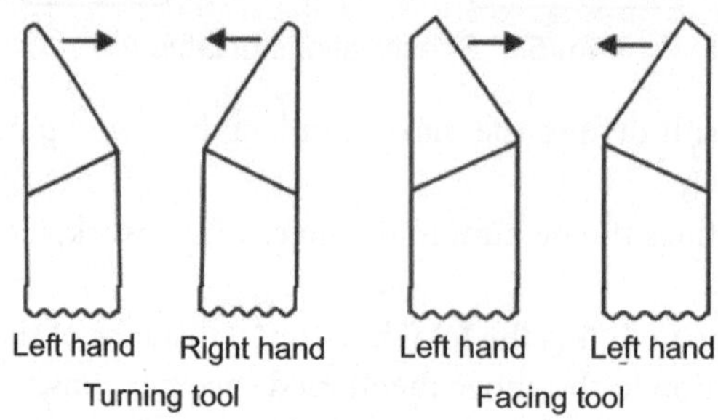

Fig. 5.4: Lathe cutting tools

5.3.4 Operations Performed on Lathe

Lathe is used to generate large number of surfaces with help of different settings and accessories. Common operations that may be performed on lathe are discussed below.

Facing: It is the operation in which ends of the workpiece are machined to make flat and the length is reduced as shown in Fig. 5.5. The cutting tool moves perpendicular to the axis of rotating workpiece and excess material is removed in the form of chips.

Turning: It is the process in which the workpiece is clamped in a rotating chuck, and the tool is fed in the direction parallel to the axis of rotating workpiece as shown in Fig. 5.6. The diameter of the workpiece is reduced throughout the length by removing the material in the form of chips. This is mostly used to produce cylindrical surfaces and is known as straight turning.

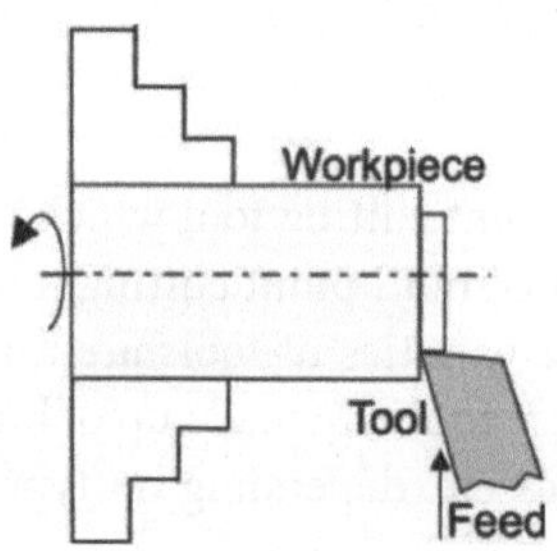

Fig. 5.5: Facing

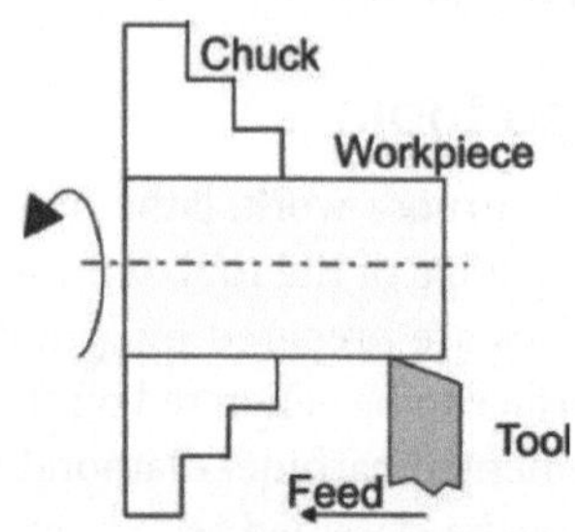

Fig. 5.6: Turning

Step turning: When the diameter of workpiece is reduced in steps to obtain different diameters is called step turning (Fig. 5.7)

Taper turning: When the diameter of workpiece is reduced in angle to obtain conical surfaces is called taper turning (Fig. 5.8). Taper turning on lathe can be carried out using following method;

- Compound slide method
- Tailstock off setting method
- By using taper turning attachment
- By using form tools

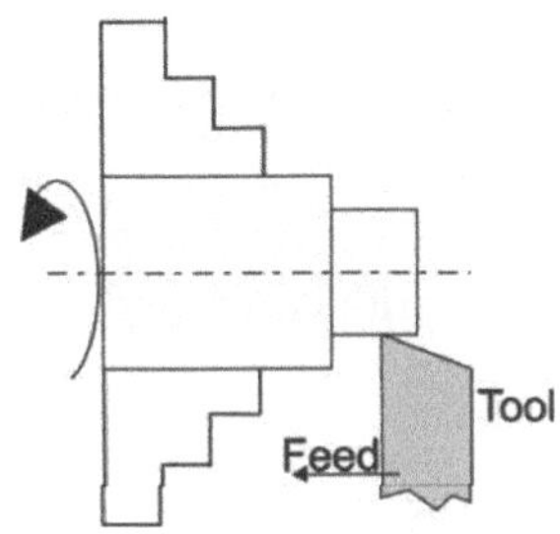

Fig. 5.7: Step turning

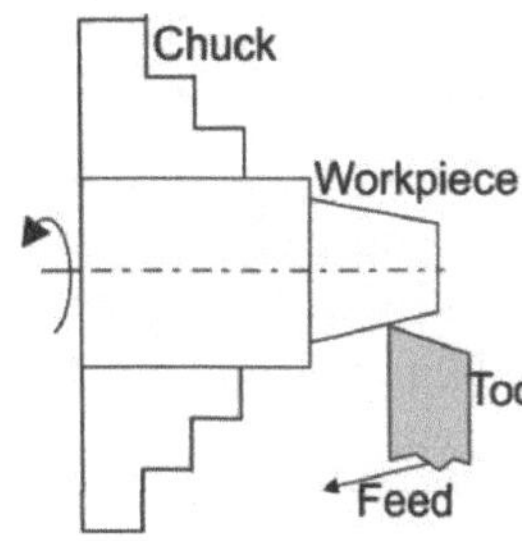

Fig. 5.8: Taper turning

Parting: In parting operation, the parting tool is clamped on tool post and is fed perpendicular to the axis of rotating workpiece to remove a piece of fixed length from the workpiece (Fig. 5.9).

Knurling: In this operation, a knurling tool having required impressions is pressed perpendicular to the workpiece until the pattern is formed on the workpiece as illustrated in Fig. 5.10. During knurling, material is not removed, but it deforms the top surface to provide gripping surface like watch turners.

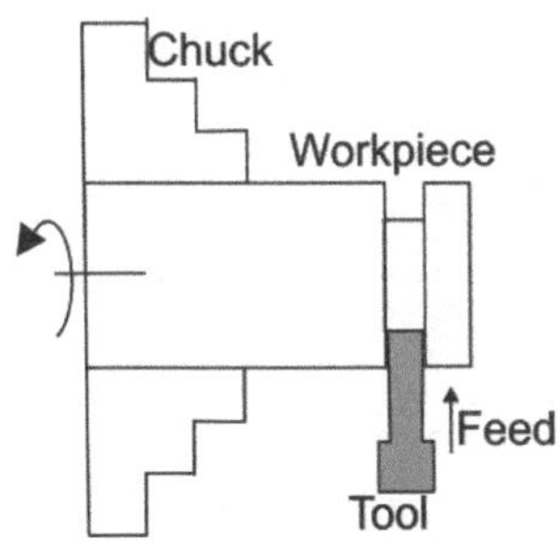

Fig. 5.9: Parting

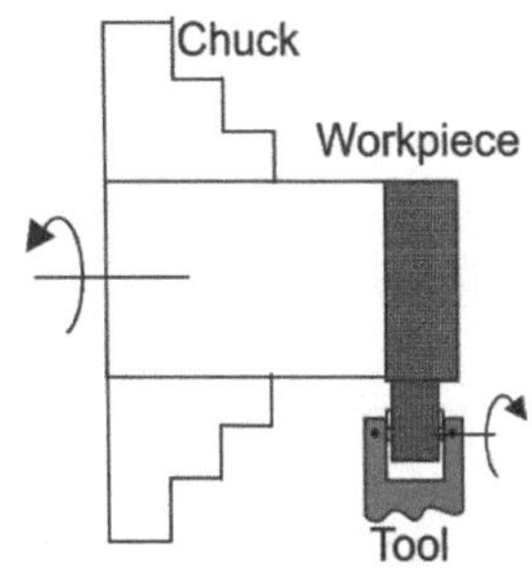

Fig. 5.10: Knurling

Chamfering: In this operation, the cutting tool is kept at small angle to the workpiece and moves perpendicular to the workpiece. The edges of the workpiece become taper to avoid sharp edges for safe handing as shown in Fig. 5.11. It also protects the end of workpiece from damage and to easily pass the nut on the threaded bolt.

Thread cutting: Cutting of helical grooves or threads on the outer surface or inner surface is called external or internal thread cutting respectively. The tool is moved longitudinal to the rotating workpiece and feed should be equal to pitch of the thread as shown in Fig. 5.12. Automatic motion to the carriage is provided with the lead screw. A pair of change gears are used to drives the lead screw and half nut, and by rotating the handle the depth of cut can be varied.

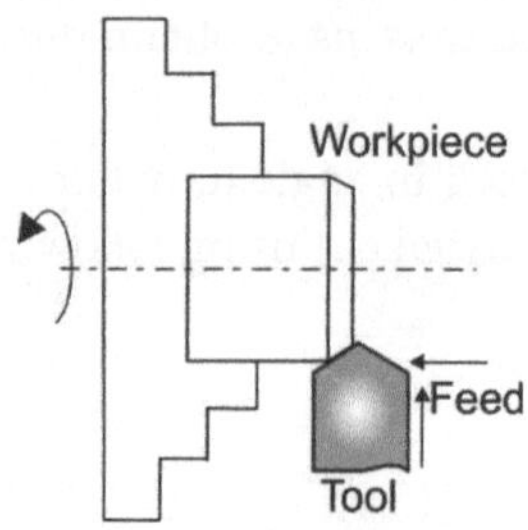

Fig. 5.11: Chamfering

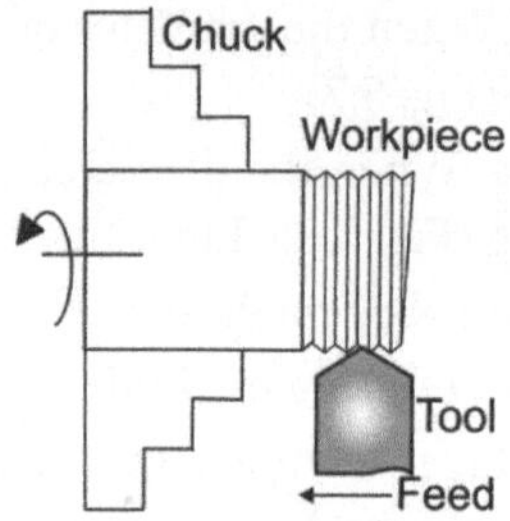

Fig. 5.12: Thread cutting

Drilling: Drilling produces a cylindrical hole in the workpiece. The workpiece is rotated with chuck and a drill bit held in the tailstock is feed into the workpiece (Fig. 5.13)

Boring: It is the operation of enlarging and turning already existing holes produced on drilling, casting and punching in the workpiece. The workpiece is rotated with chuck and a boring tool mounted on tool post is feed into the workpiece (Fig. 5.14).

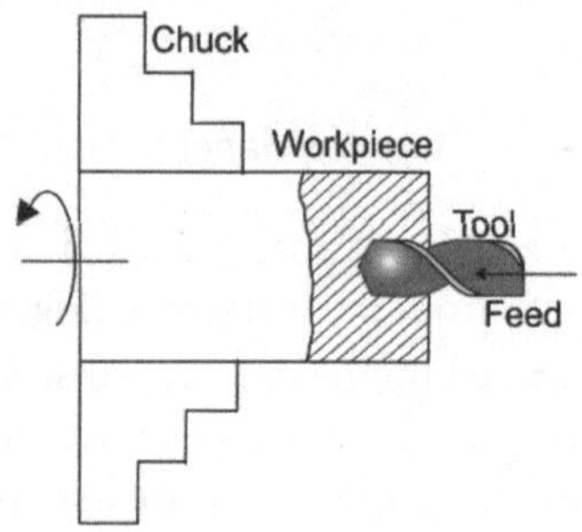

Fig. 5.13: Drilling

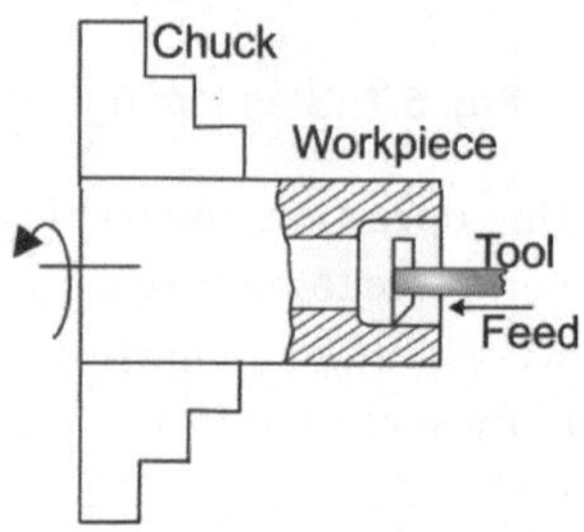

Fig. 5.14: Boring

Form turning: It is the process of generating any irregular shape on the workpiece with the help of forming tool. It does not remove the material, but forms the shape by pressing forming tool perpendicular to the workpiece (Fig. 5.15).

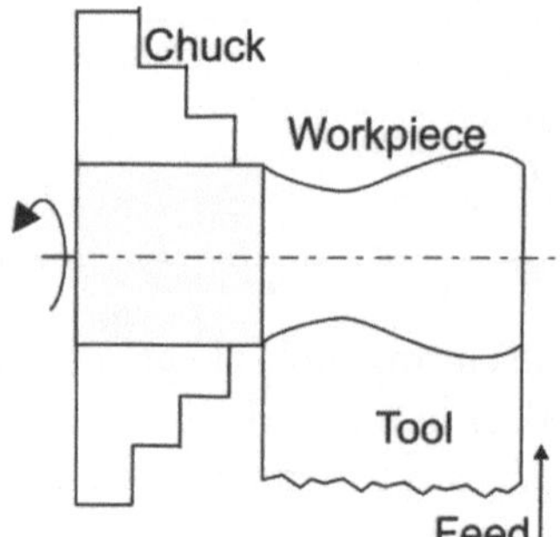

Fig. 5.15: Form turning

Scan to know more about Capstan and Turrent lathe

5.3.5 Types of Lathes

Lathes are manufactured with different types and sizes, from a small bench lathe to gigantic type of lathe. The different types are;

Engine or centre lathe: It is a common type of lathe used in machine shops. The main parts are bed, headstock, tailstock, carriage, gear box, lead screw and feed rod. The cutting tool may be fed in cross and longitudinal direction of the workpiece with the help of carriage. Multiple speeds of spindle can be achieved by a suitable mechanism. Depending on the power transmission mechanism, there may be belt and gear driven lathe. Speed variations in belt driven lathes are obtained by shifting the belt to different steps of cone type pulley, whereas in gear type, gear ratio is varied with speed lever.

Bench lathe: It is a small lathe mounted on a bench and is used for accurate and precision work, like production of gauges and punches. It performs almost all the operations which a large lathe can perform.

Speed lathe: It is also called as wood working lathe, may be of bench type or have legs fitted with the bed. It consists of headstock, tailstock and adjustable slide to support the tool. It has no gear box, carriage and lead screw, and no feed mechanism. Different speeds are obtained by cone pulley attached with variable speed motor. It is used for wood working, polishing, centring and metal spinning.

Tool room lathe: Tool room lathe is similar to an engine lathe, but have smaller bed length. It is provided with extra attachments required for more accurate and precision type of work on tool, gauges, dies and other smaller parts. The headstock is gear drive type carrying wide range of speed and feed.

Capstan and turret lathe: The centre lathe being a versatile machine is used in all the machine shops, but is not suitable for mass production. They have more idle time for changing and setting of the tool and for the setting of workpiece and their measurements. Also, skilled labour is required to operate the machine. Therefore, capstan and turret lathe form an important group useful for mass production. These are in actual a semi-automatic type of machine, used to perform a wide range of operations. It has a hexagonal block mounted in place of tailstock to fix six tools and two tool posts on cross slide to fix four tools on each. Two or more tools can operate simultaneously without resetting of tools and workpiece. This allows a quick change over in tooling and cutting operations and reduces production time. It also eliminates requirement of skilled operator. The feed of tool is controlled by feed stops. These lathes are useful for operations involving better repeatability.

Automatic lathe: Automatic lathes are well designed to automatically feed the tool to work and withdrawn after the cycle is completed. All cutting operations are performed in a sequence and controlled automatically. These lathes are used in mass production of similar parts, without participation of operator. Only bar stock is loaded manually. In semi-automatic types of machines, loading and unloading of new workpiece, starting of machine and inspection is made by operator, whereas all the operations are performed automatically.

5.3.6 Special Lathe Machines

There are many standard lathes to carry out a variety of works. But, to perform some specific operations, special lathes are designed. They can conveniently handle definite class of work, and proved to be effective and efficient.

Precision lathe: This lathe is used for precision turning of previously turned rough workpiece.

Crankshaft lathe: Specially designed and provided with all the attachments for tuning very long crankshafts.

Screw cutting lathe: It is operated through cams for the mass production of screws.

Wheel lathe: This lathe is used for turning of the locomotive wheels.

Vertical lathe: It has vertical column used for turning and boring of large rotating parts such as flywheel and gear blanks.

5.4 SHAPER MACHINE TOOL

It is a reciprocating type of machine tool used to produce flat surface which may be vertical, horizontal and inclined, grooves and slots. It is versatile machine holds a single point cutting tools in reciprocating ram and a work piece is clamped in a fixed table to remove a large amount of material. It has greater flexibility in holding workpiece, quick adjustments and use of tools. It is commonly found in all machine shops for general purpose works. In planer machine, workpiece is clamped on a table which reciprocates under stationary tool. The slotter is a vertical type of shaper where tool reciprocates vertically and cuts the workpiece rigidly held in table.

5.4.1 Working Principle of Shaper

In case of shaper, the excess material is removed from the surface with the help of single point cutting tool held in ram that reciprocates over the stationary workpiece held on table of the machine as shown in Fig. 5.16.

The workpiece is fed in small increments at right angles to the direction of the cutting tool at the end of the return stroke. The ram reciprocates back and forth, and the cutting is accomplished in forward stroke. No cutting takes place during return stroke and is called as idle stroke. Therefore, the tool should move slower to remove the material during forward stroke, whereas tool should move faster during idle return stroke, to minimize machining time. This reciprocating motion of ram can be achieved through quick return mechanism. Shaper is not economical for mass production because the cutting speeds are not high due to design constraints. Shaper find its application in tool room, die and jig making works and repair works.

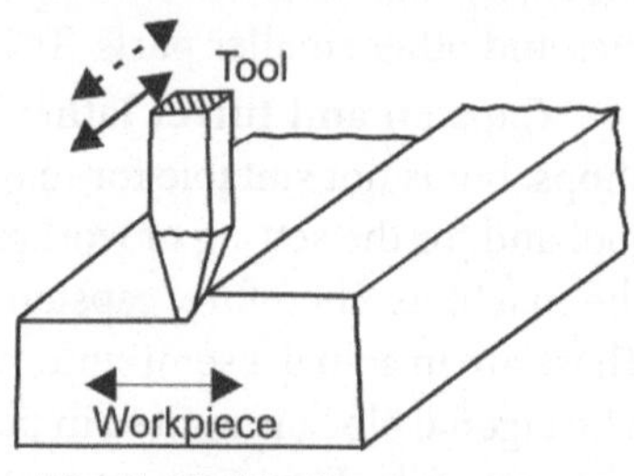

Fig. 5.16: Working principle of shaper machine

5.4.2 Specification of Shaper

Shaper machines are available in different sizes, which are specified on the basis of number of parameters. The following elements must be specified to identify the capabilities of shaper.

- Maximum length of ram stroke
- Maximum distance between table and ram.
- Size of table.
- Maximum horizontal travel of table.
- Maximum vertical travel of table.

5.4.3 Main Parts of Shaper

The various parts of shaper are shown in Fig. 5.17.

Base: It is heavy and robust cast iron body to support all other parts of the machine mounted over it. It resists the vibrations and carries the entire load of the machine and the forces generated during machining over the work.

Column: It is a box type structure of cast iron mounted at the base of the machine. It houses the operating mechanism and electrical circuits. At the top of column guideways are mounted for smooth functioning of ram during forward and return stroke.

Table: It is a cast iron body having box type structure having 'T' slots at the surface and at sides. It holds and supports the workpiece during cutting and provides the feed to workpiece in upward, downward and sideward directions.

Cross rail: It is made up of cast iron and mounted in front of the vertical guideways of the column. It carries mechanism for up and down movement of the table to accommodate different thickness of workpiece and cross transverse of the table with the help of elevating screw.

Ram: It reciprocates on the guideways at the top of the column. It carries tool head in the clapper box and mechanism for the adjustments of the stroke length.

Tool head: The tool head slides in the front the ram with the help of T- bolts. It holds the single point cutting tool and imparts necessary movements. It can be swivelled by 0° to 90° in a vertical plane. The tool head can be given up and down movements by hand feed for vertical cuts on the workpiece.

Vice: The shaper vice is made of cast iron and has robust construction of slide and jaws in comparison to ordinary machine vice. It has long and deep jaws and have maximum opening to accommodate heavy and long workpieces. The jaws are made of special steel and precisely made to assure firm grip, and long life.

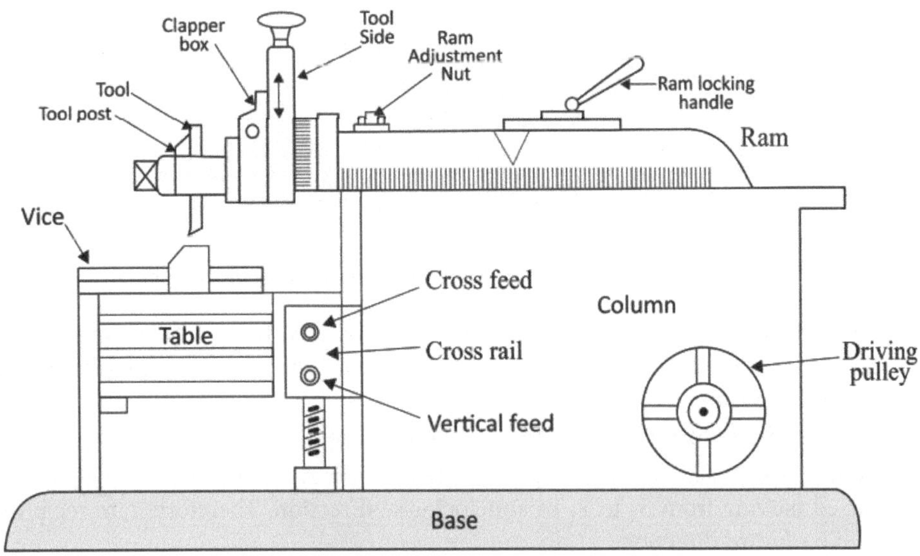

Fig. 5.17: Schematic block diagram of shaper

5.4.4 Quick Return Mechanism

A quick return mechanism is applied to convert rotary motion of crank into reciprocating motion of ram in shaping machines, which is an inversion of single slider crank chain. This mechanism indicates that the return/backward stoke is faster than the forward stroke that helps to reduce total machining time. Therefore, time ratio in quick return mechanism is the ratio of change in input displacement during the working stroke to the change in return stroke. This mechanism finds application in shaper machine, rotary internal combustion engine, power saw and slotting machine.

It basically has a motor to drive the gear which carries a crank that has uniform rotary motion. The motor controls the rpm of gear. The crank is connected to the slider, which slides freely in slotted bar in straight line path. It converts the rotary motion of crank into oscillatory motion of slotted bar about pivot point (O). A connecting rod is attached at the end of slotted bar to convert the oscillatory motion of the slotted bar into the reciprocating motion, hence ram reciprocates with tool to machine the workpiece as shown in Fig. 5.18.

5.4.4.1 Working of Quick Return Mechanism

It can be observed from Fig. 5.18 that the length of the link 'OC' is fixed, and the crank is pivoted to end 'C' and rotates with uniform angular velocity. The slider which is connected with the crank also rotates with crank. When slider starts reciprocating in slotted bar, the slotted bar pivoted at 'O' makes oscillatory motion in circular arc from point S_1 to S_2. The connecting rod attached to slotted bar and ram, converts the oscillatory motion into reciprocating motion.

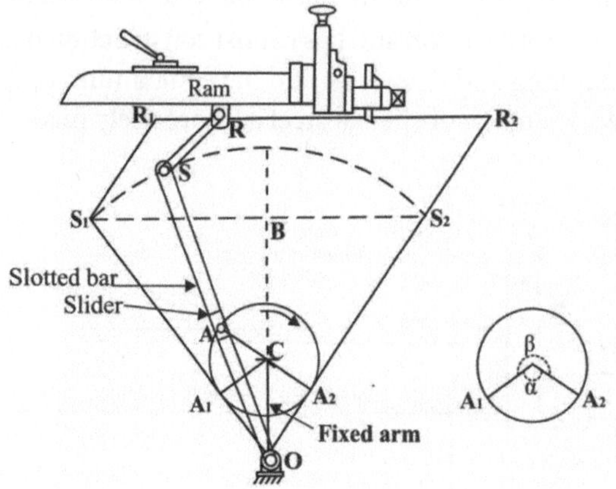

Fig. 5.18: Schematic block diagram of quick return mechanism

During forward stroke, as the crank rotates clockwise from A_1 to A_2 with an angle β, the end S of slotted bar started moving from S_1 to S_2 in clockwise direction. Therefore, ram reciprocates from Point R_1 to R_2 in forward direction.

Time of the forward stroke = Time taken to make an angle β

During return stroke, as the crank rotates clockwise from A_2 to A_1 with an angle α, the end P of slotted bar started moving from S_2 to S_1 in anticlockwise direction. Therefore, ram reciprocates from Point R_2 to R_1 in backward direction.

Time of the return stroke = Time taken to make an angle α

$$\frac{Time\ of\ forward\ stroke}{Time\ of\ return\ stroke} = \frac{\beta}{\alpha} = \frac{\beta}{360 - \beta} \qquad 5.1$$

The speed of return stroke is about 1.5 times the forward stroke.

The length of the travel must be slightly more than the actual length of the workpiece, which allows tool block of the clapper box to swing back to its position for machining. The length of stroke may be changed by varying the crank radius, larger the radius longer will be the ram stroke.

5.5 PLANER MACHINE TOOL

When the workpiece to be machined is large and heavy, and could not be clamped in shaper, then planing machine (Planer) is used. The planer produces flat surfaces and slots. The workpiece is fastened to the work table and has reciprocating motion, whereas tool remains stationary during machining. The tool is fed in horizontal or in vertical direction. The shaper works opposite to planer, where workpiece is stationary and tool reciprocates. In planer machine workpiece is mounted on table and gets support during complete machining process, therefore no work or tool deflection takes place.

5.5.1 Working Principle of Planer

The working principle of machining through planer is shown in Fig. 5.19. The workpiece is rigidly clamped to the T slots work table with special bolts. The cutting tool is vertically held in tool post mounted on the cross rail, and tool may be moved vertically up and down or in horizontal direction.

The work table with workpiece reciprocates past the vertically held stationary tool. Multiple workpiece may be clamped in line on the table to machine them simultaneously. Also, multiple tools may be applied to machine the workpiece.

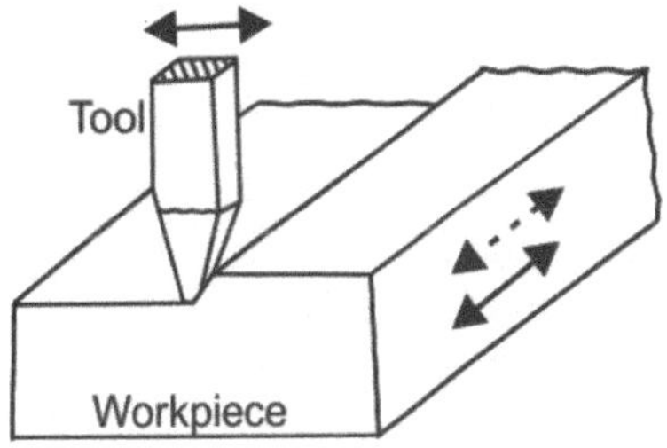

Fig. 5.19: Working principle of planer machine

5.5.2 Parts of Planer

The main parts of planer are shown in Fig. 5.20 and are discusses below:

Bed: Bed is a rigid, very large and heavy structure of cast iron as foundation and carries the column and all the other parts of machine. The length of the bed is almost twice the length of table. The bed is also provided with V shape guideways to support and slide the table along with pressure lubrication system. Beds are provided with hydraulic bumpers at the end to stop the table from over running.

Table: It is a heavy box type structure of cast iron with ribs to provide strength and it has a precisely machined top. It supports the workpiece with bolts and T slots on the top surface. The workpiece and table reciprocates along the guideways of the bed. It carries adjustable stoppers to reverse the motion at the end of each stroke.

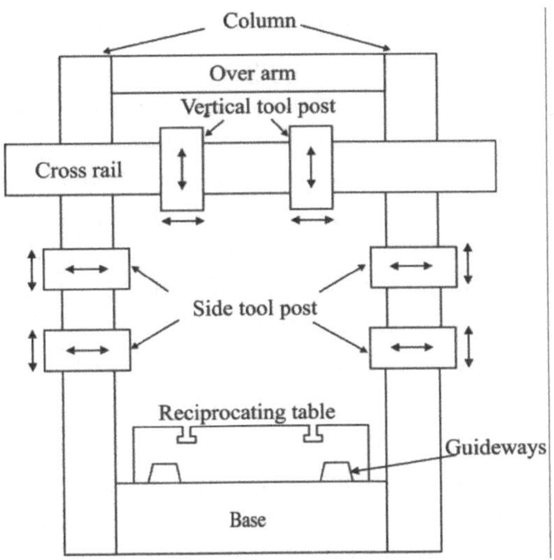

Fig. 5.20: Schematic block diagram of planer

Column: These are also called *housing* and has a rigid box like vertical structure made up of cast iron placed on each side of the bed and rigidly fastened to it. It has vertical guideways to support and provide vertically up and down movement of cross rail. It carries mechanism for power transmission to the upper parts of the machine.

Cross rail: It is a heavy box type structure mounted horizontally between two vertical columns of the machine. It also provides strength to the column structure. It can slide up and down on V or flat ways on column with the help of manual or power operated elevating screws. Cross rail may be locked in any desired position with clamps. It should be ensured that cross rail is parallel to the bed for having accurately machine flat surfaces. It carries tool heads for vertical and cross feed of the tools by means of feed rod.

Tool head: Two tools heads are mounted on cross rail and two on column with saddle which slides along the cross rail and column. All the four tool heads are independent and may be operated simultaneously to save machining time. Tools on the cross rail can be moved up and down and can be given cross feed along the rail. Similarly, the tools on column can be moved up and down and can be fed horizontally into the workpiece. The tool heads can be swivelled through 70° on either side from normal position. The clapper box is also provided to avoid scratching during idle/backward stroke.

5.5.3 Specification of Planer

Planer machines are available in different sizes and the following elements must be specified to identify the capabilities of planer.

- Maximum horizontal distance between the two vertical columns.
- Maximum vertical distance between cross rail and the table top.
- Length of table.
- Maximum travel of the table.
- Length of the bed.

5.6 MILLING MACHINE

Milling is a machining process where the material is removed from the workpiece by feeding it against the rotary milling cutters with single point or multipoint edges. Milling process is used to machine flat or curved surfaces with high precision and accuracy such as slots, gears, contours and circular profiles etc. The workpiece is clamped to the work table directly or with suitable fixtures and the cutter rotating at high speed removes the material at faster rate because of multi point cutting edges.

Milling machine is a versatile machine tool available in all machine shops in industries. It rotates the milling cutter mounted on spindle or arbor and automatically feeds the workpiece clamped at the table in desired direction and removes the material in form of chips. The machine can hold single or multiple cutters at the same time working simultaneously to reduce machining time. The higher production rate with close dimensional accuracy of the machine has replaced shaper, planer and slotter.

5.6.1 Working Principle

Milling is a power driven machine tool that removes the material by feeding the workpiece against milling cutter rotating at high speed. The surfaces are machined in single or multiple passes of the work. The milling process may be carried out by the following two methods:

1. **Up milling:** In up or conventional milling, the workpiece is fed in direction opposite of that in which cutter rotates. The material is removed in form of chips, whose thickness is minimum at the start of cut which gradually increases and reaches maximum at the end of cut. As a result of

opposing motion, it is difficult to apply coolant in cutting zone, and the chips get accumulated. These chips when carried over with cutter leaves marks on milled surface, giving poor surface finish. The working principle of up milling process is shown in Fig. 5.21.

Also, as the cutting forces are directed upwards, the cutter tends to lift the workpiece from the table. Therefore, fixtures are required to ensure the rigid clamping of workpiece.

2. **Down milling:** In down or climb milling, the workpiece is fed in same direction in which cutter rotates. As the cutter rotates, the chip thickness is maximum at the start of cut, which reaches minimum at the end. Therefore, it is easy to apply coolant and dispose of the chips and these chips do not scratch over milled surfaces. This gives a better surface finish and is used for finishing operation such as cutting, slotting and grooving. As the cutter rotates, the cutting force helps to hold the workpiece, this lowers the clamping force. The working principle of down milling process is shown in Fig. 5.22.

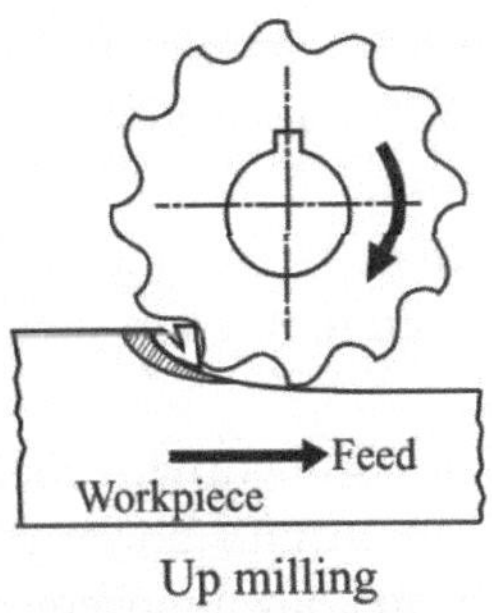

Fig. 5.21: Working principle of up milling

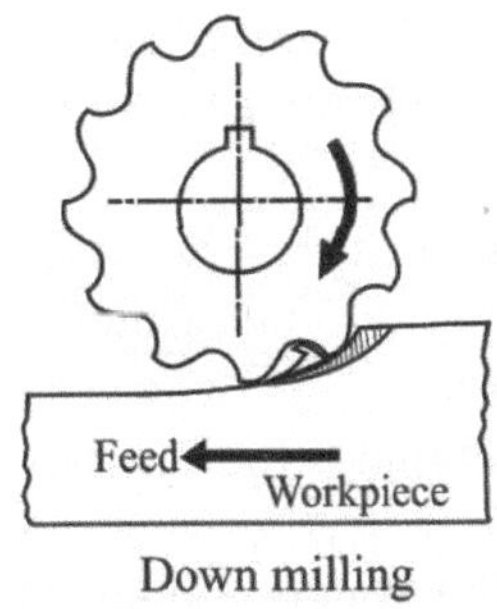

Fig. 5.22: Working principle down milling

5.6.2 Types of Milling Machine

Milling machines are designed in variety of types and sizes to fulfil the functions of cutting and these are generally classified as mentioned below:

1. **Column and Knee type milling machine:** These are general purpose machines having single spindle. Column supports the spindle, which carries the cutter, whereas knee supports worktable. Column type machines are;
   - Horizontal milling machine
   - Universal milling machine
   - Vertical milling machine
   - Omniversal milling machine
2. **Fixed bed type milling machine:** These machines are rigid and larger in size than the column and knee type machine. They carry single or multiple spindles. These machines are;
   - Simplex milling machine
   - Duplex milling machine
   - Triplex milling machine
3. **Planer milling machine:** These are used for heavy jobs and carries maximum of four tool heads.

4. **Production milling machine:**
   - Rotary table type
   - Drum type
   - Tracer controlled type
5. **Special purpose milling machine:**
   - Pantograph milling machine
   - Profile milling machine
   - Planetary milling machine

5.6.3 Horizontal Type Milling Machine

Colum and knee type milling machine is a very robust and most commonly used machine which is usually found in all machine shops. Facing and peripheral milling cutters can be mounted on spindle located horizontally to remove the material from workpiece which is normally held in a vice on table. The main parts of horizontal milling machines are shown in Fig. 5.23 and are discussed as follows;

Base: It is a heavy and rigid casting made up of cast iron acts as the foundation and carries all other parts of the machine. It absorbs the shock and vibrations generated due to cutting forces. Base may also be made hollow to store cutting fluid. It carries column and an elevating screw to support and move the knee up and down.

Column: Column is the vertical supporting frame which houses the driving mechanisms such as gears, belt drives, pulleys, driving motor and various controls. It transmits power to the spindle and the work table. The front portion of column has guideways for up and down movement of knee. It also supports the overarm at the top which is extended outwards in front of the machine.

Overarm: Overarm is mounted at the top of column which is extended in front to support the arbor and spindle. The over arm can be adjusted horizontally to position the cutters and can be fixed in any position.

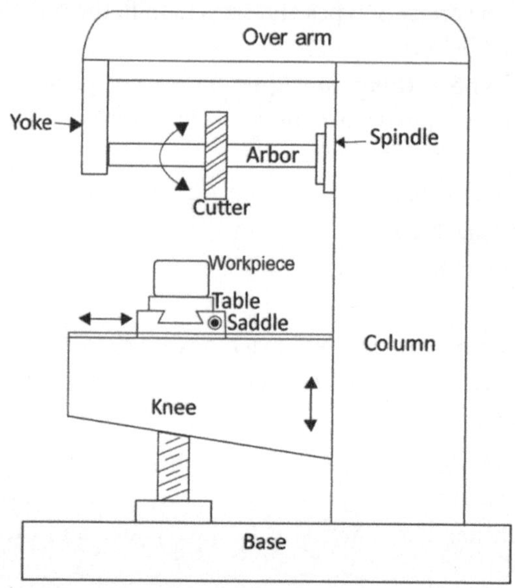

Fig. 5.23: Schematic view of horizontal milling machine

Knee: Knee is the major part of the machine which supports the table saddle, fixtures and other accessories. It is placed in front of the column which moves up and down on guideways, and height is adjusted with elevating screw. At the top, knee has slideways for the saddle which provides cross feed to the table.

Saddle: Saddle is mounted on the top of knee to support and carry the table. It is manual or power operated to provide cross feed to the table.

Table: It is mounted at the top of the saddle in guideways and can travel longitudinally in the horizontal plane. The table has T-slots to fix the workpiece or the fixtures with special T bolts. The table can be moved up and down by elevating screw and horizontally by engaging a nut on the saddle with lead screw under the table.

Spindle: The upper portion of the column contains the spindle which revolves in bearing receives power from motor and driving mechanism in the column. The front end of the spindle has tapered holes to fit the arbor to which cutters are attached.

Arbor: Arbor is a shaft tapered at one end to fit into the spindle, and other end is supported with over arm. It carries the rotating milling cutters.

5.6.4 Specifications of Milling Machine

Size of milling is represented in many ways such as;

- Size of work table
- Maximum horizontal and vertical travel of table
- Number of spindle speeds
- Number of feeds

5.6.5 Operations Performed on Milling Machine

A variety of operations may be performed on workpiece to obtain desired shapes and sizes. These are discussed below;

Plain or slab milling: It is a process employed to machine flat and horizontal surfaces which is parallel to the axis of the plain or slab milling cutter. The cutter is fixed into arbor which rotates, whereas workpiece is clamped on table and moved upward according to depth of cut (Fig. 5.24).

Face milling: It is a process used to machine flat surfaces with the help of face milling cutter rotated at an axis perpendicular to the workpiece surface (Fig. 5.25).

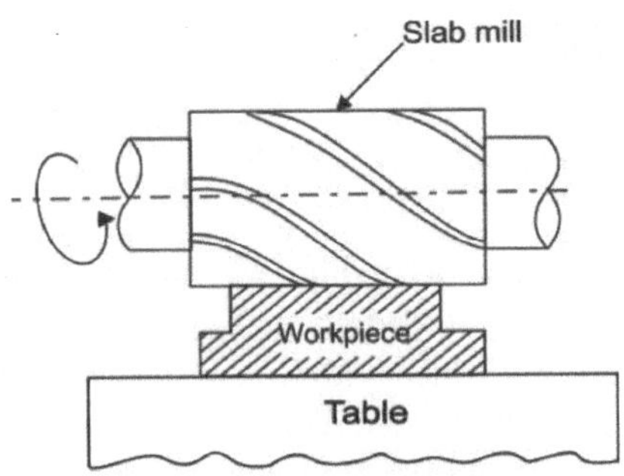

Fig. 5.24: Plain milling

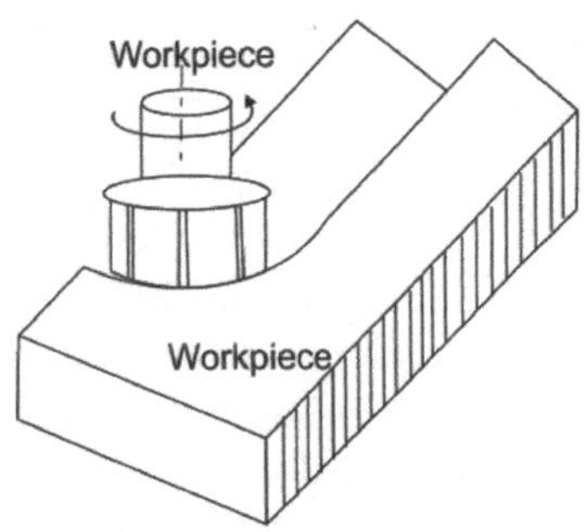

Fig. 5.25: Face milling

Angular milling: It is a process employed to produce angular surfaces other than right angle to the axis of rotating cutter. The angular grooves or V- blocks may be produced with single or double angle cutters depending on the surface to be machined (Fig. 5.26).

Form milling: It is a process employed to produce irregular shaped surfaces using form cutters. The shape of cutting teeth on cutters conforms to the shape to be produced which may be concave, convex or any other irregular surface (Fig. 5.27).

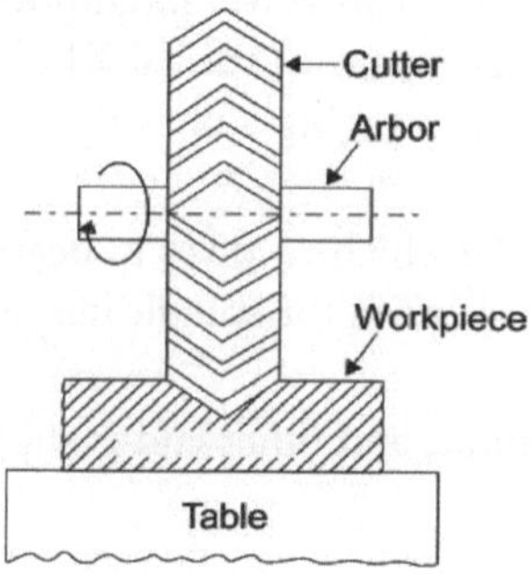

Fig. 5.26: Angular milling

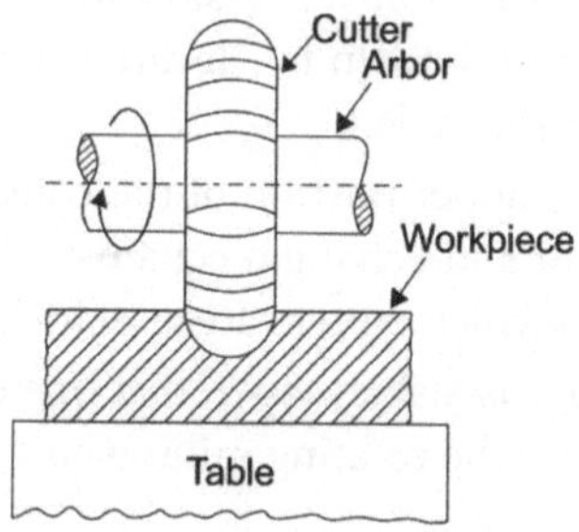

Fig. 5.27: Form milling

Straddle milling: It is a process in which a pair of side milling cutters is employed to produce two parallel surfaces simultaneously on the workpiece. The gap between two cutters may be varied with spacers (Fig. 5.28).

End milling: It is a process used for machining flat surfaces that may be horizontal, vertical or inclined with reference to the top surface of table cutters. The vertical milling machine is more suitable for producing slots, grooves and keyways in the workpiece (Fig. 5.29).

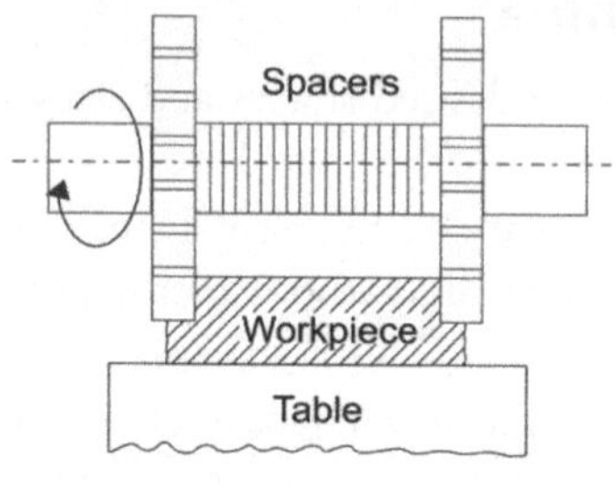

Fig. 5.28: Straddle milling

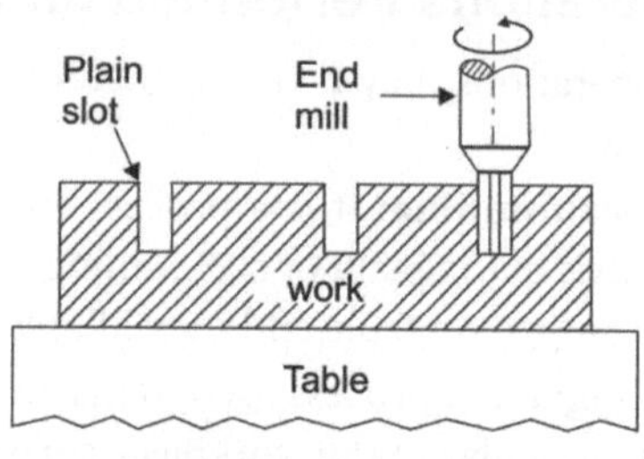

Fig. 5.29: End milling

Gang milling: It is the process of machining several surfaces on a workpiece simultaneously with the use of more than two cutters mounted on arbor. The cutters may have same or different diameters to machine flat horizontal and vertical surfaces by feeding table against cutters. This type of machining process is used for repetitive work to save machining time (Fig. 5.30).

Key way, groove and T- slot milling: The keyways, grooves and slots of different shape and sizes can be produced through milling machine cutters. Keyways are grooves milled on shafts and spindles with side or end milling cutters. Woodruff key produces seat which is machined with woodruff key seat cutter, and sunk key seat is machined with end mill cutter. V- grooves are machined with single or double angle cutters depending on the surface to be machined. Plain slots are cut with plain milling cutters or slitting saw or side milling cutters, whereas T- slots are cut in two or three steps with special cutters designed to generate required slots (Fig. 5.31).

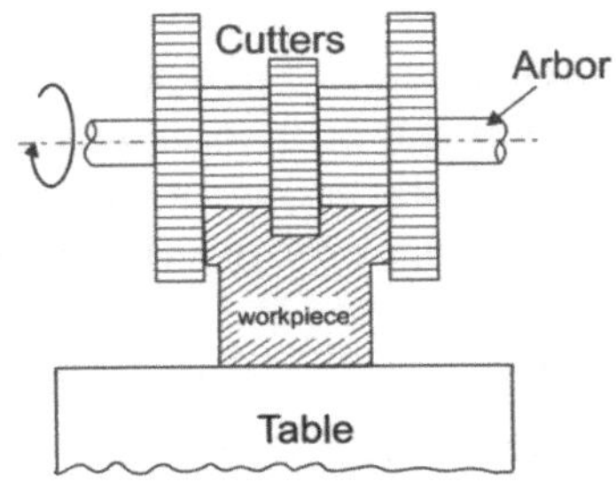

Fig. 5.30: Gang milling

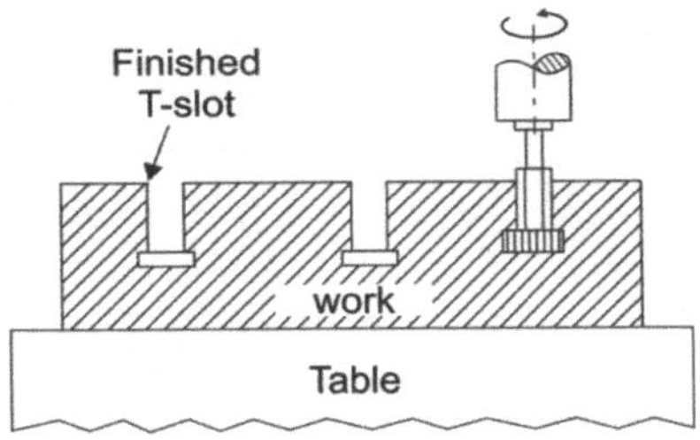

Fig. 5.31: T-slot milling

Gear cutting: It is a process of cutting gear teeth of different profiles on milling machine using form relieved or end mill cutters. Equi-spaced teeth are cut on periphery of gear blank by holding it on indexing mechanism or universal dividing head, which divides the periphery equally into required numbers. The cutter profile fits into the tooth space of the gear blank (Fig. 5.32).

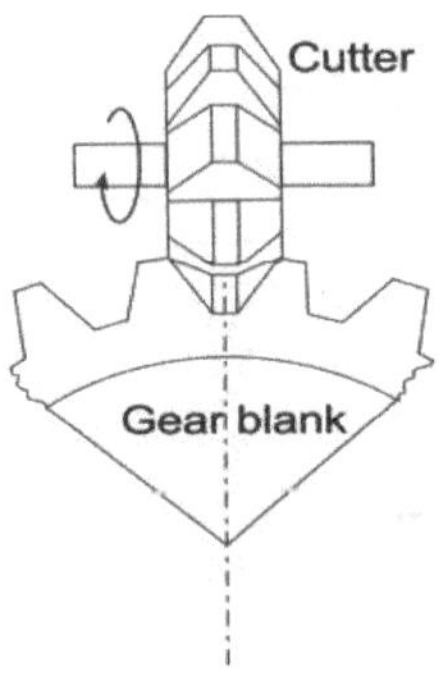

Fig. 5.32: Gear cutting

5.6.6 Milling Cutters

Milling cutters are single or multi point cutting tools used in milling machine to produce flat surfaces. Cutters are made up of high carbon steel, high speed steel, stellite, ceramics and sintered carbide tip. Milling cutters may have straight teeth parallel to the axis of rotation or have helical shape. The helix angle may be right or left hand type which decides the rotation of the cutter for machining operation. These cutters are classified on the basis of teeth profile, and purpose or use of the cutter. Some of the common milling cutters are mentioned below and discussed in milling operations.

- Plain milling cutters
- Side milling cutters
- End milling cutters
- Form milling cutters
- Face milling cutters
- Angle milling cutters
- Key milling cutters
- Slot milling cutters

Scan to know more about milling cutter

5.7 DRILLING OPERATION

When the tool is pressed in workpiece, material is removed in form of curled chips that comes out of the helical groves called flute cut on cylindrical surface of drill bit. Chips may be in form of long spirals or short flakes depending on the material and the processing parameters. Proper coolant may be applied to remove the heat generated due to friction during machining and to obtain a better surface finish of hole and increased tool life.

The holes produced by drilling are always larger than the drill diameter and internal surface of hole is usually rough. Therefore, drilling is always followed by *reaming* operation to obtain better dimensional accuracy and improved surface finish with the help of a tool called *reamer*. Boring is an operation to enlarge an already existing hole, may be casted, previously drilled or punched.

5.7.1 Working Principle

The operation of drilling can be performed with power driven machine tool called drilling machine, which holds the drill bit in the spindle rotating at high speed. The rotating drill bit is manually fed inside the workpiece as shown in Fig. 5.33. The workpiece is mounted on work table with the help of vice. The material is removed in the form of spiral chips which comes out through passage called *flute* (Fig 5.42).

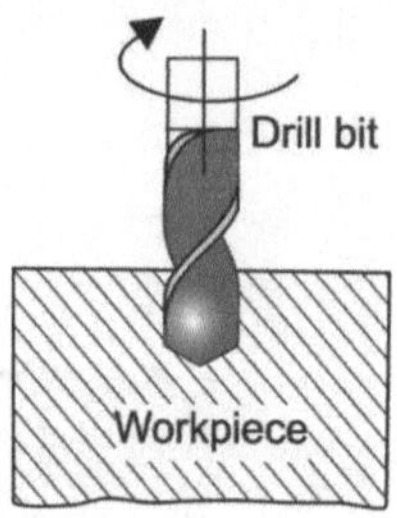

Fig. 5.33: Working principle of drilling machine

5.7.2 Parts of Drilling Machine

The drilling machine is used to drill holes by feeding the rotating drill bit along the axis of rotation. The common parts of drilling machine are shown in Fig. 5.34 and are discussed below;

Feed mechanism: Drill machine has electric motor, pulley and belt drive to transfer the power to rotate the drill spindle at different speeds. The drill head moves up and down by motors and has rack and pinion arrangement to convert rotary motion to linear motion.

Base: It is made of cast iron and acts as foundation of the machine which carries all the equipment and their weight. The base may have slots to support the workpiece and the vertical column is placed at one end of the base.

Column: Column is a cast iron structure mounted at the base and it supports the vertical arm and table. The radial arm moves in clockwise and anticlockwise direction. A sliding table is mounted on the column so that it can be moved up and down according to the requirement with help of rack and pinion arrangement on column.

Table: Table is adjustable type made up of cast iron with slots, so the vice or the workpiece can be clamped over it. The table may be either circular or rectangular in shape. The table can be moved up and down and swing horizontally depending on the workpiece requirements.

Arm: arm is a rigid cast iron structure that acts as housing for the driving and feed mechanism. It also carries drill head. In some of the machines, arm is provided with guideways to slide the drill head.

Drill head: it is mounted on the arm and carries feed and driving mechanism for spindle. Belt and pulley drive is used to transfer the power from motor to the drill head and to obtain different speeds.

Spindle: It is made up of high carbon chromium steel or alloy steel. It rotates and moves up and down in sleeve. It holds the drill chuck, which is made of alloy steel and holds the drill bit.

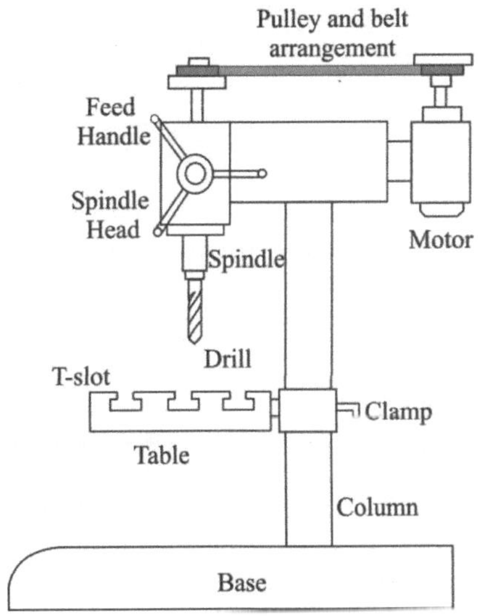

Fig. 5.34: Plain drilling machine

5.7.3 Specifications of Drilling Machine

Drilling machines are available in different sizes, which are specified on the basis of number of parameters. The following elements must be specified to identify the capabilities of drilling machine.

- Size of the table
- Maximum size of drill bit that machine can hold.
- Maximum size of workpiece that can be fixed on table.
- Maximum size of hole that can be drilled.
- Power, spindle speed and feed of machine.

5.7.4 Types of Drilling Machine

Drilling machines are manufactured in different sizes and varieties and they may be classified on the basis of features and different types of work they do. The different types are;

- Bench type drilling machine
- Portable type drilling machine
- Sensitive drilling machine
- Upright drilling machine
- Radial drilling machine
- Gang drilling machine

- Deep hole drilling machine
- Multiple drilling machine
- Automatic drilling machine

Out of these all, radial drilling machine is most common and found in almost every machine shops.

The radial drilling machine is used to drill holes in large and heavier parts. The rotating drill is moved towards desired location in stationary workpiece. The machine consists of heavy cast iron base which carries a vertical column. The column supports a radial arm on which drill head is mounted. This arm can move vertically up and down on the column through a separate motor and can be locked at desired position with clamping lever. The arm can also swing around the column to locate the drill bit for drilling at particular location. Depending on the possible movements, these machines may be classified as;

- Plain radial drilling machine which may drill in vertical plane and may swing along the arm.
- Semi-universal radial drilling machine in addition to above movements may swing spindle head around horizontal axis.
- Universal drilling machine have provision to rotate the arm at desired angle along the horizontal axis in addition to possible movements of semi- universal machine.

5.7.5 Operations Performed on Drilling Machine

The followings are the operations that can performed on drilling machine as discussed below:

Drilling: The main operation of machine is drilling in which a circular hole is made with the help of rotating tool called drill bit is shown in Fig. 5.33.

Boring: Boring is an operation of enlarging an already existing hole to achieve required size and surface finish. It makes use of adjustable tool with single cutting edge (Fig. 5.35).

Reaming: It is an operation of finishing the hole to achieve accurate size and fine surface finish. The accuracy of $\pm$ 0.005 mm can be achieved. The process makes use of multipoint cutting tool called reamers which rotates in an existing hole (Fig. 5.36).

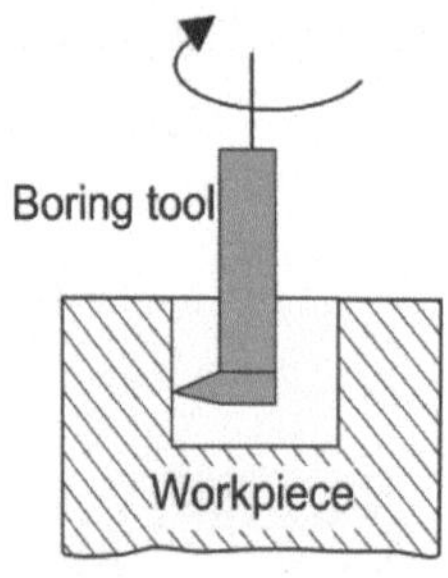

Fig. 5.35: Boring

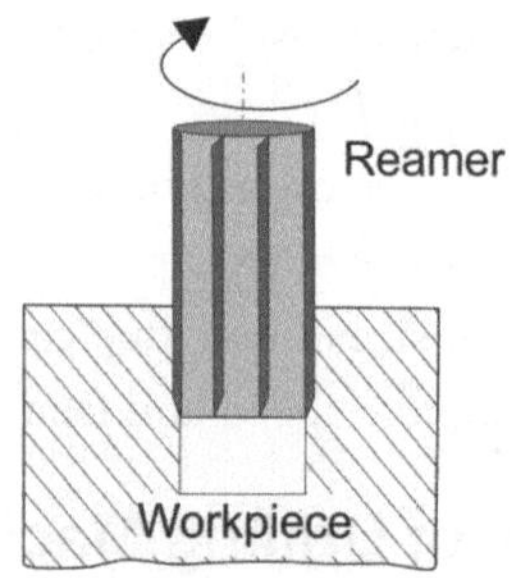

Fig. 5.36: Reaming

Counter boring: The operation is performed for enlarging a hole for a particular length. It forms a recess or shoulder to an existing hole. The counter boring tool is provided with a cylindrical projection at the end called pilot having diameter equal to the hole to maintain alignment (Fig. 5.37). The counter boring is performed to accommodate grooved nut and round head bolts.

Counter sinking: The operation is performed for enlarging the end of the hole to provide conical shape for a particular length. This allows to seat the countersunk head of screws, so that their top surface is levelled with the top surface of the workpiece (Fig. 5.38).

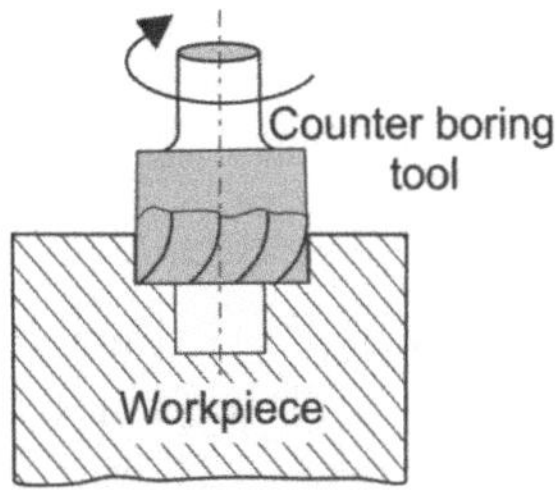

Fig. 5.37: Counter boring

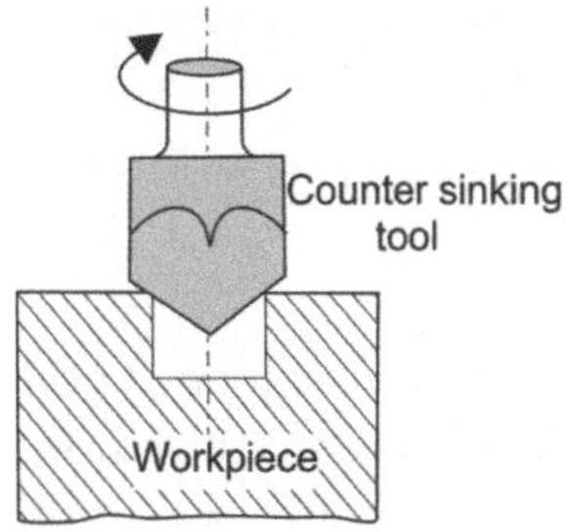

Fig. 5.38: Counter sinking

Tapping: The operation is performed to cut internal threads in the existing hole with the help of tool called tap. The tap is a cutting tool with hardened threads on the periphery (Fig. 5.39).

Trepanning: It is an operation of generating a hole by removing the material along the circumference of the hollow cutting tool (Fig. 5.40).

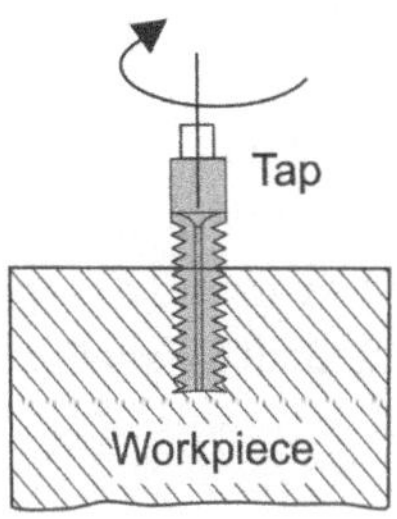

Fig. 5.39: Tapping

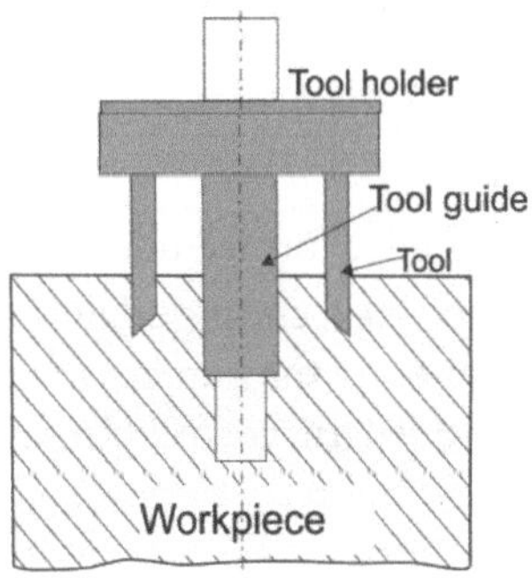

Fig. 5.40: Trepanning

Spot facing: It is the operation of squaring and finishing surface around and at the end of the hole to seat the nut or head of screw. It is performed with counter boring tool or spot facing tool (Fig. 5.41).

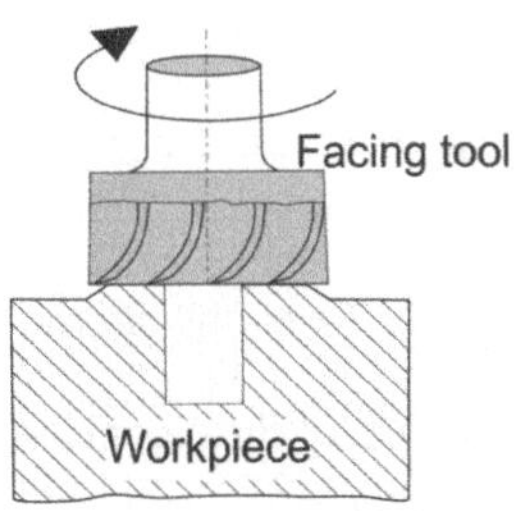

Fig. 5.41: Spot facing

5.7.6 Types of Drills

Drill is a cutting tool used to make a hole having one or more cutting edge. There are varieties of drills available, and are as follows;

- Flat drill or spade drill
- Straight drill
- Centre drill

- Twist drill
- Core drill
- Parallel shank or taper shank drill
- Masonry drill
- D-bits

Out of all the above mentioned drills, twist drill is most commonly used in drilling machines. Fig. 5.42 shows twist drill geometry and different parts of twist drill. Twist drills are made of high speed steel and available in various sizes. Drills with parallel shank are available upto 12.7 mm diameters, and beyond that drills have tapered shank with Morse taper. They have three main parts drill point or dead centre, shank and body.

Axis: It is the longitudinal centre line along which the whole body is concentric.

Body: Body is the main cutting part which extends from extreme cutting point upto neck of the drill.

Body clearance: It is the portion of the body surface reduced in diameter to provide diameter clearance.

Dead centre: It is also called chisel edge formed by the intersection of flanks. It coincides with the axis of the drill.

Heel: It is an edge formed due to intersection of body clearance and the flute of the drill.

Lip: It is also called cutting edge of the drill formed by the intersection of flank and face. Number of lips formed depends on the number of face and flank. But, in general twist drill has two lips and both lips should have equal length, correct clearance and equal inclination with the drill axis, which is about 59°.

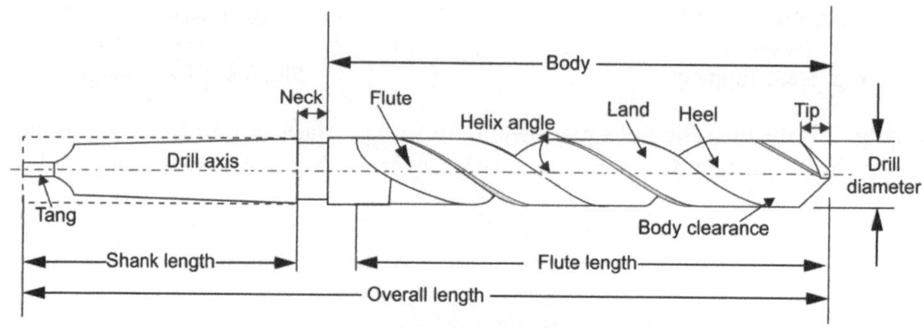

Fig. 5.42: Twist drill geometry

Point: It is a cone shaped surface at the end of flute consisting of dead centre, lips and flank.

Lip clearance: It is that part of conical surface of point, which is grounded to generate relief near to the cutting edge.

Face: It is the curved surface of the flute adjacent to lip on which the chips after cutting impinges and slides upward.

Land: It is the flat narrow surface that runs along the flute of the drill on the leading edge.

Web: It is the central metal portion of the drill that separates the flutes. It goes throughout the entire length on the drill between the flutes. The thickness gradually increases from the tip side towards shank side. It provides strength and rigidity to the drill.

Flank: It is the curved surface that extends behind the lip to flute.

Flute: The helical grooves in body of the drill are called as flute which forms cutting edges on the point. It provides the passage for the curled chips to exit and allows the cutting fluid to enter towards cutting edges. Generally, drills have two flutes.

Neck: It is the reduced diameter between the body and shank of drill. All the required particulars of the drill are engraved on it.

Shank: It is cylindrical or tapered portion of the drill by which it is gripped in the drill chuck and drives the drill. Taper shank is used in bigger size drills.

Tang: The rectangular cross-section at the end of tapered shank to fit the drill in the socket, sleeve or spindle to ensure positive drive. Drift is also provided to drive the drill out of sleeve or spindle.

5.8 GRINDING MACHINE

Grinding is a material removal process applied to all kind of materials with the help of abrasive action of tool called grinding wheel rotating at high speed. Grinding produces high quality surface with precision machining.

It is a finishing process where material removal rate is very low in form of small chips at high cutting speed to achieve close dimensional accuracy which is difficult with other machining processes.

The grinding machine or grinder is machine tool used in machine shops where abrasive wheels are attached to the tool post and workpiece is mounted on table. The power is supplied to the wheel with electric motor and the speed may be changed according to the work. The details of grinding machine with various parts are shown in in Fig. 5.43.

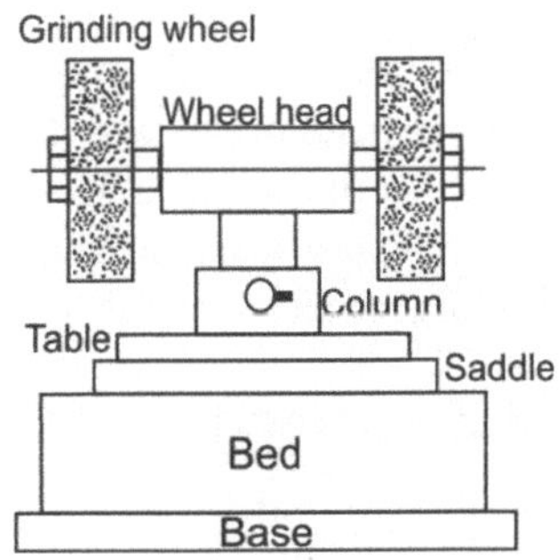

Fig. 5.43: Details of grinding machine

5.8.1 Mechanism

The workpiece is fed against the rotating wheel and removes unwanted material. The rubbing action of sharp edges of abrasive grains in grinding wheel and workpiece removes the material. Each grain in wheel acts as cutting edge which removes material in form of tiny chips from the workpiece as shown in Fig. 5.44.

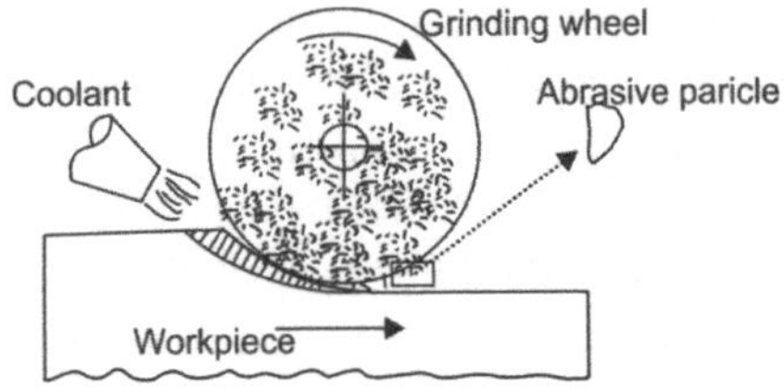

Fig. 5.44: Material removal in grinding

The common grinding operation is surface grinding, and chip cross section is similar to up milling process. The grinding may be rough or precision type grinding.

Rough grinding: It is the method of removing excess material such as fins, sharp corners and burrs from the surfaces of casting, forging and welded joints.

Precision grinding: It is the method of removing excess material which is hard to be machined with conventional machining tools or for producing surface with high dimensional accuracy and tolerances.

5.8.2 Grinding Wheel

Grinding wheel or Abrasives wheel acts as multi point cutting tool is made from thousands of hard tiny particles called as abrasive particles having many sharp cutting edges. These particles are randomly oriented and bonded together with bonding material which holds the particles when formed into different shapes called abrasive tools. The most common shape of grinding wheel is the disc wheel. The wheel should have sufficient porosity to provide space for the chips produced during grinding and for the easy flow of coolant. This prevents the adherence of chips with wheel. Abrasives are also applied in the form of paste and powders.

The abrasive particles at the surface of wheel exposed to actual cutting operation are called active particles. With continuous use of wheel, the sharp edges of active particles wear out and become blunt. On further grinding, the number of worn out particles increases which increases the friction and results in increased force on abrasive particles. When the cutting force becomes significantly high, the particles get fractured or break away from the wheel. These fractured particles generate new cutting edges, and when particles break away from the wheel, new particles are exposed and become active particles. This shows that the grinding wheel has self-sharpening characteristics. Therefore, the bond strength indicates *the maximum force an abrasive particle can withstand is an important performance characterise* of grinding wheel.

5.8.3 Selection of Grinding Wheels

The abrasive particles are cutting elements and must have sharp cutting points, high hot hardness, chemical stability and wear resistance. The wheels may get fractured when the force is increased and called fragile. The performance of grinding wheel is affected by various wheel parameters as discussed below.

1. **Size and shape of wheel**

   To achieve desired surface, it is important to select proper grinding wheel. Shape and size refers to outside diameter, spindle hole, width of the wheel. Wheels are available in variety of shapes such as disc wheels, cup wheel surface grinders, tool and cutter grinders, thin wheels for slitting and cutting purpose.

2. **Types of abrasives**

   Material used in cutting are called *abrasive,* which may be natural or artificial. Diamond has high chemical resistance and low coefficient of thermal expansion.

   Natural abrasive is sandstone, diamond, corundum and emery. Except diamond, all other natural abrasives are obsolete. Artificial abrasives are aluminium oxide, silicon carbide, cubic boron nitride and boron carbide. Aluminium oxide wheels are tough and shock resistant used for grinding of harder, tough and stronger materials such as carbon steel, high speed steel, alloy steel and tough bronze. Silicon carbide wheels are hard and sharp but not as tough as aluminium oxide. These are used for grinding of low tensile strength materials such as grey cast iron, soft bronze, copper etc. Boron carbide is harder than silicon carbide but not as hard as diamond, used as stick to dress grinding wheels. It may also be used in powder form for lapping operation. Cubic boron nitride is less reactive material used for grinding of high speed steel cutters, grinding of punches and dies, stainless steel etc.

3. **Grain size**

   It indicates the size of abrasive particles denoted by number. Higher the number, smaller is the grain size. It affects the surface finish of the material. Number between 10 to 24 indicates coarse grains, which gives faster material removal but poor surface finish, and is suitable for soft and ductile materials. Number between 70 to 180 indicates fine grains, which gives low material removal but high surface finish, and is suitable for hard and brittle materials.

4. **Type of bond**

Bond is the material which holds the abrasive particles together under the action of cutting forces. These are organic or inorganic materials. The hardness or grade of bond materials is represented by the letters which are discussed below:

- **Vitrified Bond:** Vitrified bonds are most widely used in the industry, made up of feldspar, quartz etc. These are strong, rigid and porous, and not affected by oils, acids, and water. The structure of wheel is uniform and is brittle with lack of resistance to mechanical and thermal impacts. The wheel size is up to 90 cm and is represented by letter 'V'. These are also called ceramic bonds.
- **Silicate Bond:** These wheels with silicate bonds are made by mixing sodium silicate with the abrasive particles. These bonds are not as strong as vitrified bonds, and abrasive particle gets dislodged easily. The grinding wheels with silicate bonds are used in operations where less heat is generated. The wheels are made of diameter above 90 cm, and are represented by letter 'S'.
- **Rubber Bond:** Rubber bonded wheels are made by mixing crude rubber, sulphur, and the abrasive grains together. The mixture is rolled into sheets, and then cut into circles. It is the heated under pressure to vulcanize the rubber. The wheels made are hard and tough and can be used as cutting saws for parting off operations. The rubber bonded wheels are used for fine finishing or polishing work, and are represented by letter 'R'.
- **Shellac Bond:** These wheels with shellac bonds are made by mixing shellac with the abrasive particles in a heated mixer. These are lesser used in industries and have wheel thickness of 3 mm or less. Such bonds are used for very high surface finish on paper mill rolls camshaft and cutlery, and are represented by letter 'E'.
- **Resinoid Bond:** Resinoid bonded wheels are made up of powdered synthetic thermosetting resin materials used as phenol formaldehyde mixed with abrasive particles and pressed with heat. The wheels with resinoid bonds are also called as organic wheels, and are represented with letter 'B'. The bond has more strength and elasticity than vitrified bonds. These are used for rough and heavy duty grinding of castings and parting off stocks.

5. **Grade or hardness**

Grade of the grinding wheel depends on the type, structure and amount of bonding material. It represents the hardness of the bond material and is called the grade of grinding wheel. It is the strength with which the abrasive grains are held together in the bond. More the bonding material, harder will be the grinding wheel. A harder wheel has strong bond which does not release the abrasives easily. A soft wheel has low content of bonding material and requires small force to release the grains. Soft wheels are preferred for grinding of harder materials and hard wheels for soft materials. Grade of the wheel is designated as soft, medium and hard wheels, represented by letters from A to Z in order of increasing hardness.

6. **Structure**

The structure of grinding wheel represents the porosity or the spacing between grains. It may also be represented as number of cutting edges per unit area of the wheel. The structure of the bond material in a wheel varies from about 10% to 30% of its total volume. Structure of the bond depends upon this percentage.

When the percentage of abrasive particles is more, it is called dense structure denoted by number from 1 to 4, used for cutting and snagging. Whereas, wheel with lesser percentage of abrasive particles are called as open structure denoted by number from 9 to 15 used for tool and cutter grinding. A medium structure is denoted by number from 5 to 8 and is commonly used.

5.8.4 Grinding machine operations

The following operations that are carried out on grinding machines are discussed below:

1. **Surface grinding**

   It is an operation of generating flat surfaces by removal of material with higher precision. The surface may be generated with horizontal spindle or vertical spindle type grinding wheel as shown in Fig. 5.45 and Fig. 5.46.

   The operation is similar as planer, shaper and milling, but accuracy is high. The machinery for surface grinding may be classified as;

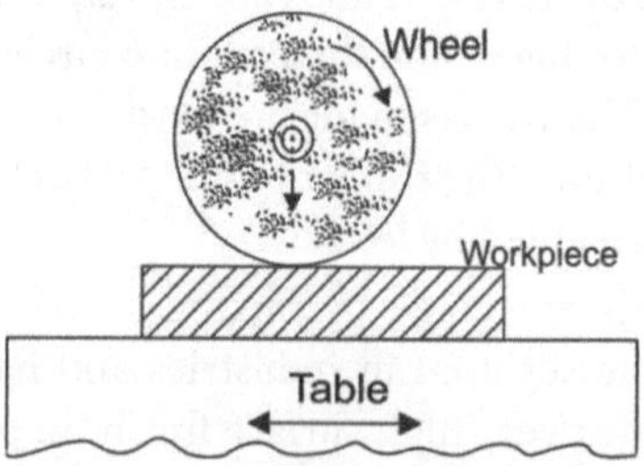

Fig. 5.45: Horizontal spindle surface grinding

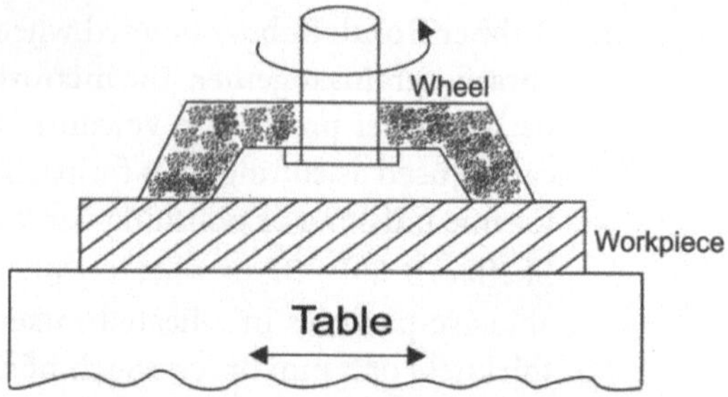

Fig. 5.46: Vertical spindle surface grinding

   (a) **Reciprocating table and horizontal spindle type**

   This is the most common surface grinding operation, where the peripheral surface of grinding wheel is used for finishing work. Surfaces grinding is one of the most common grinding operations. The workpiece is fixed on a magnetic chuck placed on table and wheel on horizontal spindle. Grinding occurs as the table reciprocates longitudinally and on completion of each stroke it feeds laterally as shown in Fig 5.47. The pump may be provided for circulation of coolant between wheel and the workpiece.

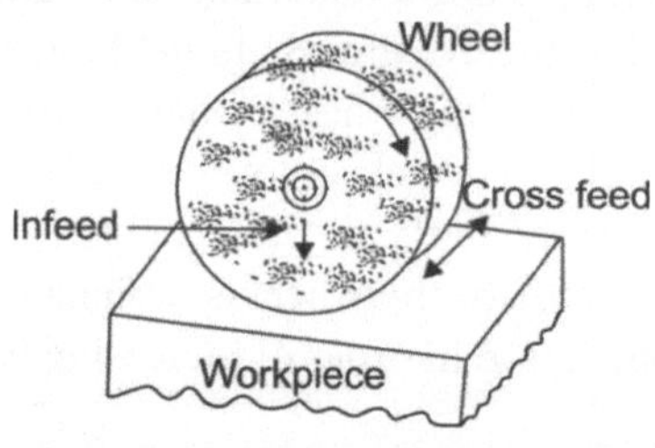

Fig. 5.47: Reciprocating table and horizontal spindle type

(b) **Reciprocating table and vertical spindle type**

It removes material faster than horizontal type and has higher precision. This grinding operation is similar to that of face milling, and the surface of a cup shaped grinding wheel finishes the workpiece over its entire width as shown in Fig. 5.48.

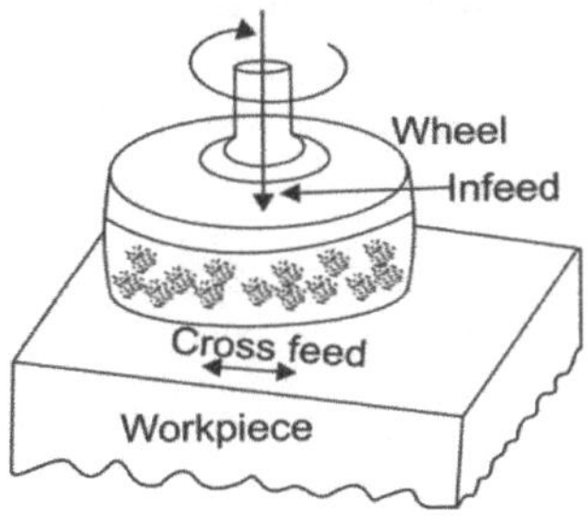

Fig. 5.48: Reciprocating table and vertical spindle type

(c) **Rotary table and horizontal spindle type**

The machine consists of horizontal wheel spindle which reciprocates from the edge to the centre of table similar to shaper arm. It has a magnetic chuck on rotary table, which can be moved up and down with provision of tilting to produce concave, convex or tapered surface as shown in Fig. 5.49. The machine is not widely used as it has a limitation to accommodate large workpieces.

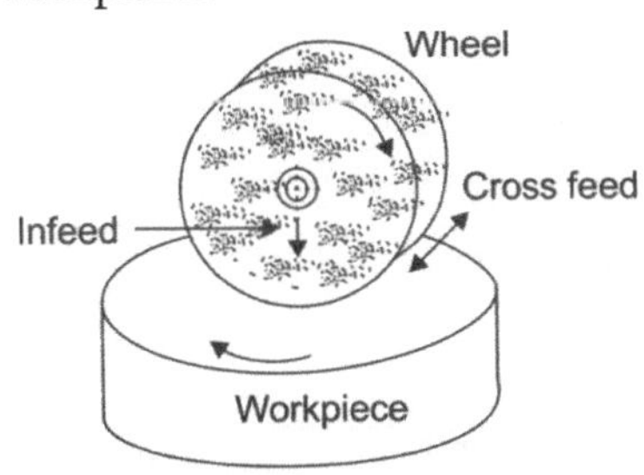

Fig. 5.49: Rotary table and horizontal spindle type

(d) **Rotary table and vertical spindle type**

This machine is used for grinding of small workpiece in large number. It has vertical wheel spindle provided with arrangement to move the wheel up and down. Both, the wheel and workpiece rotates and feed into each other. A cup shaped grinding wheel finishes the workpiece over its entire surface. The wheel axis should be exactly perpendicular for finishing, and can be tilted for rough grinding (Fig. 5.50).

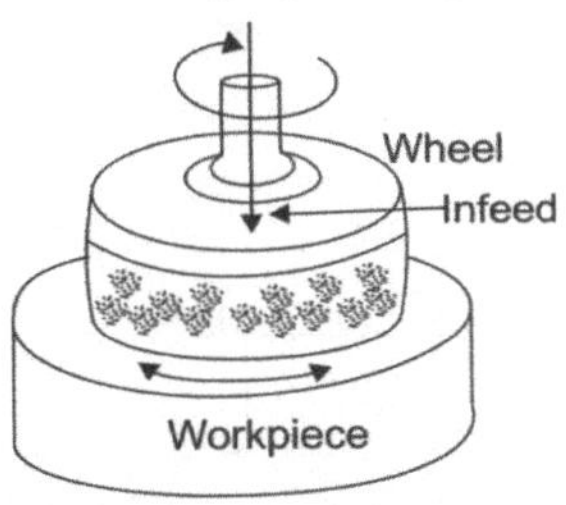

Fig. 5.50: Rotary table and vertical spindle type

2. **Cylindrical Grinding**

In cylindrical grinding, the rotating workpiece is mounted between two centres. The rotating workpiece is traversed at different feeds across the face of rotary wheels to finish entire length of the workpiece. At the end of each stroke, the workpiece is fed into the wheel equal to the depth of cut and produces long cylinders of circular profiles with better surface finish (Fig. 5.51). In case, where the width of the wheel face is more or equal to the length of workpiece, the wheel is fed into workpiece with no traversing movement of wheel or workpiece and is termed as *plunge grinding*.

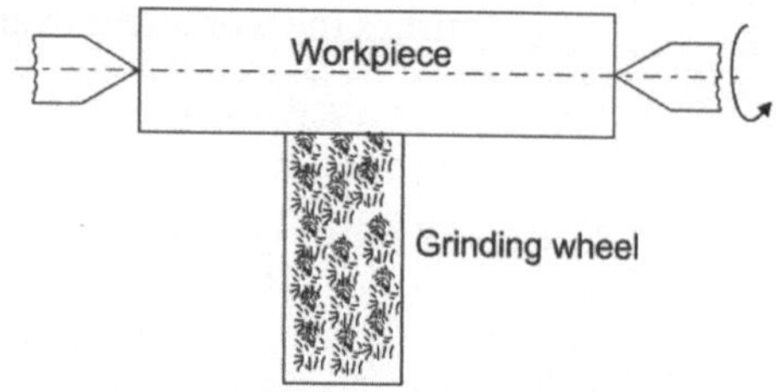

Fig. 5.51: Cylindrical grinding

(a) **External and internal cylindrical grinding**

External grinding produces straight or tapered surfaces by removing excess material form the outer surface of the cylindrical workpiece as shown in Fig. 5.52. In internal grinding, the internal bores of gears, bushes and bearing races are grinded with small wheel mounted on long spindle. The workpiece is held in stationary slide, which is mounted on a slide. The axis of the rotating wheel moves along the circular path of the hole in the workpiece. The wheel may reciprocate and the slide on which workpiece is mounted may also move backward and forward direction (Fig. 5.53).

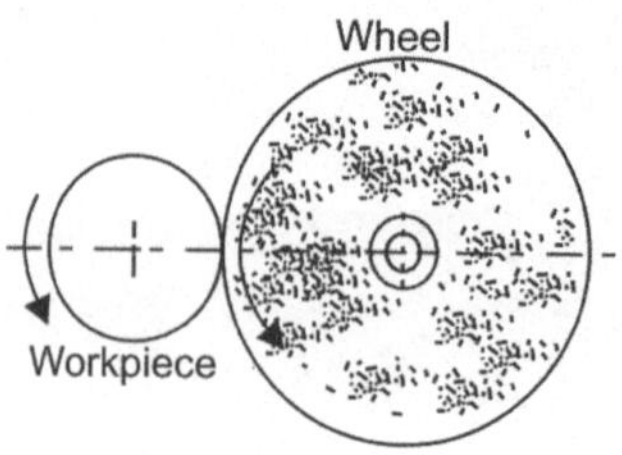

Fig. 5.52: External grinding

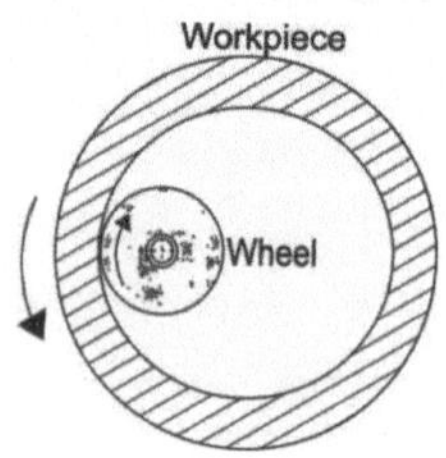

Fig. 5.53: Internal grinding

3. **Thread grinding**

It is performed on cylindrical grinders where grinding wheels are specially dressed to match the shape of the thread. The grinding wheel having the shape of the thread and abrasive particles on the periphery rotates at its axis and touches the surface of the workpiece. The particles act as cutting tools and cut the thread on workpiece (Fig. 5.54). This process produces accurate threads with very fine surface finish in harder materials such as ball screw mechanisms.

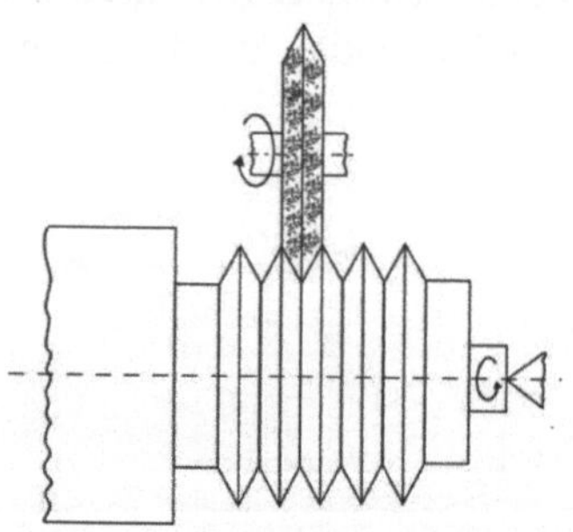

Fig. 5.54: Thread grinding

4. **Centreless Grinding Operation**

It is the grinding process where the workpiece is supported on work rest, not on centres or chucks and passed between two wheels. The grinding wheel runs at high speed and controlling wheel at low speed in same direction. The two wheels are parallel, and controlling wheel is inclined at 5-10 degree angle to provide longitudinal motion to workpiece past the grinding wheel as illustrated in Fig. 5.55.

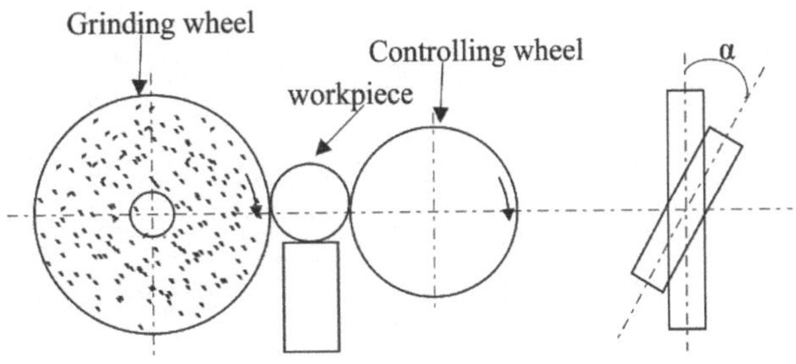

Fig. 5.55: Centreless grinding process.

The centre of workpiece is slightly above center-line of both the wheels. The grinding wheel forces the workpiece against the work rest and controlling wheel. The controlling wheel is a rubber bonded wheel with wider faces, which rotates the workpiece with same rotational speed. This enables auto feed of workpiece through the wheels and provides smooth surface. There are three ways to feed the workpiece in centreless grinding.

Through Feed: This arrangement is used for straight cylindrical workpieces. It has guides on each side to steer the workpiece between the grinding wheels, where workpiece passes from one side and comes out on the other side.

In-Feed: This type of grinding is used for workpieces that has obstructions or shoulders which prevents it to travel past the grinding wheel and after grinding the workpiece is withdrawn. It does not have guides, but stopper on the far end.

End-Feed: This type of grinding is used for the tapered workpieces. Both wheels have required shape and the work is feed against it.

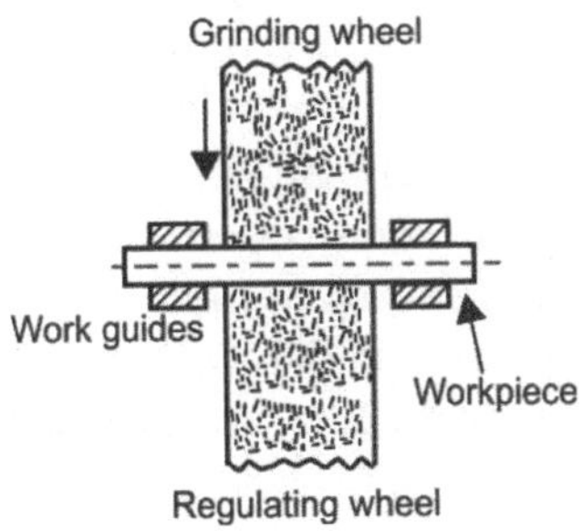

Fig. 5.56: Through feed internal grinding

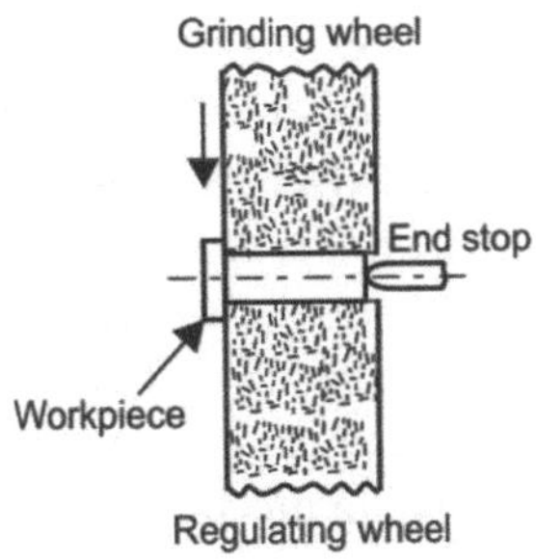

Fig. 5.57: In-feed internal grinding

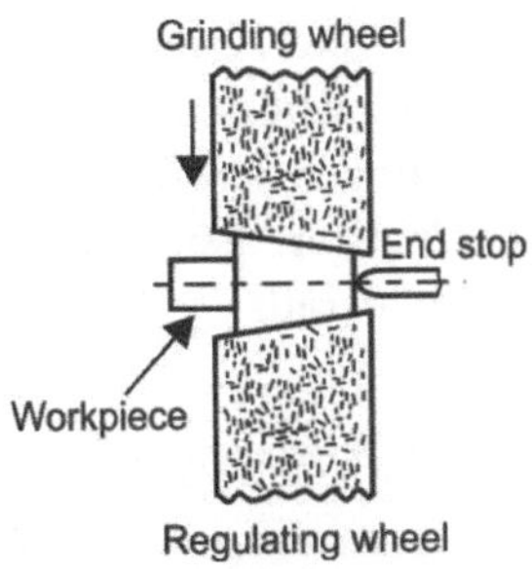

Fig. 5.58: End- feed internal grinding

5. **Form Grinding:** In form grinding, the grinding wheel is dressed in the same shape as required in final workpiece. As the workpiece is passes through the grinding wheel, it acquires the shape formed in grinding wheel (Fig. 5.59).

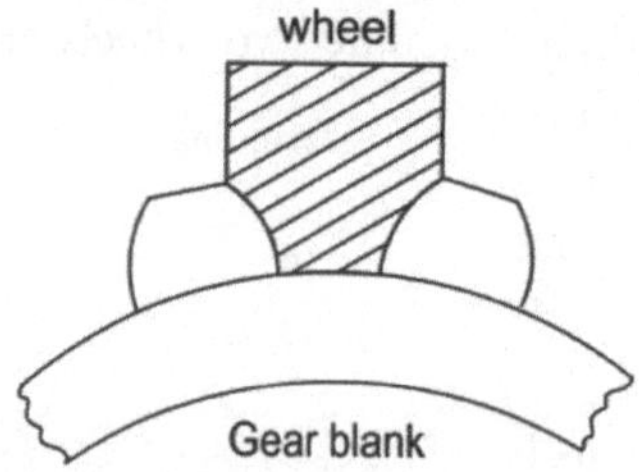

Fig. 5.59: Form grinding.

5.8.5 Balancing, Dressing and Truing

5.8.5.1 Balancing

Grinding wheels are fragile tools, which when revolved at high speed may get cracked due to unbalanced centrifugal forces. When the wheels are out of balance, it may lead to excessive vibrations, rapid wheel wear chatter, damage of bearing and finally poor surface finish. Therefore, proper mounting and balancing is required, when machining for higher accuracy. Balancing can be achieved by adding weight in the form of lead to the light side.

5.8.5.2 Dressing

With the continuous use and wear, the abrasive particles become dull. When the wheel is hard, the grains are not released by the bond and cutting surface become rounded and shiny called as glazing.

When grinding wheel is used to grind material softer than the wheel, or forcing the cut too deep, then chips of the materials get welded in the space between grains is called as loading.

The glazed and loaded wheel does not grind the material properly and reduces the cutting capacity. Grinding wheels are therefore dressed to restores the sharpness of abrasive particles. The dressing is the process of generating sharp cutting edges on abrasive grains at the surface of a grinding wheel to ensure that cutting edges are exposed and able to grind the workpiece material. Dressing also removes the residue of material left after grinding as well as dulled abrasive grains. This can be carried out with single point diamond mounted in metal holder.

5.8.5.3 Truing

The Truing is similar to dressing operation by which grinding wheel gains original shape. This corrects the uneven wear of wheel face to maintain efficient cutting. Truing is carried out when new wheel is mounted or when wheel has been in use for longer time, to correct the face and sides of the wheel. The truing and dressing of rotating grinding wheels are generally performed with a diamond tool which removes the abrasive particles and the bond material.

UNIT SUMMARY

- Machine tools hold and support the cutting tool or workpiece and remove the unwanted material from the workpiece in order to produce desired geometry. The material is removed in the form of chips.
- Machine tools are classified on the basis of machining method, type and size of object, surface generated or purpose, accuracy required and methodology adopted for their control. They are classified as general purpose and special purpose machine tools.
- Lathe is a versatile machine tool found in all manufacturing industries. They are of many types for different processing requirements. Operations that may be performed on lathe are; turning, taper turning, facing, knurling, thread cutting, drilling etc. by using attachments.
- Shaper is a reciprocating type of machine tool where tool reciprocates over the stationary workpiece to produce be vertical, horizontal and inclined flat surfaces, and works on quick return mechanism.
- In planer tool remains stationary whereas workpiece reciprocating during machining, due to this, large sized parts can be machined.
- Milling has rotating multipoint cutter mounted on arbor and automatically feeds the workpiece to machine flat or curved surfaces with more precision such as; slots, gears, contours and circular profiles etc. Milling is classified as up and down milling based on direction of movement of workpiece and rotation of cutter.
- Drilling is a process to make circular holes in solid material with the use of multi point rotating tool which has sharp twisted edges around cylindrical tool that rotates inside the material.
- Boring is process to enlarge an already existing hole, whereas reaming is finishing operation to provide close dimensional tolerances.
- Grinding makes use of bonded abrasive particles in form of grinding wheel to remove small amount of material to generate fine surface finish with higher dimensional accuracy. Grinding wheels are characterized by shape, size and type of abrasive particles, type of bond, grade and hardness, and structure.
- Dressing and truing are the necessary operations to be carried out on grinding wheel to generate sharp cutting edges and to generate required form of wheel.

EXERCISE

Multiple Choice Questions

1. Lathe bed is usually made up of ....
   - (a) Stainless steel
   - (b) Mild steel
   - (c) Cast Lathe
   - (d) All of he above
2. Lathe centers are provided with the following standard taper.
   - (a) Morse
   - (b) British
   - (c) Metric
   - (d) Sharpe

3. A 3-jaw chuck in lathe is used to clamp
   (a) Cylndrical workpiece to locate the axis of rotation
   (b) Eccentric workpiece to locate the axis of rotation
   (c) Square workpiece to locate the axis of rotation
   (d) Any type workpiece to locate the axis of rotation
4. In lathe the carriage and tail stock are guided on......
   (a) Same guideways
   (b) Different guideways
   (c) Any of the above
   (d) Not guided on guideways
5. Capacity of the lathe is expressed as
   (a) Tool post size and lathe travel
   (b) Swing and distance between centres
   (c) Horse power and chuck diameter
   (d) Bed lengtha and spindle speed
6. Facing operation in a lathe is used for producing
   (a) Cylindrical surface (b) Plane surface
   (c) Tapered surface (d) Hole
7. For accurate cutting, the cutting tool in a lathe should be set
   (a) Slightly below the axis of the workpiece so that no rubbing takes lace
   (b) Exactly at the centre of the axis of the workpiece
   (c) Slightly above the axis of the workpiece so that no rubbing takes lace
   (d) Anywhere, since the workpiece is actually rotating
8. Knurling operation in a lathe is used for producing
   (a) Plane surface (b) Cylindrical surface
   (c) Tapered surface (d) Serrated surface
9. Which of the machine tool operates with quick returm mechanism
   (a) Lathe (b) Shaper
   (c) Planer (d) Milling
10. In a shaper, the metal is removed during
    (a) Forward stroke
    (b) Return stroke
    (c) Both forward and return stroke
    (d) Neither the forward nor the return stroke
11. In shaper machine tool
    (a) The tool is stationary and work reciprocates
    (b) Work is stationary and tool reciprocates
    (c) The tool moves over stationary work
    (d) The tool moves over reciprocating work

12. The operation performed on a shaper is
(a) Machining horizontal surface (b) Machining vertical surface
(c) Machining angular surface (d) All of the above

13. Which of the following act as housing for an operating mechanism in the shaper?
(a) Base (b) Column
(c) Cross rail (d) Table

14. In planer machine tool
(a) The tool is stationary and work reciprocates
(b) Work is stationary and tool reciprocates
(c) The tool moves over stationary work
(d) The tool moves over reciprocating work

15. Lathe cutting tool can be classified as right hand and left hand according to....
(a) Method of using the tool (b) Method of holding the tool
(c) Method of applying feed (d) None of the mentioned

16. Planer is ....shaper in size
(a) Larger than (b) Smaller than
(b) Equal to (d) All of the above

17. The cutting tool moves in vertical reciprocationg motion in which machine tool
(a) Shaper (b) Planer
(c) Slotter (d) Vertical lathe

18. The type of tool used on lathe, shaper, and planer is
(a) Single-point cutting tool (b) Two-point cutting tool
(c) Three-point cutting tool (d) Multi-point cutting tool

19. During milling, the cutter usually moves
(a) Perpendicular to its axis (b) Parallel to its axis
(c) Inclined to its axis (d) Any of the above

20. In ....., the cutting action occurs primarily at the end corners of the milling cutters
(a) Face milling (b) Peripheral milling
(c) Both a and b (d) None of the above

21. Slab milling can be performed more effectively by ______ milling machine.
(a) Horizontal (b) Vertical
(c) All of the above (d) None of the above

22. When workpiece is feed in the same direction as that of the cutter at the point of contact, this milling operation is called as
(a) Down milling (b) Up milling
(c) Slot milling (d) Slab milling

23. In ....., operation, the chip thickness is minimum at the beginning of the he cut and reaches maximum at the en of cut
(a) Conventional milling (b) Climb milling
(c) Face milling (d) End milling

24. Milling cutters may be made of______
    (a) High speed steel
    (b) Cemented tipped
    (c) Super high speed steel
    (d) All of the mentioned
25. Which of the following processes results in the best accuracy of the hole made
    (a) Drilling
    (b) Reaming
    (c) Boring
    (d) Broaching
26. Grooves providede on the entire length of the body odf a twist drill are called
    (a) Webs
    (b) Lips
    (b) Flutes
    (d) Margins
27. Trepanning is a process of
    (a) Tapping internal screw threads
    (b) Producing holes by removing material along the circumference of the tool
    (c) Tapping external threads
    (d) Smooting and squaring the surfa e aound hole for a nut
28. ....is a machining oeration which makes us eo f a multi point cutting tool to produce , ingternal therads
    (a) Tapping
    (b) Trepaning
    (c) Boring
    (d) Drilling
29. Removing dull grains in order to make grinding wheel sharp is known as
    (a) Loading
    (b) Glazing
    (c) Dressing
    (d) Trueing
30. Operation done to make periphery of grinding wheel concentric with its axis to recover its lost shape is known as
    (a) Loading
    (b) Glazing
    (c) Dressing
    (d) Trueing
31. A dense structure of a grinding wheel is used for
    (a) Hard materials
    (b) Brittle materials
    (c) Finishing cuts
    (d) All of the above
32. The grade of a grinding wheel depends upon
    (a) The hardness of the material being ground
    (b) Speed of wheel and work
    (c) Condition of grinding machine
    (d) All of the above
33. In centreless grinding, workplace center will be
    (a) Above the line joining the two-wheel centers
    (b) Below the line joining the two-wheel centers
    (c) On the line joining the two-wheel centers
    (d) At the intersection of the line joining the wheel centers with the workplace plane

34. The hardness of a grinding wheel is specified by
    (a) Brinell hardness number
    (b) Rockwell hardness number
    (c) Vickers pyramid number
    (d) Letter of alphabet

Answers for Multiple Choice Questions							
1. (c)	2. (b)	3. (a)	4. (b)	5. (b)	6. (b)	7. (b)	8. (d)
9. (b)	10. (a)	11. (b)	12. (d)	13. (b)	14. (a)	15. (c)	16. (a)
17. (c)	18. (a)	19. (a)	20. (a)	21. (a)	22. (a)	23. (a)	24. (d)
25. (b)	26. (c)	27. (b)	28. (a)	29. (c)	30. (d)	31. (d)	32. (d)
33. (a)	34. (d)						

Short and Long Answer Type

1. What is a machine tool? What are their functions?
2. Define with neat sketch the working principle of lathe.
3. How is the size of lathe specified?
4. List the various methods of taper turning.
5. What are the main parts of lathe?
6. Explain the various operations performed on lathe
7. List the various types of cutting tools used n lathe.
8. What are the various types of lathe?
9. With the help of neat sketch, explain the working principle of shaper.
10. How is a shaper machine specified?
11. Discuss in brief the quick return mechanism of shaper.
12. Explain the principle parts of shaper.
13. Explain the working principle of planer.
14. Explain the fundamental difference between mechanism of shaper and planer.
15. How the size of planer is specified?
16. Explain the working principle of milling with the help of neat sketch.
17. Differentiate between up and down milling.
18. Explain with the help of neat sketch, the horizontal milling machine
19. Explain briefly the following operations performed on milling machine:
    (i) Straddle milling
    (ii) Form milling
    (iii) Face milling
    (iv) Gang milling
    (v) End milling
    (vi) Angular milling
20. Define drilling and functions of drilling machine.
21. Explain the principle parts of drilling machine.
22. Explain briefly the following operations performed on drilling machine:
    (i) Reaming
    (ii) Boring

(iii) Tapping (iv) Trepanning
(v) Countersinking (vi) Counterboring

23. With the help of neat sketch, explain the details of twist drill.
24. Define grinding and its advantages over other machining processes.
25. Explain the principle of grinding, rough and precision grinding.
26. What is grinding wheel?
27. How are the abrasives selected for grinding process.
28. Explain any two types of the following grinding machines:
    (i) Reciprocating table vertical spindle (ii) Reciprocating table horizontal spindle
    (iii) Rotary table vertical spindle (iv) Rotary table horizontal spindle
29. On what parameters the selection of grinding wheel is made?
30. Describe the following;
    (i) Balancing (ii) Dressing
    (iii) Glazing (iv) Truing
31. What is the purpose of bond? Name and explain different bonding materials.
32. Explain briefly; Silicon carbide (SiC) and Aluminium oxide Al_2O_3.

PRACTICAL

1. Study of lathe machine and prepare a job as per given drawing.
2. Study of quick return mechanism in shaper machine.
3. Study of milling machine and prepare a job as per given drawing.
4. Study of drilling machine and prepare a job as per given drawing.
   Study of simple grinding machine.

KNOW MORE

- Machining centres are multi-functional CNC machine that allows flexibility and can mill, drill, bore, tap, and perform various other works all without changing the attachment of the work piece. It has an automatic tool changer and a table that clamps the workpiece in place.
- Multi spindle machines have more than one spindle (four, six or eight), where the workpiece can be clamped. Therefore, multiple workpieces can be machined simultaneously.
- Lathe accessories are used to hold and support the workpiece and to hold the cutting tool such as; chucks, face plate, lathe dogs, rests, lathe mandrels, jigs and fixtures and lathe centres.
- Lathe attachments are not mandatory elements, but used to increase the production and efficiency, such as; stops, grinding attachments, milling attachments, taper turning attachments and copying attachments.
- To cut gear teeth, dividing head which is an important attachment is used to divide the periphery of the workpiece into equal number. The indexing plate of the dividing head has number of holes and crank pin. This crank drives the spindle and the live centre through a worm gear.

- Special set-ups such as cam milling can be used for machining special surfaces.
- Super finishing is a fine material removal process that develops very high finishes on the surfaces already been generated by other type of finishing or surface modification process.
- Honing is a super finishing process using abrasive sticks for internal cylindrical surfaces, whereas lapping makes use of loose abrasive particles on flat surfaces.

REFERENCES AND SUGGESTED READINGS

1. A. Ghosh and A. K. Mallik. Manufacturing Science, Affiliated East West Press, Delhi, India, 2019
2. P. N. Rao. Manufacturing Technology (Vol. 2), McGraw Hill, New Delhi, India, 2015.
3. M.P. Poonia and S.C. Sharma. Basic Mechanical Engineering – Khanna Publishing House, New Delhi, India, 2005.
4. S. Kalpakjian and S. R. Schmid. Manufacturing Processes for Engineering Materials, Pearson Education, New Delhi, India, 2009
5. B. S. Raghuvanshi. Workshop Technology (Vol. 2), Dhanpat Rai and Sons, New Delhi, India, 2003.
6. M. P. Groover. Fundamentals of Modern Manufacturing: Materials, Processes and systems, John Wiley and Sons, USA, 2010.
7. R. K. Rajput. Manufacturing Technology, Laxmi Publications, New Delhi, India, 2007.

REFERENCES FOR FURTHER LEARNING

1. W. D. Callister, Material Science and Engineering: An Introduction, John Wiley & Sons, New York, 2007.
2. S. K. Hajra Choudhary and A. K. Hajra Choudhary, Workshop Technology: Volume I & II, Media Promoters and Publishers, Mumbai, 1989.
3. V. Raghvan, Material Science and Engineering, Prentice Hall India Pvt. Ltd., New Delhi, 2015.
4. E P Degarmo, J T Black and R A Kohser, Materials and Process in Manufacturing, John Wiley & Sons, New York, 2003.
5. J. S. Campbell, Principles of Manufacturing Materials and Processes, Tata McGraw Hill, New Delhi, 1974.
6. P. C. Sharma. A Text Book of Production Technology, S. Chand & Company Pvt. Ltd., New Delhi, 1996.
7. S. C. Arora, S. Domkundwar, A Course In Power Plant Engineering, Dhanpat Rai & Sons, Delhi, 1991.
8. M. L. Mathur and F. S. Mehta, Thermal Engineering, Jain Brothers, New Delhi, 1998.
9. C. P. Kothandaraman, Fundamentals of Heat and Mass Transfer, New Age International Pvt. Ltd. Publishers, New Delhi, 2006.
10. C. P. Arora, Thermodynamics, Tata McGraw Hill Publishing Company Ltd., New Delhi, 1998.
11. R. S. Khurmi and J. K. Gupta, Textbook of Refrigeration and Air Conditioning, S. Chand Publishing, New Delhi, 2008.
12. E. Rathakrishnan, Fundamentals of Engineering Thermodynamics, Prentice Hall of India Pvt. Ltd., New Delhi, 2006.
13. R. K. Rajput, Internal Combustion Engines, Laxmi Publications Pvt. Ltd. New Delhi, 2005.
14. H. N. Gupta, Fundamentals Of Internal Combustion Engines, Prentice Hall of India Pvt. Ltd., New Delhi, 2012.

CO AND PO ATTAINMENT TABLE

Course outcomes (COs) for this course can be mapped with the programme outcomes (POs) after the completion of the course and a correlation can be made for the attainment of POs to analyze the gap. After proper analysis of the gap in the attainment of POs necessary measures can be taken to overcome the gaps.

Table for CO and PO attainment

Course Outcomes	ATTAINMENT OF PROGRAMME OUTCOMES (1- Weak Correlation; 2- Medium Correlation; 3- Strong Correlation)											
	PO-1	PO-2	PO-3	PO-4	PO-5	PO-6	PO-7	PO-8	PO-9	PO-10	PO-11	PO-12
CO-1												
CO-2												
CO-3												
CO-4												
CO-5												
CO-6												

The data filled in the above table can be used for gap analysis.

INDEX